Forest Pathology and Plant Health

Special Issue Editors

Matteo Garbelotto
Paolo Gonthier

MDPI • Basel • Beijing • Wuhan • Barcelona • Belgrade

MDPI

Special Issue Editors
Matteo Garbelotto
University of California-Berkley
USA

Paolo Gonthier
University of Torino
Italy

Editorial Office
MDPI AG
St. Alban-Anlage 66
Basel, Switzerland

This edition is a reprint of the Special Issue published online in the open access journal *Forests* (ISSN 1999-4907) from 2016–2017 (available at: http://www.mdpi.com/journal/forests/special_issues/pathology_health).

For citation purposes, cite each article independently as indicated on the article page online and as indicated below:

Author 1, Author 2. Article title. *Journal Name*. **Year**. Article number/page range.

First Edition 2018

ISBN 978-3-03842-671-4 (Pbk)
ISBN 978-3-03842-672-2 (PDF)

Cover photo courtesy of Matteo Garbelotto

Table of Contents

About the Special Issue Editors

Matteo Garbelotto is Adjunct Full Professor in ESPM (Environmental Science, Policy, and Management) at UC Berkeley and is the Statewide Forest Pathologist of the entire UC System. He has taught several classes on California forest diseases and has worked extensively in the Sierra Nevada, the Cascades, the Transverse Ranges of Southern California, and throughout the California Coast Range. His interests have led him to conduct research in Asia, Oceania, Mesoamerica, Europe, and the entire Mediterranean Basin. He is a recognized authority on root diseases as well as on forest Phytophthoras. His field of expertise is primarily on the evolutionary processes leading to biological invasions and on approaches to uncover pathways of global movement of microbes. He also is active in the field of biodiversity and has worked on large-scale DNA barcoding studies including the Biocode project. He has published close to 200 scientific publications and has been a pioneer in the field of molecular diagnostic of plant pathogens, but currently he is recognized for his genomic and Citizen Science projects. He has advised the US, European Union, Canadian, New Zealand, British, French, Swiss, Spanish, South African governments on several policy issues regarding the introduction and regulation of plant pathogens and is currently a member-at-large of the European Food Safety Authority. He has been recognized twice by the International Society of Arboriculture as the most relevant scientist of the year. Because of his work on Sudden Oak Death he received a proclamation by the California State Assembly, and has been declared oak savior by the City and County of San Francisco. He has received the unsung hero award by San Francisco Tomorrow, and recently he has received the US Western Extension Directors Association Award of Excellence. Matteo is a Fulbright Scholar and has two Masters and a PhD from UC Berkeley in Plant Pathology. Besides being a faculty member at UC Berkeley, he has been a visiting scientist at the Smithsonian Tropical Research Institute and a visiting professor at the University of Turin. He has been a visiting scientist at the Museum of Natural History in Venice (Italy) where he still holds an honorary curator position for its extensive fungal collection.

Paolo Gonthier is Associate Professor at the Department of Agricultural, Forest and Food Sciences of the University of Turin, where is in charge of teaching 'Forest Pathology' at the Degree Course in 'Forest and Environmental Sciences'. At the same University, he is also professor of the PhD Course in Biological Sciences and Applied Biotechnologies. He is currently Senior Editor of the Journal of Plant Pathology, Associate Editor of the European Journal of Plant Pathology, and member of the Editorial Board of Forestry. His research activity spans a broad range of aspects of the Heterobasidion root and butt rot of conifers, including impact, population biology, epidemiology and control of pathogens. Over the last 15 years, he has been studying the invasion biology and the epidemiology of the North American *H. irregulare* introduced into Italy. He was appointed as member of the Expert Working Group (EWG) for Pest Risk Analysis of the European and Mediterranean Plant Protection Organisation (EPPO), and of the Forest Fungal Pathogens categorisation Committee of the European Food Safety Authority (EFSA).

Preface to "Forest Pathology and Plant Health"

Every year, a number of new forest pathosystems are discovered as the result of introduction of alien pathogens, host shifts and jumps, hybridization and recombination among pathogens, etc. Disease outbreaks may also be favored by climate change and forest management. The mechanisms driving the resurgence of native pathogens and the invasion of alien ones need to be better understood in order to draft sustainable control strategies. In this special issue, modeling, population biology, and experimental studies are featured with the aim of providing insights on the epidemiology and invasiveness of emergent forest pathogens by contrasting different scenarios dealing with varying pathogen and host population sizes, evolvable genetics, changing phenotypes and phenologies, landscape fragmentation, occurrence of disturbances, management practices, etc. In summary, this special issue focuses on how variability in hosts, pathogens, or ecology may affect the emergence of new threats to plant species. The three elements: pathogen, host, and environment are the well-known basic elements of the plant disease triangle (PDT). The PDT is as old as the field of modern Plant Pathology, and postulates that any plant disease is the outcome of the interaction between Pathogen, Host, and the Environment. Recently, the need has emerged to study not just how the three elements of the PDT directly influence disease, but to focus on how they indirectly affect one another, consequently modifying the final outcome. Of course, anthropogenic effects need to be thrown into the mix as well.

The special issue includes 14 papers. The first is a mini-review by Garbelotto and Gonthier discussing the need for research focusing on complex interactions and on disturbances. This is followed by a much more exhaustive review on the subject by Cobb and Metz. The next four papers describe how anthropogenic effects (e.g. shorter rotation times in an article by Soularue et al., and stoking levels in a study by Munck et al.), climate change (in an article on the root pathogen *Armillaria* by Kubiak et al.), and environmental or topographic factors (Lione et al.) affect the virulence and the persistence of emerging pathogens. A seventh paper by Panzavolta and colleagues provides an interesting framework to study the correlation between stressful environmental conditions with higher susceptibility of trees to both insects and pathogens. One of the interesting unexpected conclusions of the study is the synergistic (i.e. more than additive) effect of stress in increasing fungal spread by increasing vectoring of pathogens by insects.

The following four papers focus on host variability and disease. Prospero and Cleary provide a well-structured review on the subject focusing on the effects of host variability on invasive pathogens. Ruiz Silva et al. discuss the importance of both genetic and phenological resistance against an important emergent pathogen, while Chieppa et al. experimentally uncover intraspecific competition among pine genotypes cryptically driven by susceptibility to infection by a vascular pathogen. Finally, the paper by De Urbina et al. describes how multiple diseases emerging on the same host may preferentially attack different host populations. The paper also emphasizes through an experimental study how hard it may be to manage these emergent diseases.

The last group of three papers provides solid evidence about the human role in the global movement of pathogens (Mehl et al.), and about additional human roles in enhancing the establishment of invasive exotic pathogens (Danti and Della Rocca, Ploetz et al.). The article by Ploetz et al. on Laurel Wilt also brings to the forefront the complex issue of a disease that affects both agricultural and natural forest settings.

In summary, as Garbelotto and Gonthier conclude in the first paper of the issue: "This special issue contains 13 [additional] articles that we hope will be thought-provoking in more than one way. They include widely different approaches, scales, and technical methodologies, and they well represent the cutting edge of contemporary Forest Pathology. The expectation of this special issue was to represent a range of approaches currently employed to study variability in tree diseases. We hope the reader will agree that this expectation has been met, and we hope he/she will concur that in the process of compiling this issue, we may have put together an excellent textbook for an advanced class in Forest Pathology"

Matteo Garbelotto and Paolo Gonthier

Special Issue Editors

forests

MDPI

Editorial

Variability and Disturbances as Key Factors in Forest Pathology and Plant Health Studies

Matteo Garbelotto [1],* and Paolo Gonthier [2]

[1]   Department of Environmental Science, Policy and Management, University of California at Berkeley, 54 Mulford Hall, Berkeley, CA 94720, USA
[2]   Department of Agricultural, Forest and Food Sciences, University of Torino, Largo P. Braccini 2, I-10095 Grugliasco, Italy; paolo.gonthier@unito.it
*   Correspondence: matteog@berkeley.edu; Tel.: +1-510-6434282

Received: 13 November 2017; Accepted: 14 November 2017; Published: 15 November 2017

Abstract: The plant disease triangle (PDT) is as old as the field of modern plant pathology, and it postulates that any plant disease is the outcome of the interaction between a pathogen, a host, and the environment. Recently, the need has emerged to study not only how the three elements of the PDT directly influence disease, but to focus on how they indirectly affect one another, consequently modifying the final outcome. It is also essential to structure such analyses within three major external frameworks provided by landscape level disturbances, climate change, and anthropogenic effects. The studies included in this issue cover a wide range of topics using an equally varied list of approaches, and they showcase the important role these indirect and often non-linear processes have on the development of forest diseases.

Keywords: biological invasions; climate; disease triangle; epidemiology; forest; Geographic Information System; modeling; variability

A SCOPUS (https://www.scopus.com/) search using the key words "forest pathogen", "invasive", and "variability" reveals a recent reborn interest in the concepts of variability and disturbances as major drivers of infectious forest diseases (Figure 1). Although it is still convenient to partition such variability according to the three main elements of the plant disease triangle (PDT), that is, pathogen, host, and environment [1], our interest is spiked not so much by the study of the individual variables per se, but rather by their dynamic interaction. Advancements in computational and statistical approaches provide a solid framework to focus on those effects that may have been previously discarded or considered marginal because of being too difficult to measure using standard passive analytical approaches [2,3]. These advancements allow us to compute the outcomes of multiple interactions with greater confidence than in the past, and they have provided a considerable push to cross-over across fields. Additionally, this renewed interest in the disease triangle is occurring in a broader framework provided by the awareness of the importance of both anthropogenic and climate change effects [4,5]. It should be noted that the disease triangle may be used to predict epidemiological outcomes not only in plant health, but also in public health, both in local and global communities [6]. The main aim of this special issue was to focus on disturbances and variability as important factors determining the final outcome of forest diseases.

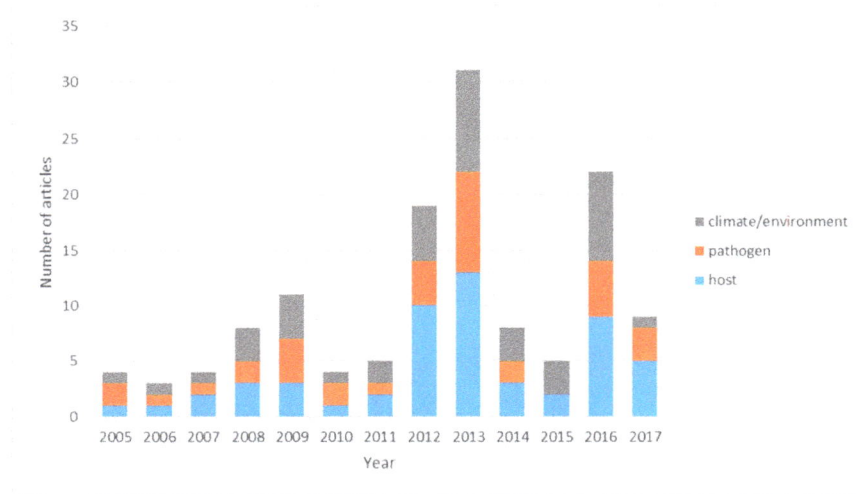

Figure 1. Number of articles retrieved on SCOPUS (https://www.scopus.com/) using the search terms "forest pathogen", "invasive" and "variability" (as of 27 July 2017) partitioned according to the three main elements of the plant disease triangle.

"Landscape-scale disturbances such as wind, fire, or land use can (i) modify the impacts of a disease, or (ii) can be influenced by disease in a manner which increases or decreases the ecological impacts of these disturbances" [7]. The above quote clearly summarizes the two-way interconnectivity between heterogeneity at the landscape level and diseases. However, it is not only the heterogeneity of the landscape that drives the epidemiology and the final outcome of forest diseases. In fact, variability in the environment, as well as in host and pathogen populations, can also have a profound impact on the spread and impacts of forest diseases. This special issue attempts to summarize some of the knowledge on this broad and novel aspect of forest pathology while providing some provoking case studies investigating several different aspects of this variability.

We have long known that hosts and pathogens are in a constant arms race through which resistance and virulence are in continuous evolution, and it has been repeatedly postulated that shorter generation times will accelerate the process [8,9]. In this issue, Soularue et al. [10] through a convincing model show that humans can also play a role in the co-evolutionary arms race, by shortening the rotation time of plantations and thus accelerating the evolution of virulence in the pathogen. This information is novel because it goes well beyond the known effects that humans can have on disease severity by altering ecosystems. Such more widely studied effects are also described in this special issue, for example, the dense monoculture of young trees and off-site plantings caused by the use of exotic species, leading to an increased susceptibility to both exotic and native emergent pathogens [11,12]. At the same time, emergent diseases are far from being in a stationary phase; climate change is currently affecting several pathosystems, especially where pathogens and/or hosts may be at the fringes of their natural or naturalized range [13]. For instance, Kubiak et al. [14] point out that increasing temperatures will allow the root rot pathogens *Armillaria* spp. to grow all year round and to decay wood more effectively, thus significantly enhancing their spread rate and pathogenicity.

Likewise, the changing climate may negatively affect the physiology of native trees, making them more attractive to insects and more susceptible to diseases caused by endophytic fungi that have turned into pathogens [15]. In addition, in such a predicament, contagion may be further compounded by the fact that the rates of vectoring of such fungi by insects also increase, as a result of a greater frequency of encounters between fungi and insects in weakened plants [15].

This is a new world we are living in: a globalized world, a changing world, and a world that requires new approaches to maximize the return of scientific investigation. The application of landscape ecology approaches [7] is greatly enhancing our insights into non-linear processes [16]. Likewise, the use of crowdsourced data provides an opportunity to generate datasets of an inconceivable scale until recently. Lione et al. [17] used crowdsourced data to uncover the non-linear progression of emergent diseases. In this paper, the authors identify precise environmental and topographic conditions that result in a reversion of infection status (from positive to negative) for the invasive and destructive forest disease Sudden Oak Death in California.

Often, even when the variability of the landscape, environment, and main ecological parameters are all embedded in our research, it is arduous to include variability of the host response in our studies. Most of the literature in this area in fact has focused on gene-for-gene resistance, but the effects of other types of resistance have been less widely studied [18]. The paper on the Eucalyptus rust *Austropuccinia psidii* [19] provides solid evidence of a phenological type of resistance present in older leaves, while even more complex is the report by Chieppa et al. [20] suggesting that genetically inherited susceptibility to a vascular fungus results in great susceptibility to changes in water availability. While interspecific competition driven by pathogens is well known [21], the example by Chieppa et al. [20] is a classic example of virtually unstudied intraspecific competition cryptically driven by a pathogen.

Even if this special issue purposefully does not focus primarily on diseases caused by an exotic organism, it would be impossible not to include this topic in an issue on the effect of variability on the epidemiology and impacts of forest diseases. Variability certainly is the issue when studying most exotic pathosystems; in fact, how can exotic pathogens be so successful in novel environments despite their limited genetic variability? There is no a single answer to this question, and it is our belief that simply invoking a lack of coevolution does a disservice to our learning of the complex mechanisms driving the invasion by exotic pathogens. In this issue, several hints are provided regarding factors other than the lack of coevolution to explain successful invasions by pathogens. In the case of cypress canker, the use of the artificially created hybrid Leyland cypress has significantly increased the severity of outbreaks, even where the causal agent is native (e.g., in California) [22]. In the case of laurel wilt, the high susceptibility of cultivated avocados has accelerated the spread of the disease in natural ecosystems [23]. On top of this, and unfortunately, the exotic laurel wilt pathogen, introduced in conjunction with the introduction of an exotic ambrosia beetle, has been picked up by multiple native beetles, thus immediately broadening its host range. Finally, exotic diseases can often emerge because humans have provided an abundance of exotic hosts. Such hosts, being exotic, generally can be regarded as planted off-site. This is the case for *Austropuccinia psidii* on Eucalyptus planted in South America [19], for *Lecanosticta acicola* on *Pinus radiata* grown in Spain [12], and likely for the many cultivated hosts of *Lasiodiplodia theobromae* [24].

This special issue contains 13 articles that we hope will be thought-provoking in more than a single way. The articles include widely different approaches, scales, and technical methodologies, and they well represent the cutting edge of contemporary forest pathology. The expectation of this special issue was to represent a range of approaches currently employed to study variability in tree diseases. We hope the reader will agree that this expectation has been met, and we hope he/she will concur that in the process of compiling this issue, we may have put together an excellent textbook for an advanced class in forest pathology.

Acknowledgments: The authors of this paper and co-editors of the special issue "Forest Pathology and Plant Health" are grateful to all the authors and reviewers of the special issue.

Author Contributions: M.G. and P.G. contributed equally to this article.

Conflicts of Interest: The authors declare no conflict of interest. The founding sponsors had no role in the design of the study; in the collection, analyses, or interpretation of data; in the writing of the manuscript; or in the decision to publish the results.

References

1. Agrios, G. *Plant Pathology*, 5th ed.; Academic Press: New York, NY, USA, 2005.
2. Yun, L.; Zeger, S.L. On the equivalence of case-crossover and time series methods in environmental epidemiology. *Biostatistics* **2006**, *8*, 337–344.
3. Roche, B.; Guégan, J.-F.; Bousquet, F. Multi-agent systems in epidemiology: A first step for computational biology in the study of vector-borne disease transmission. *BMC Bioinform.* **2008**, *9*, 435. [CrossRef] [PubMed]
4. Desprez-Loustau, M.-L.; Aguayo, J.; Dutech, C.; Hayden, K.J.; Husson, C.; Jakushkin, B.; Marçais, B.; Piou, D.; Robin, C.; Vacher, C. An evolutionary ecology perspective to address forest pathology challenges of today and tomorrow. *Ann. For. Sci.* **2016**, *73*, 45–67. [CrossRef]
5. Pautasso, M.; Doring, T.; Garbelotto, M.; Pellis, L.; Jeger, M. Impacts of climate change on plant diseases—Opinions and trends. *Eur. J. Plant Pathol.* **2012**, *133*, 295–313. [CrossRef]
6. Scholthof, K.G. The disease triangle: Pathogens, the environment and society. *Nat. Rev. Microbiol.* **2007**, *5*, 152–156. [CrossRef] [PubMed]
7. Cobb, R.C.; Metz, M.R. Tree diseases as a cause and consequence of interacting forest disturbances. *Forests* **2017**, *8*, 147. [CrossRef]
8. Parker, I.M.; Gilbert, G.S. The evolutionary ecology of novel plant-pathogen interactions. *Annu. Rev. Ecol. Evol. Syst.* **2004**, *35*, 675–700. [CrossRef]
9. Oliva, J.; Boberg, J.; Hopkins, A.J.M.; Stenlid, J. Concepts of epidemiology of forest diseases. In *Infectious Forest Diseases*; Gonthier, P., Nicolotti, G., Eds.; CABI: Wallingford, UK, 2013; pp. 1–28.
10. Soularue, J.-P.; Robin, C.; Desprez-Loustau, M.-L.; Dutech, C. Short rotations in forest plantations accelerate virulence evolution in root-rot pathogenic fungi. *Forests* **2017**, *8*, 205. [CrossRef]
11. Munck, I.A.; Luther, T.; Wyka, S.; Keirstead, D.; McCracken, K.; Ostrofsky, W.; Searles, W.; Lombard, K.; Weimer, J.; Allen, B. Soil and stocking effects on Caliciopsis canker of *Pinus strobus* L. *Forests* **2016**, *7*, 269. [CrossRef]
12. Ortíz de Urbina, E.; Mesanza, N.; Aragonés, A.; Raposo, R.; Elvira-Recuenco, M.; Boqué, R.; Patten, C.; Aitken, J.; Iturritxa, E. Emerging needle blight diseases in Atlantic *Pinus* ecosystems of Spain. *Forests* **2017**, *8*, 18. [CrossRef]
13. Thomas, C.D. Climate, climate change and range boundaries. *Div. Distr.* **2010**, *16*, 488–495. [CrossRef]
14. Kubiak, K.; Żółciak, A.; Damszel, M.; Lech, P.; Sierota, Z. *Armillaria* pathogenesis under climate changes. *Forests* **2017**, *8*, 100. [CrossRef]
15. Panzavolta, T.; Panichi, A.; Bracalini, M.; Croci, F.; Ginetti, B.; Ragazzi, A.; Tiberi, R.; Moricca, S. Dispersal and propagule pressure of *Botryosphaeriaceae* species in a declining oak stand is affected by insect vectors. *Forests* **2017**, *8*, 228. [CrossRef]
16. Holdenrieder, O.; Pautasso, M.; Weisberg, P.J.; Lonsdale, D. Tree diseases and landscape processes: The challenge of landscape pathology. *Trends Ecol. Evol.* **2004**, *19*, 446–452.
17. Lione, G.; Gonthier, P.; Garbelotto, M. Environmental factors driving the recovery of bay laurels from *Phytophthora ramorum* infections: An application of numerical ecology to citizen science. *Forests* **2017**, *8*, 293. [CrossRef]
18. Laine, A.L.; Burdon, J.J.; Dodds, P.N.; Thrall, P.H. Spatial variation in disease resistance: From molecules to metapopulations. *J. Ecol.* **2011**, *99*, 96–112. [CrossRef] [PubMed]
19. Ruiz Silva, R.; Costa da Silva, A.; Antônio Rodella, R.; Eduardo Serrão, J.; Cola Zanuncio, J.; Luiz Furtado, E. Pre-infection stages of *Austropuccinia psidii* in the epidermis of eucalyptus hybrid leaves with different resistance levels. *Forests* **2017**, *8*, 362. [CrossRef]
20. Chieppa, J.; Eckhardt, L.; Chappelka, A. Simulated summer rainfall variability effects on loblolly pine (*Pinus taeda*) seedling physiology and susceptibility to root-infecting ophiostomatoid fungi. *Forests* **2017**, *8*, 104. [CrossRef]
21. Gilbert, G.S. Evolutionary ecology of plant diseases in natural ecosystems. *Annu. Rev. Phytopathol.* **2002**, *40*, 13–43. [CrossRef] [PubMed]
22. Danti, R.; Della Rocca, G. Epidemiological history of Cypress Canker Disease in source and invasion sites. *Forests* **2017**, *8*, 121. [CrossRef]

23. Ploetz, R.C.; Kendra, P.E.; Choudhury, R.A.; Rollins, J.A.; Campbell, A.; Garrett, K.; Hughes, M.; Dreaden, T. Laurel wilt in natural and agricultural ecosystems: understanding the drivers and scales of complex pathosystems. *Forests* **2017**, *8*, 48. [CrossRef]
24. Mehl, J.; Wingfield, M.J.; Roux, J.; Slippers, B. Invasive everywhere? Phylogeographic analysis of the globally distributed tree pathogen *Lasiodiplodia theobromae*. *Forests* **2017**, *8*, 145. [CrossRef]

forests

MDPI

Review

Tree Diseases as a Cause and Consequence of Interacting Forest Disturbances

Richard C. Cobb [1,*] and Margaret R. Metz [2]

[1]   Department of Plant Pathology, University of California, Davis, One Shields Ave, Davis, CA 95616, USA
[2]   Department of Biology, Lewis & Clark College, 0615 S.W. Palatine Hill Road MSC 53,
     Portland, OR 97219, USA; mmetz@lclark.edu
*   Correspondence: rccobb@ucdavis.edu; Tel.: +1-530-754-9894

Academic Editors: Matteo Garbelotto and Paolo Gonthier
Received: 15 March 2017; Accepted: 27 April 2017; Published: 28 April 2017

Abstract: The disease triangle is a basic and highly flexible tool used extensively in forest pathology. By linking host, pathogen, and environmental factors, the model provides etiological insights into disease emergence. Landscape ecology, as a field, focuses on spatially heterogeneous environments and is most often employed to understand the dynamics of relatively large areas such as those including multiple ecosystems (a landscape) or regions (multiple landscapes). Landscape ecology is increasingly focused on the role of co-occurring, overlapping, or interacting disturbances in shaping spatial heterogeneity as well as understanding how disturbance interactions mediate ecological impacts. Forest diseases can result in severe landscape-level mortality which could influence a range of other landscape-level disturbances including fire, wind impacts, and land use among others. However, apart from a few important exceptions, these disturbance-disease interactions are not well studied. We unite aspects of forest pathology with landscape ecology by applying the disease-triangle approach from the perspective of a spatially heterogeneous environment. At the landscape-scale, disturbances such as fire, insect outbreak, wind, and other events can be components of the environmental 'arm' of the disease triangle, meaning that a rich base of forest pathology can be leveraged to understand how disturbances are likely to impact diseases. Reciprocal interactions between disease and disturbance are poorly studied but landscape ecology has developed tools that can identify how they affect the dynamics of ecosystems and landscapes.

Keywords: forest pathogens; disease triangle; landscape ecology; disturbance interactions; fire; insect outbreak

1. Introduction

Forest diseases occur in an intricate environmental context that is a reflection of long host lifespan and fixed location. In contrast to other foci of disease ecology, environmental influences on forest health have been recognized as important factors influencing disease in individual trees since the emergence of forest pathology as a topic of scientific inquiry [1]. Environmental factors act on pathogens and tree hosts independently and, as a consequence, disease incidence may increase or decrease due to environmental variability. This environmental variation can include year-to-year climate variation (temperature, precipitation, humidity), environmental pollutants (ozone, acid deposition), and a range of biophysical or biotic disturbances such as fire, wind, herbivory or insect outbreak [2,3]. A substantial body of research has developed to elucidate environmental influences on both pathogens and their tree hosts, encompassing a diversity of environmental stresses and effects at scales ranging from cellular to landscape [4,5]. Integrating dynamics of both host and pathogen simultaneously with environmental effects on disease remains challenging, and yet is essential to understanding how forest diseases are likely to change in the future [6].

The disease triangle is a scale-free, flexible, and general model that has been used extensively to determine the etiology of many plant diseases and, more recently, zoonotic diseases. The model is a visualization of host, pathogen, and environment in a tripartite dynamic interaction framework that has been especially useful when applied at the individual-to-stand level. However, this application is in some contrast to the scale of the most problematic disease outbreaks which tend to occur across landscapes or regions [7–9].

The field of landscape ecology focuses on identifying mechanistic processes that lead to landscape-level patterns and the implications of cross-scale interactions in driving these patterns. Scale mismatch among mechanisms that underlie an emergent condition—including disease—is not a problem unique to forest pathology. Landscape ecology has developed understanding of the causes, consequences, and patterns of ecological changes that occur over large spatial scales such as plant community shifts resulting from fire, biotic agents including insects and disease, or land-use patterns [10–12]. Landscape ecology applies to a range of spatial extents, but is often employed to understand changes in collections of heterogeneous ecosystems (a landscape) or regions (a collection of landscapes). The field has developed a range of spatial analysis tools that pair with remote sensing technologies; these efforts have improved understanding of the causes and consequences of spatial patterns as well as increasingly accurate resource inventories.

Forest health researchers have applied landscape ecology approaches to gain insight into disease drivers acting at the spatial extent of a landscape or region [13,14]. These efforts have produced maps of spatiotemporal patterns of disease risk with great management application [10,15]. These advances are also important in that they apply epidemiological theory to large spatial extents although, at the landscape extent, forest pathology has tended to examine disease in isolation from other dynamics and processes. At the same time, understanding ecological dynamics at landscape-to-regional scales is increasingly focused on the role of interactions among disturbances which alter spatial structure and variation [16,17]. We suggest that this is a potential nexus of the two fields that can improve understanding of landscape dynamics as well as the mechanistic basis of disease emergence at broad spatial extents.

Disease forecasting and prediction of the resulting ecological impacts have obvious value to managers and policy makers. Shaping landscape-level structure, such as fuel levels, age-class, or species distribution, demands substantial economic investments. Disturbance events, including but not limited to disease, can challenge these goals and incur great ecological or management costs. Increasing interest within landscape ecology has focused on the overlap and potential interactions of landscape-level disturbances. Global environmental change, including altered fire regimes, changes in regional climate and weather, as well as increased introduction of exotic organisms, are all potential drivers of disturbance interactions that could increase the ecological impacts of these events [16–18]. Landscape-scale models and empirical studies of disease impacts are relatively infrequent compared to fire and land use [16–18]. At the same time, cellular-to-population level studies of disease illustrate considerable variation in intensity and underlying causes that emerge from the components and interactions encompassing the disease triangle. These insights can be leveraged into predictive epidemiological models that account for host susceptibility and transmission, characteristics that are distinct biologically from risk or contagion as applied in models of fire or land-use change. Put another way, disease is an emergent phenomenon with inherent mechanistic differences from other landscape-scale disturbances. Yet diseases are likely to affect, and be affected by, other landscape-level disturbances in ecologically important ways.

This paper examines linkages between forest diseases and landscape-scale disturbances to understand how these distinct events may interact to affect each other. Landscape ecology and forest pathology are notable for strong research foundations that suggest disease–disturbance interactions are likely to shape ecosystems. Yet, very few empirical examples of these interactions have been undertaken in spite of a strong emphasis within both fields on the role of environmental factors. This suggests potential for improving prediction of disease and disturbance impacts through an assessment of the strengths and research needs in each field. Although pathogens and insects have important differences in how they affect tree health, we examine inferences gained through the study of insect outbreak–disturbance interactions to frame hypotheses of disease–disturbance dynamics from ecosystem-to-landscape scales. Empirical research linking epidemiological process and ecosystem or landscape-level impacts of forest disease is generally lacking, so we also point to several studies aimed at addressing this knowledge gap. Landscape ecology has built metrics for quantifying landscape pattern, processing of remotely sensed data, and models of disturbance dynamics that can be placed within traditional forest pathology approaches. This structure provides a basis for rapid integration of ecosystem-to-landscape impacts and dynamics of disease into landscape models and associated theory. We link these bodies of knowledge by identifying mechanisms of interaction that are likely to determine if or when disease–disturbance interactions are ecologically important.

2. The Disease Triangle: A Primer for Landscape Ecologists

The disease triangle (Figure 1) is a valuable heuristic tool for envisioning and testing drivers of disease emergence. The model has a long history both within forest pathology and plant pathology more broadly [1]. The model frames disease as an emergent condition resulting from interactions of a pathogen with a susceptible host in suitable environmental conditions. Changes to any of the components can accelerate or dampen disease dynamics, while also altering the other components of the triangle. As a general, parsimonious, and dynamic conceptual model, the disease triangle also requires a researcher or manager to contextualize the disease system, including any unique characteristics of the focal host or pathogen.

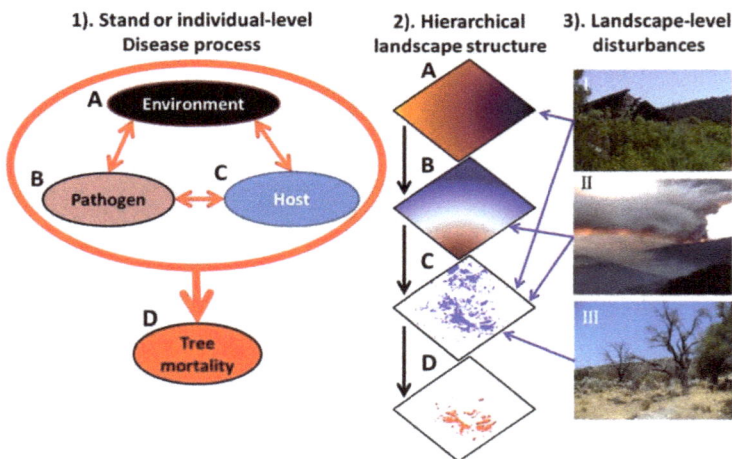

Figure 1. Linkages between forest pathology and landscape ecology with example landscape-level disturbances which can influence disease emergence and impacts. The traditional disease triangle (**1**) factors (A–C) are shown simultaneously in the spatial context of landscape ecology (**2**) along with landscape-level disturbances that are likely to interact with disease (**3**). Environment (1-A and 2-A) is shown on a gradient along with a waveform pathogen invasion process (1-B and 2-B) and realistic spatial heterogeneity of host distribution (1-C and 2-C) and mortality (1-D and 2-D). Examples of interactive disturbances are shown with arrows indicating the components of landscape structure and the disease triangle that are impacted: Invasive plants competitively inhibit forest host reestablishment following land abandonment with impacts to environmental conditions and host distribution (I—*Genista monspessulana* invasion of an old vineyard), fire can alter host and pathogen distribution (II—Soberanes Fire 2016, Big Sur—Photo credit K. Frangioso), and emergent insect outbreak-caused mortality of host populations (III—*Agrilus coxalis* mortality of coast live oak).

2.1. Pathogens

The major disease-causing pathogens of forest trees represent many branches on the tree of life, including fungi, oomycetes, or parasitic plants [3,19]. These pathogens can vary widely in their spatial distribution across a landscape. Some weak or facultative pathogens can be so widespread as to be practically ubiquitous, likely an evolutionary consequence of minimal impacts to host health or primarily saprotrophic energy acquisition [20,21]. Other more aggressive native pathogens such as *Heterobasidion* or *Dothistroma* species may be widespread within a region but only emerge as diseases following land use or climatic changes, respectively [22,23]. In contrast, diseases caused by invasive pathogens are often characterized by introduction foci that strongly control the spatial distribution of disease risk prior to equilibration of the pathogen in a new range. Introduced populations of *Phytophthora ramorum*, *P. cinnamomi*, or the beech bark disease insect-pathogen complex all share this strong signature of spatiotemporal variation in the distribution of disease risk [9,12,24].

2.2. Tree Hosts

Disease emergence or suppression may be driven by shifts in host presence or abundance. In forests, species abundance reflects a range of climate drivers, ecological processes, and cultural practices; several examples of documented or potential increases in disease with changes in host population are worth highlighting in light of these dynamics. Increased cultivation of *Hevea* species for rubber production is implicated in increased frequency and severity of South American leaf blight [19,25], a disease that rarely reaches inoculum levels needed for disease emergence under non-cultivated conditions. Increased tanoak importance in response to inadequate silvicultural investments has also been implicated in patterns of *P. ramorum* establishment risk and emergence of sudden oak death [26–28]. Afforestation in response to changes in land use or fire suppression has also been shown to increase *P. ramorum* hosts and subsequent patterns of disease emergence [29]. Vegetation changes in response to environmental and social or economic dynamics can also shift species abundances and spatial distributions with no significant changes in the represented taxa. For example, in northern forests of Quebec, human impacts reflecting changes in economic factors and silvicultural practices could influence future likelihood of insect outbreak [30]. Plant invasions can also alter host populations at broad spatial scales. In the southern hemisphere, several invasive pine species along with their mycorrhizal symbionts are aggressive invaders of native ecosystems. These invasions cause severe landscape-to-regional scale transformations of forest communities that also transform standing biomass, biogeochemical processes, and biodiversity [31,32].

Hosts also experience changes in susceptibility due to environmental responses at cellular-to-individual scales. Environmentally driven increases in host susceptibility to pathogens are commonly termed host "predisposition" to infection or disease. Tree hosts respond to physical or biotic stresses such as fire, drought, or insect and pathogen attack through a range of physical structures, chemical defenses, and biochemical pathways [4,5]. Thus, the physiological status of individual trees or entire populations can also determine the likelihood of disease emergence.

2.3. Environment

Within the context of the disease triangle, many environmental factors influence disease by limiting hosts or pathogen ranges and through direct influences on pathogens or their hosts. For example, temperature affects pathogen metabolic activity as well as host predisposition in response to environmental stress [33]. Global change-driven range expansions or contractions could make disease emergence more, or less, likely due to changes in the abundance or spatial distribution of either biological component [3,19,34]. Physical host wounds that act as a de facto surmounting of host physical defenses play an important role in infection [1,22,35,36]. Drought has been shown to cause physical damage to plant vascular tissues that may require multiple years to repair [37] and it is likely that other environmental stresses also cause multi-year changes in susceptibility.

Land management can be considered part of the environmental portion of the disease triangle, and is sometimes envisioned as a fourth component. This is warranted as land management policy can be more readily changed compared to an environmental event such as drought; policy and management can also affect environmental drivers of disease or landscape-level host distribution [3,38].

3. The Disease Triangle in the Context of Disturbance

In the absence of disease, disturbances play a critical role in structuring forest communities at local-to-landscape scales [17,30,39]. Ecological disturbances can alter any component of the disease triangle (Figure 1). This creates a dual context for disturbances: they are a component of the environmental portion of the model or also an external factor that can influence host composition or pathogen abundance. These effects in conjunction with environmental changes that determine host stresses or pathogens infection, can also create disturbance-feedbacks that will alter disease dynamics.

3.1. Environmental vs. Biotic Stress on Tree Hosts

An intricate system of plant hormone signaling has a strong influence on secondary chemical pathways and allocation of resources to repair physical damage from biotic and abiotic causes. Problematically for plant hosts, biochemical pathways that confer resistance to environmental stress can suppress those that confer resistance to insects and pathogens. The resulting hormonal interference of defensive pathways is known as plant hormonal cross-talk [4] and can be a significant factor leading to increased susceptibility (predisposition). Carbon- or nitrogen-based secondary chemistry can be important factors determining the degree of damage caused by insects or pathogens, but available carbon resources are also important for repairing tissue damage resulting from environmental stresses [5]. This suggests another resource dilemma for trees—whether to respond to environmental stress or biotic attack. Of course, physical damage, heat, and water stress may suppress defensive pathways for biotic attack via hormonal cross-talk [4] suggesting plant response to environmental factors may simply take precedence to biotic attack in some circumstances or species. It is likely that trees are sometimes caught in untenable "catch 22" traps where the host must respond to potentially overwhelming environmental or overwhelming biotic stresses and is likely to be overcome either way.

3.2. Example Disturbance Effects on Pathogens

Forest management and fire are especially important disturbances in forested landscapes that have been documented to influence disease dynamics in some circumstances. Each of these events can alter temperature and relative humidity which are particularly important controls on infection success for many important fungal pathogens (Figure 1). Tree harvesting can further alter stand characteristics that influence pathogen dispersal. For example, dwarf mistletoe dispersal is facilitated by increased canopy opening following thinning [40], and it is likely that increased air-flow and reduced tree density also increases the connectivity of susceptible hosts. Harvest and the resulting effects on substrate for pathogen survival, creation of infection pathways, and changes in host distribution or abundance have strong influences on disease emergence [14,20,22]. Increased input of below-ground organic matter, important for inoculum buildup of some root diseases, could occur following a variety of management actions [41,42]. Fire is somewhat unique in that it often causes substantial host and pathogen mortality [43] which should caution researchers against generalization without more examples. In contrast to fire impacts, fire suppression can lead to increased host density, which in turn may increase local inoculum build-up as well as alter microclimate in ways that influence pathogen establishment and the emergence of specific diseases [44].

Non-host plant invasions can also alter mutualistic host–microbial interactions, nutrient availability, and increase competition for other limiting resources such as light and water [45]. Most obviously, strong competitive interactions or competitive exclusion of host species by a novel invader could have powerful influences on disease emergence. However, this kind of shifting species dynamic is more likely when the invading plant amplifies pathogen populations and creates positive feedback between pathogen and invader [46,47]. Plant invasions can suppress or amplify pathogen populations and the relevance to disease will often depend on the interactions between a specific plant invader and pathogen.

3.3. N Deposition, an Example of Changing Environmental Conditions at Broad Spatial Extents

Environmental contamination has received less attention as environmental drivers of disease, but the spatial scales at which these events occur make them relevant to any discussion of disease–disturbance interactions. Atmospheric N deposition has potential to alter host populations by shifting species composition and dominance via changes in growth rates. N deposition has also been documented to increase pathogen attack [48,49] which may reflect mechanistic drivers from cellular-to-community scales. Studies of N-deposition on disease emergence are more common in understory or herbaceous species but whether this is due to an overall lack of study in overstory trees vs. greater susceptibility in these habitats is unclear. Increased N availability could alter a range of ecological interactions that influence tree host susceptibility such as phyllosphere host–microbial interactions and within-plant carbon allocation [33]. Species invasions, climate change, and other kinds of pollution could alter host or pathogen populations at broad spatial scales and are worth examining for disease-disturbance interactions.

3.4. Disease as Disturbance

Both native and exotic pathogens are substantial ecosystem disturbances that can cause a range of impacts with variable duration and intensities [50]. Root nematode pathogens, aerially dispersed *Phytophthora* pathogens, and native *Phellinus* root pathogens have each been shown to alter biogeochemical processes including nitrogen and carbon cycling [51–53]. Forest structural changes can be dramatic, and include shifts in size class distribution, selective removal of individuals, or changes in species composition [7,12,54,55]. The nature and extent of forest structural changes depend on specific biological characteristics of the pathogen such as the host range and virulence, as well as interactions with the biology of the tree host, including age- or size-related variation in susceptibility or host competency to transmit the pathogen. In the most extreme cases where host range is broad and many hosts suffer disease following infection, pathogens can drive a conversion from one ecosystem type to another. *Phytophthora cinnamomi* is an illustrative example, this pathogen causes the devastating disease Jarrah dieback which can convert woodland or forest to a low-statured heathland [9].

4. Disturbance Interactions: Perspectives from Landscape Ecology

Disturbance interactions are increasingly studied outside of the purview of forest pathology and the associated empirical and theoretical advances are relevant for bridging the fields (Table 1). Landscape ecology has made important advances in framing general theory and expectations regarding how a broad set of disturbances are likely to interact [16,17,56]. These efforts have incorporated differences in frequency, spatial pattern, and severity of physical and biological disturbances with the aim of assessing the relevance to landscape structure and management. Disturbance frequency, patterns, and impacts lie at the core of inquiry in landscape ecology, which stems from the importance of these events in driving spatial patterns [56]. In light of this, it is unsurprising that many of the intellectual frontiers of disturbance interactions (including but not limited to disease) are also firmly rooted within landscape ecology [13,14]. Systems at the landscape spatial scale are so often subject to variation in environmental stresses and stochastic events—such as lightning ignition sources or individual land-use actions—that understanding overlapping disturbances and their interactions is fundamental to mechanistic understanding and prediction over large areas.

Table 1. Example studies linking environmental changes and disease emergence or the interactive effects of biotic agents and landscape-level disturbance dynamics.

Biological Agent	Interacting Disturbance	Comments	Examples
Landscape-level examples			
Native insects	Fire, wind, salvage harvest	Tested for interactive effects	[11,57,58]
Invasive pathogen	Fire	Tested for interactive effects	[18,43,59,60]
Root pathogens	Wind	Focused on environmental drivers of disease	[35,61]
Root pathogens	Management, fire suppression	Etiological investigation of landscape-level disease drivers	[22]
Individual-level examples			
Insects or pathogens	Drought, salt stress, and heat	Synthesis, laboratory experiments	[4]
Pathogens	Drought	Synthesis	[5]

Discrete and intense disturbances—such as insect or disease interactions with fire—are notable for the relative ease of assessing population-level changes to vegetation. In these examples, mortality from either disturbance is rapid and easily recognized. The relative tractability of these study systems further leads to empirical and model based studies [11,56,59]. However, slowly accumulating and large-extent disturbance such as climate change and atmospheric N deposition are much more nuanced and difficult to assess. These events will have strong influences on landscape-level processes including disturbance frequency, extent, and impact but quantifying these dynamics with an empirically-backed and mechanistic framework poses many practical challenges [56].

How landscape scale dynamics affect ecological resiliency is also important for gauging the relevance of disturbance interactions to land management policy. Post-disturbance vegetation recovery rates and successional dynamics have a strong control over ecological resilience to interacting disturbances, for example, by affecting the likelihood of future disturbance events and their intensities [16]. Policy makers and managers face the dilemma that disturbance interactions are known to occur, or are highly likely, yet it is unclear how much these events will challenge local and regional management goals. This problem may be acute when multiple possible successional trajectories emerge from the local species pool and variation in disturbance impacts [17,50]. Despite important uncertainties, enough information is in hand to provide a point-of-departure for quantifying disease–disturbance interactions and incorporating these insights into land management policy.

4.1. Insect–Fire Interactions

Interactions between fire and insect outbreak have received attention among researchers due to the frequent overlap and relative importance of each disturbance [11,57,58]. Although biologically distinct from forest diseases, insect outbreaks provide insights into disturbance interactions and can help formulate expectations given changes to canopy structure, species composition, and changes in biomass distribution. Insect-driven mortality events have been dramatic and cause for alarm among land managers in fire-prone conifer forests of western North America. Many of these forests have been impacted by drought and fire suppression, which have influenced insect population dynamics and host physiological status [62,63]. In this respect, insect outbreaks are a useful model for disease in that tree mortality is driven by environmental changes including climate dynamics and land management. Severe insect-caused mortality often results in a short-term increase in highly flammable canopy fuels followed by an increase in ground fuels as dead material moves from the canopy to the forest floor. Somewhat counter to initial expectations, these mortality patterns have led to an overall decrease in fire impacts as measured by the likelihood of crown-fire, which are often the most damaging and dangerous forest fire conditions, for both bark beetle [11,57] and spruce budworm [58]. In these studies, the timing of fire and insect outbreak events was more important than the spatial overlap of each disturbance.

4.2. Disease–Disturbance Interactions

As previously noted, etiological studies have dominated most applications of the disease-triangle in forest pathology. For example, wind pressure on tree canopies can transfer a substantial amount of tension and compression force to root systems. The resulting wind-caused damage to fine roots is an important pathway of infection and subsequent creation of disease centers in balsam fir-spruce forests of the White Mountains, New Hampshire [61]. Historical management actions including fire suppression and cutting to control bark beetle outbreak have been demonstrated to influence patterns of *Heterobasidion* root disease centers in Yosemite Valley, California [22]. Historical patterns of conifer harvest have also likely influenced the landscape-level distribution of invasion risk for *P. ramorum* in coastal California and Oregon [24]. However, for each of these example systems, there has been little effort to inform prediction of future resource status, disturbance patterns, and rates (or capacity) of ecosystem recovery [17,56]. For example, wind-driven root infection decreases tree capacity to withstand wind events. Thus, wind can facilitate disease which then renders stands or ecosystems more vulnerable to wind-driven mortality [61]. Problematically, temporally delayed disease–disturbance interactions could be easily masked and result in misattribution of underlying mortality causes [50].

The few published studies focused on disease–disturbance interactions provide some insights that may help overcome this research challenge. The authors of this article along with colleagues conducted a series of studies examining interactive impacts of *P. ramorum* and fire in coastal California forests. Using a set of permanent study plots in disease-impacted forests that were surveyed before and after the 2008 Basin Fire, this work showed that fire impacts are contingent on the stages of disease progression, supporting comparisons to insect–fire interactions (compare with [57]). In our plots, fire intensity was positively related to the amount of fine canopy fuels associated with recently killed trees and decreased as these materials move from the canopy to the soil surface [18]. In contrast, fire-caused mortality of redwood, a species that is resilient to *P. ramorum* and typically also resilient to fire, increased with greater accumulation of ground fuels [59]. This pattern suggested that disease-generated fuels accumulated to the point that damage to tree root systems or cambium tissue was substantially increased. Ground fuel accumulation also impacted soil resources, losses of critical soil nutrients and carbon increased with increasing disease-related mortality [60], changes that may in turn alter post-fire vegetation succession. Although these studies are associated with an individual disease and fire (i.e., a single example), the mechanistic link between mortality, fuels, and ecosystem impacts suggests these disease–disturbance interactions could occur where other disease and fire events overlap.

5. Frontiers in the Study of Disease–Disturbance Interactions

Approaching a new study system or challenge with the most parsimonious model is often a recipe for rapid advancement. With some notable exceptions [11,17,56], most interactive disturbance studies have focused on two events. In the sudden oak death system, our sole explicit test of disease–disturbance interactions, we studied a single forest disease and a single fire. This study system spans multiple forest types which provides insight into variation in disease impacts, the importance of pre-disease species composition and structure [18,43,59,60] but ultimately provides only a partial view on the importance of interactive impacts. For example, our Big Sur study region is a fire-prone landscape that burned with a landscape-level wildfire only eight years following our initial study (the 2016 Soberanes Fire). How do repeated or subsequent disturbances modify or reflect previous disease–disturbance interactions? In the case of Big Sur and sudden oak death, *P. ramorum* was found to rapidly reinvade following the initial 2008 Basin wildfire [43], presumably impacts from the recent 2016 Soberanes wildfire reflect both the previously documented disease-fire interactions as well as an additional eight years of disease that occurred between wildfires. In Yosemite Valley, *Heterobasidion* disease centers are a serious public safety concern as several deaths have been associated with failures of infected trees [22]. This has motivated the aggressive removal of at risk trees, particularly in high-recreation use areas, as well as the greater application of prescribed burning in other parts of the valley. Yet, it is unclear how these management actions have influenced disease dynamics and furthered public safety beyond the immediate danger of tree failures. In both cases, disease–disturbance interactions could be reasonably argued to have increased or decreased ecological impacts under specific edaphic, vegetation, or management conditions.

Third and fourth-order disease–disturbance interactions remain minimally studied although these are certain to represent future management challenges. Fire-prone regions where multiple fires and disease events occur over time could be shaped by these factors. Paine et al. [56] and Johnstone et al. [17] suggested several general ecosystem trajectories resulting from disturbance interactions that tend to focus on transitions between stable ecosystem states or temporary states associated with recovery (Figure 2; examples 1 and 2). We recognize that these expectations may also fit disease–disturbance interactions. For example, disease-center formation in the White Mountains of New Hampshire resulting from wind impacts is a temporary ecosystem state [35,61]. Similarly, it has not yet been shown that sudden oak death–fire interactions will lead to long-term and permanent shifts in forest structure although it is plausible that increased tree mortality and soil nutrient loss will result in these changes. However, biotic disturbances are strongly influenced by the coincidence of environmental conditions favorable to establishment or outbreak suggesting that dynamic ecosystems should be expected in some circumstances (Figure 2; example 3). Oscillating dynamics where host populations recover and provide the basis for future outbreaks are a common pattern in many tree mortality events driven by native organisms [35,51,62]. Intensifying outbreaks resulting from climate-change-associated shifts in temperature, precipitation, and resulting host-stress appears likely given documentation of landscape-level declines in host physiological status and its impact to plant defense [3,4,37].

Hysteresis is a common challenge for much of ecology and the importance as well as the difficulty of capturing these effects is increased in studies at the landscape scale. The difficultly of scaling-up of results from the plot or stand scale where most ecological inquiry is made to the landscape scale is a common challenge in landscape ecology including the study of interactive disturbances [14,16]. An especially complex factor is that as the landscape is changed, the dynamics within the unit of study may become more dependent on processes and structure external to the measurement unit, often the plot or stand (see Meentemeyer et al. [14]). These changes may require many years to emerge as well as quantify, suggesting that surprises, often problematic to management goals, are likely to emerge without the tools to make judicious forecasts of disease–disturbance impacts. Several lines of inquiry can help prevent these problems: (1) increased efforts to disentangle three-way interactions that lead to disease at broad spatial scales. Disease-triangle applications are biased towards etiological descriptions of disease, yet disease emergence could be a secondary or contributing factor influencing tree mortality and disturbance interactions [19,37]. Understanding if disease is a cause (primary agent) or a symptom (secondary agent) of a forest health problem is essential to any response. (2) Increase focus on multi-pathogen systems. Tree mortality from one event, such as the invasion of an exotic pathogen is likely to alter the conditions for the emergence of a second pathogen. For example, native Armillaria pathogens are widespread in forests at risk to *P. ramorum* invasion and the substantial addition of buried coarse roots is likely to increase local Armillaria biomass and associated disease [8,20]. (3) Increase the scope of disease–disturbance studies to pathogens with different epidemiological dynamics, plant parts which are attacked, and impacts to host health. At present, studies of fire–insect outbreak has examined insects which attack both the bole as well as the canopy of hosts [57,58]. A broad range of pathogens attack roots and leaves which may alter the timing and duration of dead fine fuels in the canopy; previous work has shown this to be a critical factor influencing disease–fire relationships [18]. Contrasting these classes of pathogens with bole-canker pathogens, vascular wilts, and a range of other biological differences among pathogens [5] could help link pathogenicity with effects on other disturbances.

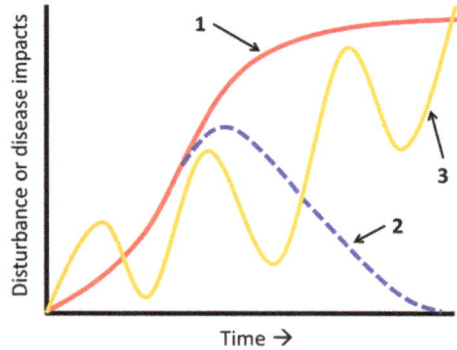

Figure 2. Several examples of disturbance and disease interactive outcomes over time. Impacts may increase and reach a new stable state (1—red); increases may gradually recover to pre-outbreak levels (2—blue dashed); or feedbacks between disease and disturbance could create fluctuating dynamics over time, here depicted as an oscillating, but increasing impact (3—yellow).

6. Conclusions

Landscape-scale disturbances such as wind, fire, or land use can (i) modify the impacts of a disease [22,61]; or (ii) can be influenced by disease in a manner which increases or decreases the ecological impacts of these disturbances [59]. In both cases, the potential to magnify the ecological impacts of either disturbance can have important management implications [16,17], but the mechanistic processes driving these impacts are quite different. The disease triangle is a valuable model for improving understanding of disease–disturbance interactions but to date, the approach has been primarily employed to study environmental influences on disease emergence. This body of work provides the mechanistic foundation for landscape ecologists to model disease emergence at broad spatial scales along with providing the basis to address associated management goals and challenges. However, the influence of disease on other disturbance processes is less well studied, and although the disease triangle is fully reciprocal, little effort has focused on disturbance changes stemming from ecosystem-to-landscape level impacts of disease. A nexus of interest and research efforts is developing between the fields [13,14]; forest pathology can provide epidemiological foundations unique to understanding disease emergence while landscape ecology can provide broad spatial extent models and empirical assessment of ecosystem and landscape dynamics. Of course, more data on ecological impacts of disease and collaboration between forest pathologists and landscape ecologists is needed to develop and test predictions of disease–disturbance interactions that will inform management decisions. However, each field is primed to take advantage of knowledge gained from the other and determine when, if, or how disease–disturbance interactions can be incorporated into land management policy and actions.

Acknowledgments: This work was funded by NSF Grant DEB EF-0622770 as part of the joint NSF-NIH Ecology of Infectious Disease program, the Gordon and Betty Moore Foundation, Cal Fire Green House Gas Reduction Fund, and the USDA Forest Service Forest Health Protection and Pacific Southwest Research Station. The authors thank David Rizzo and two anonymous reviewers for helpful comments on earlier versions of this manuscript.

Author Contributions: R.C.C. and M.R.M. conceived the research, conducted the analysis, and wrote the paper.

Conflicts of Interest: The authors declare no conflict of interest.

References

1.   Hartig, R. *Text-Book of the Diseases of Trees*; Macmillan and Company: London, UK, 1894.

2. Desprez-Loustau, M.-L.; Marçais, B.; Nageleisen, L.-M.; Piou, D.; Vannini, A. Interactive effects of drought and pathogens in forest trees. *Ann. For. Sci.* **2006**, *63*, 597–612. [CrossRef]

3. Pautasso, M.; Schlegel, M.; Holdenrieder, O. Forest health in a changing world. *Microb. Ecol.* **2014**, *69*, 826–842. [CrossRef] [PubMed]

4. Bostock, R.M.; Pye, M.F.; Roubtsova, T.V. Predisposition in plant disease: Exploiting the nexus in abiotic and biotic stress perception and response. *Annu. Rev. Phytopathol.* **2014**, *52*, 517–549. [CrossRef] [PubMed]

5. Oliva, J.; Stenlid, J.; Martínez-Vilalta, J. The effect of fungal pathogens on the water and carbon economy of trees: Implications for drought-induced mortality. *New Phytol.* **2014**, *203*, 1028–1035. [CrossRef] [PubMed]

6. Stenlid, J.; Oliva, J. Phenotypic interactions between tree hosts and invasive forest pathogens in the light of globalization and climate change. *Philos. Trans. R. Soc. B* **2016**, *371*, 20150455. [CrossRef] [PubMed]

7. Paillet, F.L. Chestnut: History and ecology of a transformed species. *J. Biogeogr.* **2003**, *29*, 1517–1530. [CrossRef]

8. Rizzo, D.M.; Garbelotto, M.; Hansen, E.M. Phytophthora ramorum: Integrative research and management of an emerging pathogen in California and Oregon forests. *Annu. Rev. Phytopathol.* **2005**, *43*, 309–335. [CrossRef] [PubMed]

9. Shearer, B.L.; Crane, C.E.; Barrett, S.; Cochrane, A. Phytophthora cinnamomi invasion, a major threatening process to conservation of flora diversity in the South-west Botanical Province of Western Australia. *Aust. J. Bot.* **2007**, *55*, 225–238. [CrossRef]

10. Orwig, D.A.; Foster, D.R.; Mausel, D.L. Landscape patterns of hemlock decline in New England due to the introduced hemlock woolly adelgid. *J. Biogeogr.* **2002**, *29*, 1475–1487. [CrossRef]

11. Kulakowski, D.; Veblen, T.T. Effect of prior disturbances on the extent and severity of wildfire in Colorado subalpine forests. *Ecology* **2007**, *88*, 759–769. [CrossRef] [PubMed]

12. Garnas, J.R.; Ayres, M.P.; Liebhold, A.M.; Evans, C. Subcontinental impacts of an invasive tree disease on forest structure and dynamics. *J. Ecol.* **2011**, *99*, 532–541. [CrossRef]

13. Holdenrieder, O.; Pautasso, M.; Weisberg, P.J.; Lonsdale, D. Tree diseases and landscape processes: The challenge of landscape pathology. *Trends Ecol. Evol.* **2004**, *19*, 446–452. [CrossRef] [PubMed]

14. Meentemeyer, R.K.; Haas, S.E.; Václavík, T. Landscape epidemiology of emerging infectious diseases in natural and human-altered ecosystems. *Annu. Rev. Phytopathol.* **2012**, *50*, 379–402. [CrossRef] [PubMed]

15. Cunniffe, N.J.; Gilligan, C.A. A theoretical framework for biological control of soil-borne plant pathogens: Identifying effective strategies. *J. Theor. Biol.* **2011**, *278*, 32–43. [CrossRef] [PubMed]

16. Buma, B. Disturbance interactions: Characterization, prediction, and the potential for cascading effects. *Ecosphere* **2015**, *6*, 70. [CrossRef]

17. Johnstone, J.F.; Allen, C.D.; Franklin, J.F.; Frelich, L.E.; Harvey, B.J.; Higuera, P.E.; Mack, M.C.; Meentemeyer, R.K.; Metz, M.R.; Perry, G.L.; et al. Changing disturbance regimes, ecological memory, and forest resilience. *Front. Ecol. Environ.* **2016**, *14*, 369–378. [CrossRef]

18. Metz, M.R.; Frangioso, K.M.; Meentemeyer, R.K.; Rizzo, D.M. Interacting disturbances: Wildfire severity affected by stage of forest disease invasion. *Ecol. Appl.* **2011**, *21*, 313–320. [CrossRef] [PubMed]

19. Desprez-Loustau, M.-L.; Aguayo, J.; Dutech, C.; Hayden, K.J.; Husson, C.; Jakushkin, B.; Marçais, B.; Piou, D.; Robin, C.; Vacher, C. An evolutionary ecology perspective to address forest pathology challenges of today and tomorrow. *Ann. For. Sci.* **2016**, *73*, 45–67. [CrossRef]

20. Baumgartner, K.; Rizzo, D.M. Ecology of *Armillaria* spp. in Mixed-Hardwood Forests of California. *Plant Dis.* **2001**, *85*, 947–951. [CrossRef]

21. Stergiopoulos, I.; Gordon, T.R. Cryptic fungal infections: The hidden agenda of plant pathogens. *Plant-Microbe Interact.* **2014**, *5*, 506. [CrossRef] [PubMed]

22. Slaughter, G.; Rizzo, D. Past forest management promoted root disease in Yosemite Valley. *Calif. Agric.* **1999**, *53*, 17–24. [CrossRef]

23. Woods, A.J.; Martín-García, J.; Bulman, L.; Vasconcelos, M.W.; Boberg, J.; La Porta, N.; Peredo, H.; Vergara, G.; Ahumada, R.; Brown, A.; et al. Dothistroma needle blight, weather and possible climatic triggers for the disease's recent emergence. *For. Pathol.* **2016**, *46*, 443–452. [CrossRef]

24. Cunniffe, N.J.; Cobb, R.C.; Meentemeyer, R.K.; Rizzo, D.M.; Gilligan, C.A. Modeling when, where, and how to manage a forest epidemic, motivated by sudden oak death in California. *Proc. Natl. Acad. Sci. USA* **2016**, *113*, 5640–5645. [CrossRef] [PubMed]

25. Lieberei, R. South American Leaf Blight of the Rubber Tree (*Hevea* spp.): New Steps in Plant Domestication using Physiological Features and Molecular Markers. *Ann. Bot.* **2007**, *100*, 1125–1142. [CrossRef] [PubMed]

26. Bowcutt, F. Tanoak target: The rise and fall of herbicide use on a common native tree. *Environ. Hist.* **2011**, *16*, 197–225. [CrossRef]
27. Cobb, R.C.; Rizzo, D.M.; Hayden, K.J.; Garbelotto, M.; Filipe, J.A.N.; Gilligan, C.A.; Dillon, W.W.; Meentemeyer, R.K.; Valachovic, Y.S.; Goheen, E.; et al. Biodiversity Conservation in the Face of Dramatic Forest Disease: An Integrated Conservation Strategy for Tanoak (*Notholithocarpus densiflorus*) Threatened by Sudden Oak Death. *Madroño* **2013**, *60*, 151–164. [CrossRef]
28. Meentemeyer, R.K.; Cunniffe, N.J.; Cook, A.R.; Filipe, J.A.N.; Hunter, R.D.; Rizzo, D.M.; Gilligan, C.A. Epidemiological modeling of invasion in heterogeneous landscapes: Spread of sudden oak death in California (1990–2030). *Ecosphere* **2011**, *2*, 1–24. [CrossRef]
29. Meentemeyer, R.K.; Rank, N.E.; Anacker, B.L.; Rizzo, D.M.; Cushman, J.H. Influence of land-cover change on the spread of an invasive forest pathogen. *Ecol. Appl.* **2008**, *18*, 159–171. [CrossRef] [PubMed]
30. Danneyrolles, V.; Arseneault, D.; Bergeron, Y. Pre-industrial landscape composition patterns and post-industrial changes at the temperate–boreal forest interface in western Quebec, Canada. *J. Veg. Sci.* **2016**, *27*, 470–481. [CrossRef]
31. Nuñez, M.A.; Horton, T.R.; Simberloff, D. Lack of belowground mutualisms hinders Pinaceae invasions. *Ecology* **2009**, *90*, 2352–2359. [CrossRef] [PubMed]
32. Dickie, I.A.; John, S.G.M.; Yeates, G.W.; Morse, C.W.; Bonner, K.I.; Orwin, K.; Peltzer, D.A. Belowground legacies of Pinus contorta invasion and removal result in multiple mechanisms of invasional meltdown. *AoB Plants* **2014**, *6*, plu056. [CrossRef] [PubMed]
33. Burdon, J.J.; Thrall, P.H.; Ericson, L. The current and future dynamics of disease in plant communities. *Annu. Rev. Phytopathol.* **2006**, *44*, 19–39. [CrossRef] [PubMed]
34. Ghelardini, L.; Pepori, A.L.; Luchi, N.; Capretti, P.; Santini, A. Drivers of emerging fungal diseases of forest trees. *For. Ecol. Manag.* **2016**, *381*, 235–246. [CrossRef]
35. Rizzo, D.M.; Harrington, T.C. Root movement and root damage of red spruce and balsam fir on subalpine sites in the White Mountains, New Hampshire. *Can. J. For. Res.* **1988**, *18*, 991–1001. [CrossRef]
36. Etheridge, D.E.; Craig, H.M. Factors influencing infection and initiation of decay by the Indian paint fungus (*Echinodontiumtinctorium*) in western hemlock. *Can. J. For. Res.* **1976**, *6*, 299–318. [CrossRef]
37. Anderegg, W.R.L.; Plavcová, L.; Anderegg, L.D.L.; Hacke, U.G.; Berry, J.A.; Field, C.B. Drought's legacy: Multiyear hydraulic deterioration underlies widespread aspen forest die-off and portends increased future risk. *Glob. Chang. Biol.* **2013**, *19*, 1188–1196. [CrossRef] [PubMed]
38. Dillon, W.W.; Meentemeyer, R.K.; Vogler, J.B.; Cobb, R.C.; Metz, M.R.; Rizzo, D.M. Range-wide threats to a foundation tree species from disturbance interactions. *Madroño* **2013**, *60*, 139–150. [CrossRef]
39. Thompson, J.R.; Carpenter, D.N.; Cogbill, C.V.; Foster, D.R. Four centuries of change in Northeastern United States forests. *PLoS ONE* **2013**, *8*, e72540. [CrossRef] [PubMed]
40. Mehl, H.K.; Mori, S.R.; Frankel, S.J.; Rizzo, D.M. Mortality and growth of dwarf mistletoe-infected red and white fir and the efficacy of thinning for reducing associated losses. *For. Pathol.* **2013**, *43*, 193–203. [CrossRef]
41. Garbelotto, M.; Gonthier, P. Biology, epidemiology, and control of heterobasidion species worldwide. *Annu. Rev. Phytopathol.* **2013**, *51*, 39–59. [CrossRef] [PubMed]
42. Rizzo, D.M.; Whiting, E.C.; Elkins, R.B. Spatial distribution of armillaria mellea in pear orchards. *Plant Dis.* **1998**, *82*, 1226–1231. [CrossRef]
43. Beh, M.M.; Metz, M.R.; Frangioso, K.M.; Rizzo, D.M. The key host for an invasive forest pathogen also facilitates the pathogen's survival of wildfire in California forests. *New Phytol.* **2012**, *196*, 1145–1154. [CrossRef] [PubMed]
44. Hawkins, A.E.; Henkel, T.W. Native forest pathogens facilitate persistence of Douglas-fir in old-growth forests of Northwestern California. *Can. J. For. Res.* **2011**, *41*, 1256–1266. [CrossRef]
45. Ehrenfeld, J.G. Ecosystem Consequences of Biological Invasions. *Annu. Rev. Ecol. Evol. Syst.* **2010**, *41*, 59–80. [CrossRef]
46. Cobb, R.C.; Meentemeyer, R.K.; Rizzo, D.M. Apparent competition in canopy trees determined by pathogen transmission rather than susceptibility. *Ecology* **2010**, *91*, 327–333. [CrossRef] [PubMed]
47. Holt, R.D.; Dobson, A.P.; Begon, M.; Bowers, R.G.; Schauber, E.M. Parasite establishment in host communities. *Ecol. Lett.* **2003**, *6*, 837–842. [CrossRef]
48. Mitchell, C.E.; Reich, P.B.; Tilman, D.; Groth, J.V. Effects of elevated CO_2, nitrogen deposition, and decreased species diversity on foliar fungal plant disease. *Glob. Chang. Biol.* **2003**, *9*, 438–451. [CrossRef]

49. Strengbom, J.; Englund, G.; Ericson, L. Experimental scale and precipitation modify effects of nitrogen addition on a plant pathogen. *J. Ecol.* **2006**, *94*, 227–233. [CrossRef]
50. Eviner, V.T.; Likens, G.E. Effects of pathogens on terrestrial ecosystem function. In *Infectious Disease Ecology. Effects of Ecosystems on Disease and Disease on Ecosystems*; Ostfeld, R.S., Keesing, F., Eviner, V.T., Eds.; Princeton University Press: Princeton, NJ, USA, 2008; pp. 260–283.
51. Matson, P.A.; Boone, R.D. Natural Disturbance and Nitrogen Mineralization: Wave-Form Dieback of Mountain Hemlock in the Oregon Cascades. *Ecology* **1984**, *65*, 1511–1516. [CrossRef]
52. Hobara, S.; Tokuchi, N.; Ohte, N.; Koba, K.; Katsuyama, M.; Kim, S.J.; Nakanishi, A. Mechanism of nitrate loss from a forested catchment following a small-scale, natural disturbance. *Can. J. For. Res.* **2001**, *31*, 1326–1335. [CrossRef]
53. Cobb, R.C.; Eviner, V.T.; Rizzo, D.M. Mortality and community changes drive sudden oak death impacts on litterfall and soil nitrogen cycling. *New Phytol.* **2013**, *200*, 422–431. [CrossRef] [PubMed]
54. Hansen, E.M.; Goheen, E.M. Phellinus weirii and other native root pathogens as determinants of forest structure and process in western North America. *Annu. Rev. Phytopathol.* **2000**, *38*, 515–539. [CrossRef] [PubMed]
55. Metz, M.R.; Frangioso, K.M.; Wickland, A.C.; Meentemeyer, R.K.; Rizzo, D.M. An emergent disease causes directional changes in forest species composition in coastal California. *Ecosphere* **2012**, *3*, 86. [CrossRef]
56. Paine, R.T.; Tegner, M.J.; Johnson, E.A. Compounded perturbations yield ecological surprises. *Ecosystems* **1998**, *1*, 535–545. [CrossRef]
57. Simard, M.; Romme, W.H.; Griffin, J.M.; Turner, M.G. Do mountain pine beetle outbreaks change the probability of active crown fire in lodgepole pine forests? *Ecol. Monogr.* **2010**, *81*, 3–24. [CrossRef]
58. James, P.M.A.; Robert, L.-E.; Wotton, B.M.; Martell, D.L.; Fleming, R.A. Lagged cumulative spruce budworm defoliation affects the risk of fire ignition in Ontario, Canada. *Ecol. Appl.* **2016**, *27*, 532–544. [CrossRef] [PubMed]
59. Metz, M.R.; Varner, J.M.; Frangioso, K.M.; Meentemeyer, R.K.; Rizzo, D.M. Unexpected redwood mortality from synergies between wildfire and an emerging infectious disease. *Ecology* **2013**, *94*, 2152–2159. [CrossRef] [PubMed]
60. Cobb, R.C.; Meentemeyer, R.K.; Rizzo, D.M. Wildfire and forest disease interaction lead to greater loss of soil nutrients and carbon. *Oecologia* **2016**, *182*, 265–276. [CrossRef] [PubMed]
61. Worrall, J.J.; Harrington, T.C. Etiology of canopy gaps in spruce–fir forests at Crawford Notch, New Hampshire. *Can. J. For. Res.* **1988**, *18*, 1463–1469. [CrossRef]
62. Negrón, J.F.; McMillin, J.D.; Anhold, J.A.; Coulson, D. Bark beetle-caused mortality in a drought-affected ponderosa pine landscape in Arizona, USA. *For. Ecol. Manag.* **2009**, *257*, 1353–1362. [CrossRef]
63. Allen, C.D.; Breshears, D.D.; McDowell, N.G. On underestimation of global vulnerability to tree mortality and forest die-off from hotter drought in the Anthropocene. *Ecosphere* **2015**, *6*, 129. [CrossRef]

![forests logo] *forests*

MDPI

Article

Short Rotations in Forest Plantations Accelerate Virulence Evolution in Root-Rot Pathogenic Fungi

Jean-Paul Soularue *, Cécile Robin, Marie-Laure Desprez-Loustau and Cyril Dutech

BIOGECO, INRA, Univ. Bordeaux, 33610 Cestas, France; cecile.robin@inra.fr (C.R.);
marie-laure.desprez-loustau@inra.fr (M.-L.D.L.); cyril.dutech@inra.fr (C.D.)
* Correspondence: jean-paul.soularue@inra.fr

Academic Editors: Matteo Garbelotto, Paolo Gonthier and Timothy A. Martin
Received: 7 April 2017; Accepted: 6 June 2017; Published: 10 June 2017

Abstract: As disease outbreaks in forest plantations are causing concern worldwide, a clear understanding of the influence of silvicultural practices on the development of epidemics is still lacking. Importantly, silvicultural practices are likely to simultaneously affect epidemiological and evolutionary dynamics of pathogen populations. We propose a genetically explicit and individual-based model of virulence evolution in a root-rot pathogenic fungus spreading across forest landscapes, taking the *Armillaria ostoyae–Pinus pinaster* pathosystem as reference. We used the model to study the effects of rotation length on the evolution of virulence and the propagation of the fungus within a forest landscape composed of even-aged stands regularly altered by clear-cutting and thinning operations. The life cycle of the fungus modeled combines asexual and sexual reproduction modes, and also includes parasitic and saprotrophic phases. Moreover, the tree susceptibility to the pathogen is primarily determined by the age of the stand. Our simulations indicated that the shortest rotation length accelerated both the evolution of virulence and the development of the epidemics, whatever the genetic variability in the initial fungal population and the asexuality rate of the fungal species.

Keywords: forestry; tree fungal pathogen; root-rot disease; *Heterobasidion annosum*; *Ganoderma boninense*; evolutionary epidemiology; quantitative host–pathogen interaction; asexuality; clonality; saprotrophism

---

1. Introduction

While the demand for forest products is still increasing [1,2], high damage levels caused by pathogens in planted forests have emphasized the urgent need for management measures containing the development of epidemics [3–5]. The design and implementation of such measures require a clear understanding of the effects of common silvicultural operations on the expansion of pathogen populations in forest plantations. However, at present this understanding is far from being achieved. One key reason is that the evolutionary potential of pathogens has been widely neglected in forestry [6]. For a long time, the evolution of pathogens has indeed been assumed to be a much slower process than the development of the epidemics they cause [7]. However, there is now evidence that pathogens can rapidly evolve [8,9], particularly within ecosystems under strong human influence [10].

Among biological traits assumed to be under strong evolutionary pressure, the virulence of pathogens, generally defined as the quantitative ability of a pathogen to induce host mortality [11,12], is the focus of considerable attention from evolutionary ecologists [12] and plant pathologists [13]. Despite still imprecise experimental characterizations, mainly based on proxies, and a complex theory including the transmission–virulence trade-off hypothesis [14], the evolution at this trait is thought to be tightly related to the spread of diseases [11,15,16]. This evolutionary perspective, together with the large spatio-temporal scales considered in forestry and the often numerous uncertainties about the biology of the pathogens, make integrative experimental assessments of the effects of common forestry practices on fungal pathogen populations hardly feasible. Overall, available experimental observations about effects of silvicultural practices on the epidemiology of forest diseases remain scarce. As a consequence, when disease outbreaks are detected in forests, practitioners have very few elements at their disposal to confidently define and implement the most relevant management operations both preserving the productivity of the stands and containing the development of pathogen populations.

Within this context, modeling has much to contribute. Interestingly, a theoretical framework focused on the integration of the co-occurring epidemiological and evolutionary processes at short time-scales has been set (e.g., [17–19]). For instance, assuming a population made of a wild and a mutant genotype showing distinct virulence and transmission rates, Berngrüber set a mathematical model to explore the transient effects of mutation on virulence evolution in a bacteriophage virus during the onset of epidemics [11]. On the same basis and for the same kind of organism, Griette et al. showed that the level of virulence determines the speed of epidemics but also that virulence is favored at the front line of an epidemic and counter-selected behind the front [16]. Besides, other general models have investigated the evolutionary consequences of differences in life cycles. For instance Bazin et al. found that a reproduction system which combines a high rate of asexuality with a non-null sexuality rate increased the success of invasion [20]. Papaïx et al. explored the effect of dispersal and environmental heterogeneity on the specialization of organisms in a meta-population context [21]. Finally, other models were set to explore the effect of spatial heterogeneity associated with management practices on the evolutionary trajectories of both pathogens and their hosts in agrosystems of annual plants. These latter models were mostly used to investigate the durability of resistance genes deployed in crops from a qualitative definition of the host–pathogen interaction relying on gene-for-gene interactions. As examples Fabre et al. identified the optimal strategies of resistance gene deployments within agrosystems infected by plant viruses [22], and Bourget et al. investigated the adaptive dynamics of pathogen populations facing a multicomponent treatment [23].

All of these theoretical studies have yielded useful information about key determinants of the evolutionary and epidemiological dynamics of pathogenic organisms, sometimes in intensively exploited ecosystems. By their diversity, they also underlined the fact that these determinants are tightly related to the biological characteristics of the pathogen, and the specificities of the ecosystem of interest. Nevertheless, all of the previous modeling studies simultaneously investigating the evolutionary and epidemiological dynamics remain very general. Moreover, the strong assumptions made on pathogen life-history are more often not relevant to tree pathogens emerging in a forest plantation. Besides, in a general way, directly transposing existing models, theory and results into different ecosystems that often constitute distinct complex systems remains a challenge [24–26]. Hence, there is a need for new simple models clearly focused on pathogens spreading in a forest plantation and using existing theory in evolutionary epidemiology to address practical questions. To be helpful these new models have to account explicitly for the following aspects: (1) the key biological characteristics of the targeted fungal pathogen such as its reproduction system; (2) the specificities of forest management like the times of thinning and clear cutting; (3) the biological specificities of the planted trees such as their long lifespan; and (4) an adequate definition of the tree–pathogen interaction.

We propose here a stochastic individual-based simulation model dedicated to the study of the interplay between common forestry practices and the evolution and the propagation of a root-rot pathogenic fungus in planted forests. Individual-based modeling is a bottom-up approach that relies on algorithmic descriptions of each individual constituting the system of interest. It provides highly customizable and integrative abstractions of complex systems, making it an ideal tool to address practical questions involving many entities interacting at different levels [27–29]. Though this kind of modeling has been fruitfully developed in evolutionary ecology (e.g., [30–37]), it remains almost absent from the fields of phytopathology and evolutionary epidemiology (however, see [21]). Unlike most of the existing models in phytopathology, our model describes host–pathogen interactions in a quantitative way, through the exploitation of the quantitative genetics theory. Hence, the model formalizes quantitatively the phenotype of the fungus, the underlying genetic architecture of the targeted trait (here the virulence), and the effects of the main processes driving the evolutionary change: mutation, recombination, dispersal or selection. The environmental conditions affecting the evolution of the trait and the expansion of the fungus are also expressed quantitatively. They essentially consist of the susceptibility of the host trees, which is determined by the management of forest plantations. Although meant to be generic, our model was built and parameterized with the *Armillaria ostoyae–Pinus pinaster* pathosystem in mind. *A. ostoyae* is one of the main root-rot pine pathogens of the northern hemisphere [38,39]. The model meets the four aforementioned conditions and particularly considers the following elements. First, because the life cycle of an organism determines its evolutionary responses to changing environmental conditions, and thus its invasive potential [9,20,40,41], the life cycle of *A. ostoyae* is explicitly described. In particular, following the *Armillaria* model, the simulated life cycle includes sexual mating, vegetative growth and saprotrophic phases, i.e., the ability to survive in dead pine material like stump or wood [38,42,43]. Second, the host–pathogen interaction is formalized quantitatively and depends not only on the virulence of the fungus but also on the susceptibility of the tree to the fungus. Here, the tree susceptibility was assumed to be determined only by the age of the tree; young pines are known to be very sensitive to *A. ostoyae*, whereas adults are more resistant [44,45]. Third, the spatio-temporal heterogeneity in environmental conditions induced by management practices [22,23] is explicitly accounted for to investigate their consequences on the evolutionary and epidemiological dynamics. Clear-cutting and thinning operations constitute a central part of our model because they simultaneously act on the spatio-temporal distributions of the resistance levels through age effect, and on the number of healthy trees in each stand.

Considering a forest made of even-aged stands in which susceptibility only depends on age class, we specifically addressed the following question: how are the evolution of virulence and the expansion of the fungal population affected by changes in rotation length? We simulated multiple scenarios, each one being characterized by rotation length, asexuality rate and level of standing genetic variability in the source fungal population. Because asexuality rate [20] and standing genetic variability [46] are known to determine the adaptive and invading potential of species [47], we anticipated that depending on the values assigned to these two parameters, the effects of rotation length would result in contrasting evolutionary and epidemiological dynamics. We further discuss our results on the basis of related theory, and, finally, we make suggestions for further investigations.

2. Materials and Methods

Our model consists of a stochastic simulation program we wrote in Python from the standard [48] and Numpy [49] libraries. We also used Matplotlib [50] library to create the figures.

2.1. Fungal Pathogen

The model is based on the biology of *A. ostoyae*, which causes important mortality in maritime pine (*Pinus pinaster*) planted forests [45,51,52]. This diploid fungus combines sexual and asexual reproduction phases. It can infect pines either through mycelial growth from the root of an infected tree to the root of another healthy tree [52], or through mycelium resulting from the fusion of two basidiospores regularly released by its fruiting bodies [53]. The combination of these two modes of infection produces disease foci that correspond to large clonal patches [54], which are genetically differentiated [53]. Moreover, *A. ostoyae* can survive for years on dead wood material as a saprotroph. During this saprotrophic phase, *A. ostoyae* still sporulates and grows clonally [38,42,43,55].

We call *individual* a fungal genotype which successfully infects and kills one tree. Hence, in this paper, the term *individual* refers both to a genotype and its location in the forest. It is important to note that a single genotype can infect multiple trees through root contact in a stand (see below), which generates multiple individuals carrying the same genotype. Taking into consideration the clonal structures characterized in maritime pine plantations [53] and in natural forests [56] infected by *A. ostoyae*, we only allow single infections to occur in our simulations, i.e., a tree can be only infected by one individual. Each individual was characterized by its virulence defined here as the quantitative ability of a genotype to infect a tree, with all fungal infections resulting in tree mortality. Considering a diploid fungal species, we modeled virulence as a fully heritable quantitative trait Z determined additively by 10 loci without dominance nor epistasis:

$$Z = \sum_{l=1}^{10} (\alpha_i + \alpha_j)_l + \epsilon$$

where α_i, α_j represent the allelic effects of the two alleles associated with locus l, and ϵ is an environmental variation following the uniform distribution $\mathcal{U}(0,1)$. In the following, the term genetic value (G) refers to the additive contribution of the alleles at the 10 loci involved in the variability of the trait Z, such as $Z = G + \epsilon$.

Each simulation starts with n genetically distinct fungal individuals in a single stand. In the present study, the genetic composition of the starting fungal population was drawn randomly from sets of 200 alleles per locus, the allelic effects following a normal distribution $\mathcal{N}(0, V_a)$ with V_a set here to 0.5. We assumed that the fungal pathogen population undergoes a number of annual evolutionary cycles. The evolutionary cycles are non-overlapping which means that all individuals simultaneously undergo all the events constituting a cycle. However, unlike most of the studies based on individual-based models in evolutionary ecology (e.g., [20,36]), the generations overlap, i.e., "parents" are not removed from the fungal population after having produced offspring through asexual growth or sexual reproduction. During one evolutionary cycle, two types of propagules are generated by each fungal individual. Clonal propagules, also called "clones", are the result of asexual growth. As a consequence of clonal growth several trees can be infected by the same genotype in a stand. On the other hand, efficient sexual spores that truly deposite and germinate in stands contribute to generating, after mating, sexual propagules, i.e., "zygotes". Following [20], the quantity of clonal propagules and sexual spores produced by each fungal individual is constant and determined a priori from a fecundity f and an asexuality rate c varying between 0 and 1. Hence, at each evolutionary cycle, each fungal individual produces $n_c = c \times f$ clonal propagules and $n_s = (1 - c) \times f \times p$ efficient sexual spores that will contribute to generate a total of n_s/p sexual propagules after dispersal, germination and mating, n_s/p being rounded to the nearest lower integer when it is an odd number. In the present study, the simulated fungal species had a constant fecundity of 5 as in [20], and was diploid ($p = 2$).

Each evolutionary cycle (Figure 1) starts with gametogenesis during which the fungal individuals produce haploid sexual spores through meiosis and recombination. Assuming that the loci are located on separated chromosomes, recombination occurs at a rate of 0.5 per locus. Then, dispersal of sexual spores occurs at the same time than clonal growth, simultaneously in all stands. In what follows we may use the term asexual reproduction to refer to clonal growth. While the sexual spores emitted by the individuals of a stand can reach the neighboring stands according to the stepping-stone dispersal model [57], the clones produced remain local, i.e., in the same stand. We set the proportion of the efficient sexual spores produced in a stand that germinate in neighboring stands at 5%, the remaining 95% germinating in the original stand. Finally, the dispersal step ends with the generation in each stand of the sexual propagules through the random fusion of pairs of the efficient spores that is deposited in the stands. Then, density regulation operates to determine which propagules (sexual or clonal) are kept in the stands and will attempt to infect healthy trees, when the fungal carrying capacity of a stand is exceeded, namely when the number of healthy trees is lower than the number of propagules. This step of density regulation simply consists in a random draw of individuals independent from their virulence level. At the end of the density regulation step, any propagule that cannot attempt an infection is automatically removed from the stand.

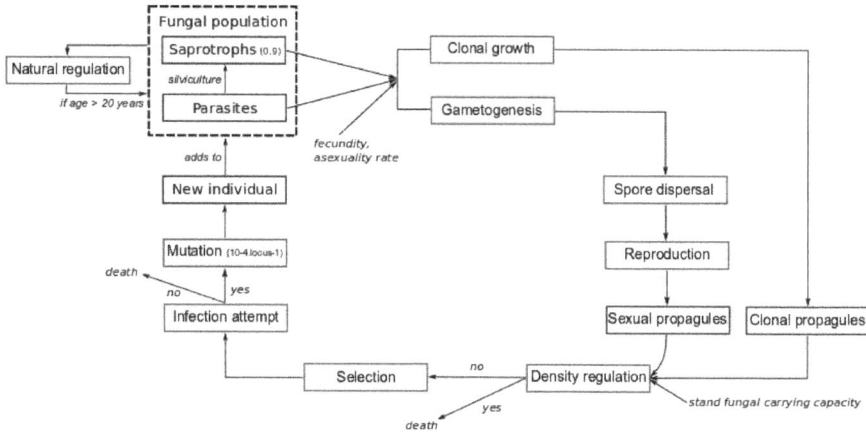

Figure 1. Simulated evolutionary cycle. Each green square represents an evolutionary step. Red squares indicate different evolutionary states of the fungus in our simulations. The dashed square represents the fungal population, i.e., a set of fungal individuals at the parasitic or saprotrophic stage. Mutation and survival in dead material as saprotrophs occur randomly according to the rates specified.

During the next step of selection, the propagules remaining after density regulation experience directional selection [58] towards a phenotypic optimum Z_{opt}. Here the fitness value [59] assigned to each propagule weights the probability of infection success (see below) which further determines the survival and the ability to reproduce. For a trait Z, a fitness value W is computed from:

$$W(Z) = \exp\left(\frac{-(Z - Z_{opt})^2}{\omega}\right)$$

for Z values comprised in $]-\infty, Z_{opt}]$, otherwise the maximal fitness value of 1 is systematically assigned to the trait values exceeding the Z_{opt} value. According to preliminary sensitivity analysis, we specified here uniformly in all stands a strong selection pressure through a ω value of 20 and a phenotypic optimum Z_{opt} of 10, a value much greater than the mean trait value of the initial fungal populations close to 0 (see [20,32,35]). The resistance R of a stand is included in $[0, 1]$, and $R = 1 - S$, where S is the mean susceptibility of the stand to the pathogen. The outcome of each infection attempt (success or failure) by a propagule i in a stand x is drawn from a binomial law weighted by a probability of success $\rho_s = W(Z_i) \times S_x$. When a propagule successfully infects a tree, a new individual is created, otherwise the propagule is removed from the stand. Here, for the sake of simplicity, each successful infection leads unavoidably to host death. This simplification is consistent with earlier results indicating an important vulnerability of maritime pine to *A. ostoyae* [44]. At the mutation step, the genotype of each individual mutates at a rate of 10^{-4} per locus as in [20] according to the K-allele model [60]: when a mutation occurs, a new allele is drawn randomly among the 200 possible alleles. Given the high variance of allelic effects ($V_a = 0.5$), mutation is likely to generate substantial genetic variability during simulations. Due to the saprotrophism ability of the fungal pathogen [45], when an infected tree is cut (see next sub-paragraph), the fungal individual established in this tree can persist on dead substrate in the stand with a probability s, where s designates the saprotrophic ability. The saprotrophs remaining in stumps of cut trees continue to disperse both asexually and sexually until their age reaches the maximal age A_{max}, and thus contribute to the production of propagules each year.

Finally, a natural regulation of individuals according to their age occurs, those of age greater than A_{max} are removed from the landscape. Based on [61], we set a life time A_{max} of each fungal individual to 20 years. At the forest scale, this value allows clonal structures composed of several individuals at the parasitic or saprotrophic stage to persist during the whole simulation process, which is in agreement with the very long life-span of *Armillaria* organisms (e.g., [56]). Figure 1 summarizes the modeled evolutionary cycle and the different possible living states of the pathogenic fungus.

2.2. Forest Plantation and Silvicultural Operations

The modeled forest landscape was composed of 49 even-aged stands arranged according to a 7×7 square. We considered three age classes in the tree population: young, adult and mature, associated with distinct carrying capacities, and contrasted levels of susceptibility to the fungal pathogen. Using a year as unit of time, the three age classes were arbitrarily delimited by years 15 and 32. Thereby, trees younger than 15 years were considered as young, trees between 15 and 32 years were considered as adult, and trees older than 32 years were viewed as mature. The carrying capacities associated with each age-class were 1000, 500 and 300 for young, adult and mature stands, respectively. The corresponding resistance values were arbitrarily set at 0.3, 0.8, 0.6 following earlier observations reporting that juvenile trees are particularly sensitive to fungal pathogens [44,45].

The starting point of each simulation is a randomly generated planted forest landscape with a third of stands of each age class, the maximum age being 55. During simulations, year after year, trees get older, and stands are managed through clear-cutting and thinning operations. Thinning operations occur at each age–class transition, and mainly consist in cutting the required number of trees to match the carrying capacity of the next age class. This process can however be affected by infections. Because only lethal infections are explicitly modeled here, a tree has two states: (1) "healthy" when the tree is not infected or has efficiently contained previous infection attempt(s) sufficiently early; or (2) "infected/killed" when the tree is subject to an ongoing lethal infection. Thus, we assume that symptoms of lethal infections are always visible and the forest managers make sanitary thinning; they systematically choose to cut all the infected trees during thinning. We also assume that the forest managers can decide anticipated clear-cutting operations (i.e., end of rotation) if the number of remaining trees resulting from thinning is lower than a threshold N_r. In the default case, the number of trees cut during a thinning operation is $N_c = K - K'$ where K and K' refer to the fungal carrying capacities in the lower and upper age class at each age transition, respectively. This relation holds as long as the number of infected trees N_i is lower than $K - K'$. Otherwise the number of trees to be cut N_c equals either N_i as long as $K - N_i \geq N_r$ (i.e., the number of remaining trees is greater than the threshold N_r) or K (anticipated clear-cut) when $K - N_i < N_r$. Taking as example a stand at the juvenile–adult transition, the forest managers should cut 500 trees in the general case. If fewer than 500 trees show symptoms of disease, the forest managers will cut 500 trees including all the infected trees. However, if more than 500 trees have been infected, for example 850, the forest managers will cut all the 850 infected trees. If the minimal threshold of remaining trees is 200 in an adult stand, an anticipated clear-cutting operation is triggered. In the normal case, clear-cutting is realized at regular intervals corresponding to the rotation length, all trees of a stand are cut, and new juvenile trees of age 0 are replanted immediately. In our simulations, N_r, the minimal number of trees triggering an anticipated clear-cutting operation in place of a simple thinning, was set at 150. Depending on the scenario, different rotation lengths were considered.

2.3. Simulated Scenarios

By considering two initial fungal populations, combined with two asexuality rates and three distinct rotation lengths of 15, 30 and 50 years, we simulated a total of 12 scenarios, and ran 50 independent stochastic replicates for each scenario.

The starting point of each simulation was a small fungal population of $n = 10$ individuals emerging in a single stand located near the western edge of the landscape (Figure 2). Both the initial fungal population and the forest plantation landscape were generated randomly. However, because the initial genetic composition drastically influences the evolutionary potential in populations of such limited size, the starting fungal population was randomly generated once for each level of genetic variance considered, and thus was identical among the replicates. This ensured a reasonable stochastic variability among replicates. In the default scenario, the starting fungal population was characterized by an initial mean genetic value G_0 of 0.77, and an elevated initial standing genetic variance V_0 of 4.06 that, combined with the elevated variance of mutational effect ($V_a = 0.5$), guarantees substantial adaptive and invasive potentials of the fungal population [46,47]. We also simulated additional scenarios starting from another initial fungal population showing a mean initial genetic value G_0 of 0.33 but a much lower initial genetic variance V_0 of 1.63. Several independent random generations of the population were realized to obtain two starting fungal populations showing similar G_0 values close to 0. This mean initial genetic value defined an initial important maladaptation level [35] in both cases: the distance between the phenotypic optimum ($Z_{opt} = 10$) and the mean initial genetic value G_0 was important but of the same magnitude. Two values of the asexuality rate c were compared, 0.2 and 0.8, to investigate the effects of two contrasted dispersal and reproduction strategies, either mainly sexual or mainly clonal. In the following, *fp1* refers to the mostly asexual reproductive strategy ($c = 0.8$) and *fp2* refers to the mostly sexual reproductive strategy ($c = 0.2$). Following the *Armillaria* model, we assumed that the two fungal profiles had a high saprotrophic ability s [38,42,45,55], set here to 0.9. The initial planted forest landscape (i.e., the geographical distribution of age classes in the 49 stands) was randomly generated before each run in order to eliminate the influence of the initial spatial distribution of the mean stand ages. The initial proportion of young, adult and mature stands remained however always fixed to $(1/3, 1/3, 1/3)$. Tables 1 and 2 summarize all the variable and fixed parameters simulated.

Table 1. Variable parameters.

Parameter	Value
Initial genetic variance V_0	1.63, 4.06
Mean initial genetic value G_0	0.33, 0.77
Asexuality rate c	0.2, 0.8
Rotation length in years	15, 30, 50

Table 2. Fixed parameters. *J* refers to juvenile, *A* to adult, *M* to mature.

Parameter	Value
Number of stands	49
Tree age classes in years (J, A, M)	$([0, 15], [16, 32], [32, \infty[)$
Carrying capacities K of stands (J, A, M)	$(1000, 500, 300)$
Mean resistance levels R of stands (J, A, M)	$(0.3, 0.8, 0.6)$
Initial proportions in J, A and M stands	$(1/3, 1/3, 1/3)$
Threshold N_r before anticipated clear-cut	150
Size n of the initial fungal population	10
Optimal phenotype Z_{opt}	10
Number of loci	10
Intensity of stabilizing selection ω	20
Mutation rate	$10^{-4} \times locus^{-1}$
Mutational variance V_a	$0.5 \times locus^{-1}$
Recombination rate	0.5
Fecundity f	5
Ploidy p	2
Proportion of spores dispersed among stands	5%

(a)

(b)

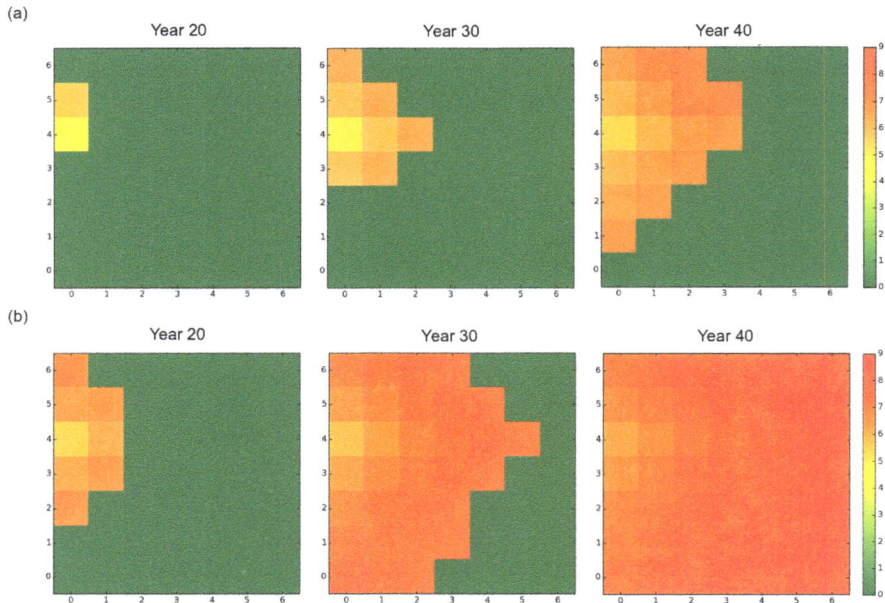

Figure 2. Spatial distribution of the mean virulence level over the forest landscape, at years 20, 30 and 40, for a rotation length of 15 years. Dark green corresponds to healthy stands. The other colors inform about the mean genetic value of the local fungal sub-population in each infected stand. (**a**) mostly clonal fungal pathogen profile *fp1* (clonality rate $c = 0.8$); (**b**) mostly sexual fungal pathogen profile *fp2* ($c = 0.2$). In both cases, the starting genetic variance G_0 was 4.06 (default scenario). The mean fungal genetic values were computed in each stand from the 50 independent replicates simulated.

3. Results

3.1. Observed Epidemiological and Evolutionary Dynamics

Initially present in a single stand located near the western edge of the landscape, the fungal population progressively expanded and finally reached all the stands without regressing, whatever the initial settings (Figures 2 and 3). The monitoring of the number of infected trees and stands indicated that the expansion of the pathogen population was very slow during about the first ten years and was limited to the original stand. Once the pathogen reached another stand, the progression of the epidemics became exponential (Figures 3 and S1). The velocity of the propagation depended on the rate of asexuality of the fungal species and the initial standing genetic variation; the lower the asexuality rate and the higher the initial genetic variation, the faster the spread of the epidemics, both at the landscape scale (Figures 2 and 3) and at the stand scale (data not shown). For example, under a rotation regime of 30 years, *fp2* needed 5 years to reach a first neighboring stand and 64 years to infect durably all the stands. By contrast *fp1* established in a first neighboring stand after 8 years and needed 85 years to infect all the stands.

(a)

(b)

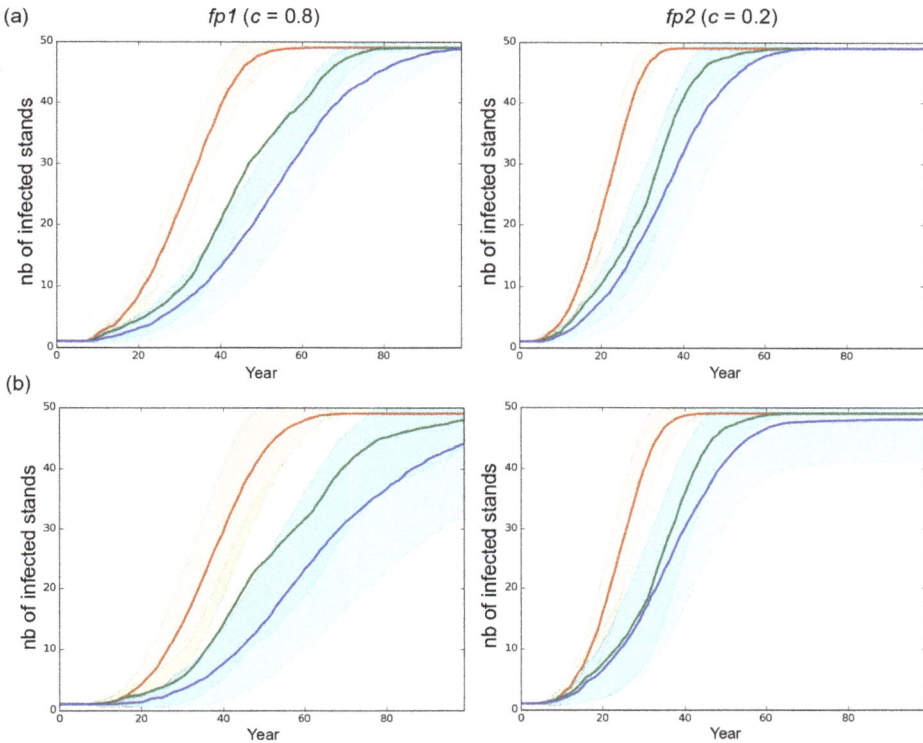

Figure 3. Effect of the rotation length on the propagation of the pathogen. (**a**) high initial genetic variance V_0 of 4.06 in the starting fungal population (default scenario); (**b**) reduced initial genetic variance V_0 of 1.63 in the starting fungal population. Left column: mostly clonal fungal profile *fp1* (clonality rate $c = 0.8$). Right column: mostly sexual fungal profile *fp2* ($c = 0.2$). Red, green and blue represent a rotation length of 15, 30 and 50 years, respectively. Each line is the mean of 50 independent replicates simulated. Each colored area represents the standard deviation of the 50 replicates.

In all scenarios, the mean level of virulence Z followed a logistic growth but never reached the optimal phenotypic value ($Z_{opt} = 10$). The faster and larger evolutionary change was observed under the mostly sexual reproduction system (*fp2*) and when the initial standing genetic variation was large ($V_0 = 4.06$). In this case, at year 100 the maximal level of virulence was 8.2 whereas it was only 7.7 for *fp1* in our simulations (Figure 4). Hence, in our simulations, the velocity of the expansion of the fungal population was related to the evolution of virulence. The large initial genetic variance set in the default scenario rapidly decreased under the joint action of selection and genetic drift during the first 20 years despite the elevated probability of mutation set in the model ($10^{-4} \times locus^{-1}$). Overall, a high sexuality rate (*fp2*) systematically: (1) maintained a higher level of genetic variance than a low sexuality rate (*fp1*) in a first phase of about 40 years; and (2) resulted in the lowest level of genetic variance at year 100 (Figure 5). It is worth noting that when V_0 was low (1.63 vs. 4.06), the level of genetic variance observed at year 100 slightly exceeded the initial level V_0, particularly in the case of the mostly asexual species *fp1* (Figure 5). Finally, a slight spatial gradient of the virulence level progressively appeared following the direction of the fungal population expansion (Figure 2).

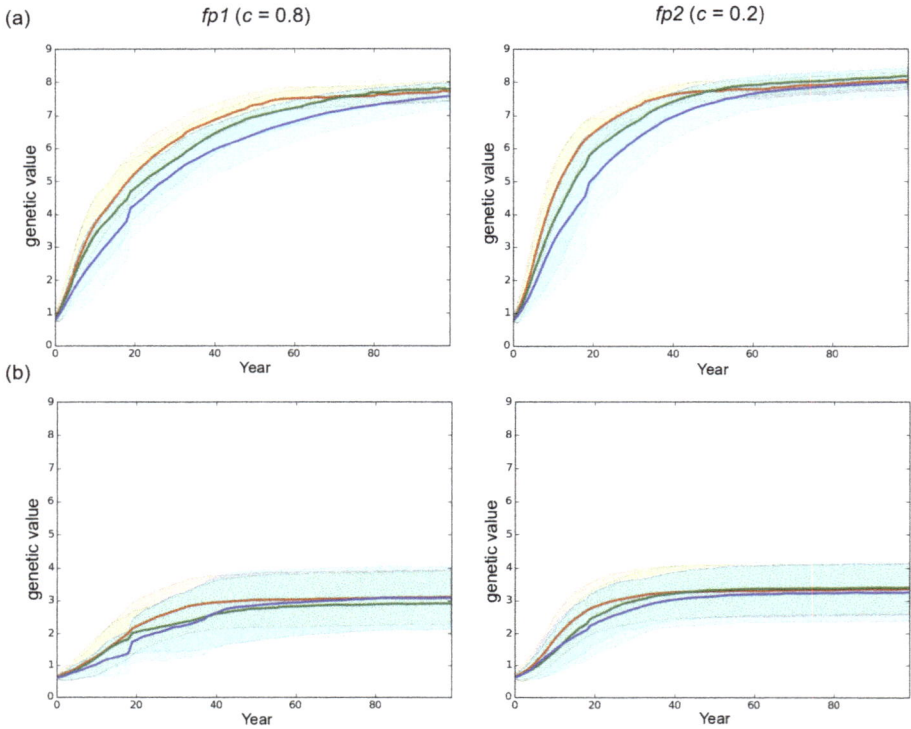

Figure 4. Effect of the rotation length on virulence evolution. The genetic value is the sum of the allelic effects at the 10 loci involved in the variability of the trait. (**a**) high initial genetic variance V_0 of 4.06 in the starting fungal population (default scenario); (**b**) reduced initial genetic variance V_0 of 1.63 in the starting fungal population. Left column: mostly clonal fungal profile *fp1* (clonality rate $c = 0.8$). Right column: mostly sexual fungal profile *fp2* ($c = 0.2$). Red, green and blue represent a rotation length of 15, 30 and 50 years, respectively. Each line is the mean of 50 independent replicates simulated. Each colored area corresponds to the standard deviation of the 50 replicates.

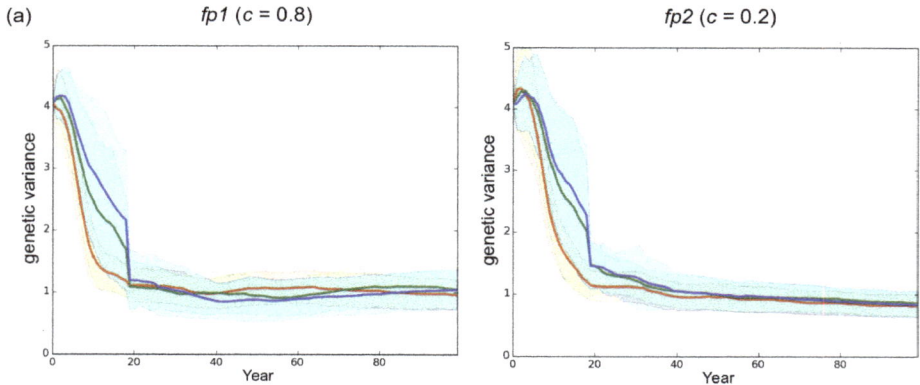

Figure 5. *Cont.*

(b)

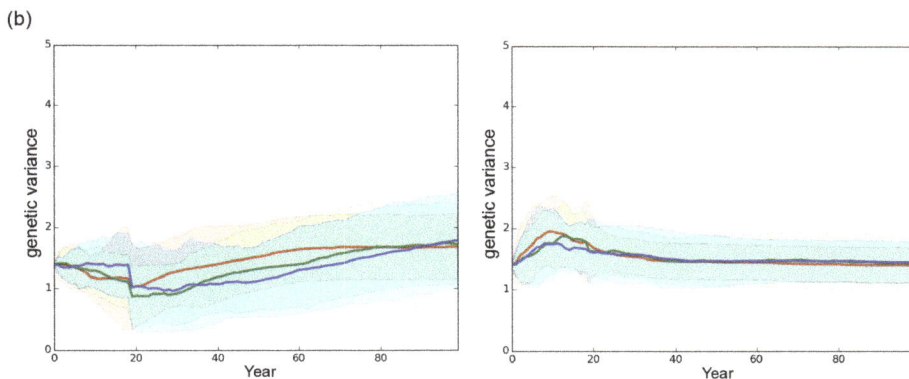

Figure 5. Evolution of the genetic variance under different rotation lengths. The genetic variance was computed at the landscape scale. (**a**) high initial genetic variance V_0 of 4.06 in the starting fungal population (default scenario); (**b**) reduced initial genetic variance V_0 of 1.63 in the starting fungal population. Left column: mostly clonal fungal profile *fp1* (clonality rate $c = 0.8$). Right column: mostly sexual fungal profile *fp2* ($c = 0.2$). Red, green and blue represent a rotation length of 15, 30 and 50 years, respectively. Each line is the mean of 50 independent replicates simulated. Each colored area corresponds to the standard deviation of the 50 replicates.

3.2. Effect of the Rotation Length on Pathogen Propagation and Virulence Evolution

A shorter rotation length clearly resulted in faster development of the epidemics (Figures 3 and S1). For instance, in our simulations the *fp2* population infected all the stands in 38, 63 and 72 years when rotation lengths of 15, 30 and 50 years were specified, respectively. Besides, decreasing the rotation length substantially accelerated the evolutionary change at virulence during a first phase which lasted 70 years in the case of *fp1* and 46 years in the case of *fp2* in the default scenario. During this first phase, the shorter the rotation length, the faster the evolution of virulence. This relationship was however transient and progressively disappeared in the subsequent years. Taking for example the case of *fp1*, virulence levels of 7.2, 6.8 and 6.2 were reached at years 50 when the rotation lengths equaled 15, 30 and 50 years, respectively, but at year 100 the corresponding virulence levels equaled 7.65, 7.7 and 7.5. Importantly, the monitoring of the number of anticipated clear-cutting operations (see model) revealed that the negative correlation between the rotation length and the evolved virulence level weakened when anticipated clear-cutting operations were triggered (Figure 6). In our simulations the mostly sexual fungal pathogen *fp2* induced the highest number of anticipated clear-cutting operations (Figure 6).

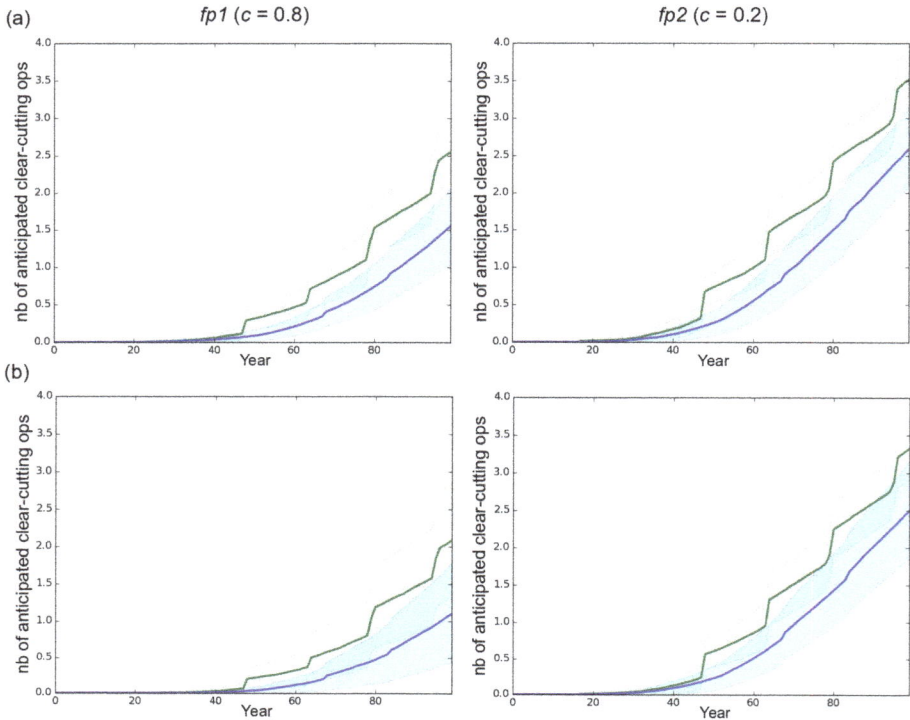

Figure 6. Number of anticipated clear-cuttings operations. In our simulations, sanitary thinnings occurred at 15 and 32 years, i.e., at the juvenile–adult and adult–mature transitions, respectively. An anticipated clear-cutting operation was triggered (end of rotation) in a stand when the number of healthy trees remaining after thinning was lower than 150. (**a**) high initial genetic variance V_0 of 4.06 in the starting fungal population (default scenario); (**b**) reduced initial genetic variance V_0 of 1.63 in the starting fungal population. Left column: mostly clonal fungal profile *fp1* (clonality rate $c = 0.8$). Right column: mostly sexual fungal profile *fp2* ($c = 0.2$). Green and blue represent a rotation length of 30 and 50 years respectively. Each line is the mean of 50 independent replicates simulated. Each colored area corresponds to the standard deviation of the 50 replicates simulated.

4. Discussion

4.1. The Effect of Silvicultural Operations on a Necrotroph Fungal Pathogen Alternating Parasitic and Saprotrophic Stages

Our simulations showed that the scenario involving the highest number of clear-cutting operations generated the largest increase in virulence and the fastest spread of epidemics. Although we purposely parameterized our simulations to allow the fungus to rapidly evolve and expand (e.g., high mutation rate, high fecundity), there were strong limitations to the evolutionary change in our simulations. First, because of the small size of the starting fungal population and the multiple founder effects accompanying its expansion, the genetic drift was likely to severely impede the adaptive response of the fungus to directional selection [62]. Second, the host death, clear-cutting and thinning rarely led to pathogen death because most of the fungal individuals remained as saprotrophs in stumps and dead wood ($s = 0.9$).

Saprotrophism facilitates the persistence of the fungal population in a stand and limits the effects of genetic drift and founder effects by massively sustaining the production of propagules that will try to infect the healthy trees available (see Model section). Nonetheless, saprotrophism is also likely to slow down the evolutionary change of the trait, by favoring the maintenance in the stands of many early individuals showing largely suboptimal virulence levels ($Z_{opt} - Z >> 0$), which dilutes the new arising virulent genotypes in the fungal population. This negative effect of saprotrophism on virulence evolution is particularly striking when many trees of a stand are previously infected by fungal individuals showing a low virulence level. In this patho-system where: (1) no coinfection can occur; and (2) the pathogen can maintain in the stand as a saprotroph (Figure 1), new infections, whatever the level of virulence of fungal individuals, can only occur when healthy trees are available, which is more likely during the early stage of epidemics. Such a process favoring early-arriving individuals over late- arriving individuals is called a priority effect [63].

Interestingly, in this context in which the pathogen virulence may be the subject of opposite pressures, one can expect the silvicultural operations to have a substantial impact on its evolution and the expansion of the pathogen population. In the case of thinning during juvenile–adult transition, the maintenance of numerous saprotrophs comes along with a rarefaction of healthy hosts (K decreased from 1000 to 500) which, at the same time, become much more resistant to the pathogen (R increased from 0.3 to 0.9). As a first consequence, the combination of a lower fungal carrying capacity with a considerable quantity of propagules largely provided by the saprotrophs is likely to induce drastic density regulations. As a second consequence, the increase in the mean resistance of the stand steadily amplifies the selection pressure exerted on virulence. A mature–adult transition induces a similar effect on the fungal carrying capacity of the stand (K decreasing from 500 to 300) but results in a moderate decrease of the resistance of the stand (R decreasing from 0.9 to 0.6). Finally, clear-cutting operations result in the replacement of resistant trees by numerous young susceptible trees, which weakens the intensity of selection during infection attempts (R falling down to 0.3 after plantation of young trees), and makes density regulation of the propagules less drastic or inexistent (K increasing from 500 or 300 to 1000). To sum up, each silvicultural operation modifies both the strength of the selection and the magnitude of the density regulation of propagules. While the strength of selection is one of the main determinant of the adaptive response of the fungal population [58], in our model the density regulation step can act against the evolution of the level of virulence in our simulation, notably when the fungal population is large and contains few virulent individuals. Let us figure a set of N propagules, N being greater than the number of healthy trees in a stand. Let us assume there are only n_v very virulent propagules among the N existing propagules, such as $n_v << N$. The step of density regulation reduces the probability of infection attempt by a very virulent genotype by a factor n_v/N. Predicting accurately how the modifications of the selection intensity and the magnitude of density regulation interact with the effects of saprotrophism in the case of thinning and clear-cutting operations requires additional investigations. Nonetheless, the fact that the scenario involving the highest number of clear-cutting operations generated the largest increase in virulence and the fastest spread of epidemics in our simulations suggests that the release of the selection pressure at each clear-cutting was largely outweighed by the reduction of: (1) the genetic drift; (2) the founder effects; and (3) the magnitude of density regulation events. Finally, although silvicultural operations induce disturbances in a pathosystem dominated by a priority effect, it is interesting to note that in our simulations, the mean virulence level was systematically greater at the front of the epidemics than in the first stands infected (Figure 2). This result confirms the predictions of [16] in a radically different context. Nonetheless the reduction in differences in the mean virulence level between the later stands infected suggests that the priority effect was progressively canceled as the epidemics developed.

4.2. The Importance of Initial Genetic Variation

A low initial genetic variability (1.63 vs. 4.06) in the emerging fungal population did not qualitatively alter the clear-cutting effect characterized: the shorter the rotation length, the faster the evolution of virulence and the development of the epidemics. However, the level of initial genetic variability influenced quantitatively both the evolutionary change of virulence and the speed of the epidemic (Figures 3 and 4). Available genetic variance strongly determines the evolutionary potential of populations (e.g., [36,64,65]) particularly in the case of an expansion across spatio-temporally heterogeneous landscapes [47,66]. Our results thus confirm and underline the crucial influence of the level of standing genetic variation in fungal pathogen populations emerging in forest plantations. Surprisingly, the simulation of a low initial genetic variance ($V_0 = 1.63$) resulted in a substantial increase in the genetic variance between year 0 and year 100 (Figure 5), the final level being almost twice greater in this case than in the default scenario ($V_0 = 4.06$) when a high asexuality rate was simulated. The very limited change in the virulence level occurring at the same time in these cases (Figure 4) indicates that the generated additive variance was not efficiently exploited by genetic adaptation, particularly in the case of the mostly asexual species *fp1*. This observation illustrates again the consequences of the priority effect and saprotrophism discussed in the sub-paragraph above. Indeed, an epidemic can develop in spite of a very limited genetic variation among the fungal individuals, provided they are sufficiently virulent to infect available hosts. Here, in the context of priority effect and saprotrophism, a small increase in genetic variance, without a visible effect on the mean virulence, corresponds to the arrival of new virulent genotypes that remained however much less frequent that the first early maladapted individuals. This suggests that the genetic variation in the fungal population can really sustains the increase in virulence only when it is sufficiently elevated during the early stage of the epidemics.

4.3. The Importance of the Reproduction System

Little is known about the ratio of sexual to asexual reproduction modes in populations of root-rot fungal pathogens of conifers such as *A. ostoyae* [53], or *Heterobasidion annosum* [40]. In this context our model provides new elements which contrast with existing theory. While the rate of asexuality did not alter the effects of clear-cutting-operations, the fungal profile with the highest sexuality rate (*fp2*, $c = 0.2$) evolved more rapidly towards the highest virulence levels and generated the fastest epidemics in our simulations. First of all, it is important to note that in our model, at constant fecundity, a higher sexuality rate comes along with more sexual spores and thus greater spore flow between stands. Moreover the number of infections in the first infected stand grew slightly faster over 100 years in the *fp2* than in the *fp1* population (data not shown), as a result of a higher mean evolved virulence level. Indeed, in an evolutionary context characterized by important genetic drift and recurrent founder effects, gene flow between stands sustains the adaptive response of the pathogen to uniform directional selection by allowing the alleles with positive effects on virulence to be rapidly exchanged among stands [36]. In addition, higher sexuality rates make new genotypes potentially more virulent to be generated by recombination [67]. In contrast, fungal populations characterized by high asexuality rates have a lower capacity to break existing linkage disequilibrium, the mutations becoming the main source of genetic variability [20]. As a consequence of recombination and gene flow, the mostly sexual reproduction system induced an increase in the genetic variance during the first evolutionary cycles in our simulations (Figure 5), this increase corresponding to infections realized by new virulent genotypes. When they are generated sufficiently early (see the priority effect discussed above), these new genotypes are more likely to infect the healthy hosts compared to the early less virulent genotypes. Consequently, new virulent individuals rapidly prevail, which leads to a rapid decrease of the genetic variance in the following years and ultimately allows for the largest evolutionary change and the fastest spread of the pathogen.

Our results differ from those of Bazin et al. (2014) [20] who found that organisms that combined a very high asexuality rate (0.95) and a low sexuality rate were the best invaders. This difference stems from the fact that in their model the between-stand dispersal ability was kept constant whatever the asexuality rate, which is not the case in our model. Hence, in our study, the mostly sexual fungal population of a stand generated more infections in neighboring stands for a given population size and virulence level. Moreover, the model proposed by Bazin et al. (2014) [20] simulates non-overlapping generations, i.e., the generated offspring systematically replaces the adults at each evolutionary cycle. In contrast, generations overlap in our model, which might also explain the observed differences. Earlier works have indeed shown that describing realistic demography by simulating non-overlapping generations can affect the evolutionary outcomes (e.g., see [68]).

4.4. Generalization and Future Directions

Interestingly, our results can be generalized to other root-rot pathogenic fungi such as the *Heterobasidion* species which cause root-rot disease in conifers. Indeed, these fungi combine sexual and asexual reproduction modes and alternate parasitic and saprotrophic stages [40]. Moreover, although we found no experimental results on the relationship between host age and host susceptibility to these fungal species, Pukkala et al. proposed a model of the spread of *H. annosum* in even-aged conifer stands which explicitly formalizes a higher mortality rate for young (i.e., small) trees exposed to the pathogen than mature (i.e., large) trees, for a given period length (see Figure 6, Equations (21) and (22) in [69]). Together with the experimental results of Lung-Escarmant and Guyon regarding *A. ostoyae* [44], this model supports the idea that young trees of smaller size are more rapidly weakened and killed by a pathogenic fungus than mature trees of greater size.

Nonetheless, further disentangling the complex effects of the artificial disturbances induced by silviculture on the evolutionary and epidemiological dynamics of pathogenic fungal populations requires consideration of a wider range in important parameters. Within the framework of a direct extension of the present study, this will require definition and simulation of multiple tree age class structures, rotation ages, and relationships between tree resistance and tree age. On the pathogen side, this will require investigation of how different levels of saprotrophic ability, sizes and genetic composition of the initial fungal population can affect the results obtained in this work. We now detail three other points that deserve to be considered in future investigations following our study. First, it is unlikely that all the stands composing a forest plantation landscape are managed according to a unique scheme. Hence, it might be interesting to assess how heterogeneity in management practices could affect the positive relationship between the number of occurrences of clear-cutting operations and the evolution of virulence. Following earlier modeling studies in phytopathology which have assessed the effect of spatio-temporal deployments of plant resistance genes on the durability of resistance (e.g., [22,23]), we could try to characterize the spatio-temporal dynamics of silvicultural operations that maximize the productivity of a forest plantation while containing the development of the epidemics. Such a study could be the opportunity to assess the effects of stump removal on the evolutionary and epidemiological dynamics of the fungal population. Second, the reproduction system simulated here included a substantial rate of selfing that was allowed in our simulations. Accounting for sexual compatibility in our simulations on the basis of dedicated loci could modify the dynamics characterized in this study. Indeed, selfing is largely viewed as a constraint to adaptation of populations, which was recently confirmed experimentally by Noel et al. (2017) [70] in snails. This point might have important consequences in our pathosystem, especially during the early evolutionary cycles which, because of the priority effect, strongly determine the evolutionary potential of the fungal population and thus the epidemic speed. Third, integrating a transmission character that is heritable and under selection could be of great help to further consider the evolution of virulence in the more realistic context of the virulence–transmission trade-off [14] within managed forest plantations.

5. Conclusions

Our study shows that frequent clear-cutting operations accelerate virulence evolution and propagation in root-rot fungal pathogens in forest plantations. More generally, this study yields insights into the evolutionary and epidemiological dynamics of a fungus combining sexual and asexual reproduction modes and showing a saprotrophic ability. Although more investigations are needed to further characterize the dynamics identified here, this work shows that genetically-explicit and individual-based models can be useful to improve our understanding of the relationship between forest management and spread of epidemics.

Supplementary Materials: The following are available online at www.mdpi.com/1999-4907/8/6/205/s1, Figure S1: total number of infected trees.

Acknowledgments: We thank Céline Meredieu, Stéphanie Mariette and Antoine Kremer for helpful discussions during model development. We thank three anonymous reviewers for their constructive comments. Computer time for this study was provided by the computing facilities MCIA (Mésocentre de Calcul Intensif Aquitain) of the Université de Bordeaux and of the Université de Pau et des Pays de l'Adour.

Author Contributions: All of the authors designed the study, identified the different scenarios to be tested, and wrote the paper. J.P.S. developed the model and realized the simulation work.

Conflicts of Interest: The authors declare no conflict of interest.

References

1. FAO. Planted forests are a vital resource for future green economies. In *Planted Forests Increasing in Importance Worldwide, Proceedings of the Third International Congress on Planted Forests, Estoril, Portugal, 15–21 May 2013*; FAO: Roma, Italy, 2013; p. 6.
2. WWF (World Wildlife Fund). *WWF's Living Forest Report: Chapter 4—Forests and Wood Products*; World Wildlife Fund: Washington, DC, USA, 2013.
3. Santini, A.; Ghelardini, L.; Pace, C.D.; Desprez-Loustau, M.L.; Capretti, P.; Chandelier, A.; Hantula, J. Biogeographical patterns and determinants of invasion by forest pathogens in Europe. *New Phytol.* **2013**, *197*, 238–250.
4. Ennos, R.A. Resilience of forests to pathogens: An evolutionary ecology perspective. *Forestry* **2015**, *88*, 41–52.
5. Wingfield, M.J.; Brockerhoff, E.G.; Wingfield, B.D.; Slippers, B. Planted forest health: The need for a global strategy. *Science* **2015**, *349*, 832–836.
6. Desprez-Loustau, M.L.; Aguayo, J.; Dutech, C.; Hayden, K.J.; Husson, C.; Jakushkin, B.; Vacher, C. An evolutionary ecology perspective to address forest pathology challenges of today and tomorrow. *Ann. For. Sci.* **2016**, *73*, 45–67.
7. Perkins, T.A. Evolutionarily labile species interactions and spatial spread of invasive species. *Am. Nat.* **2012**, *179*, E37–E54.
8. Sapoukhina, N.; Durel, C.E.; Le Cam, B. Spatial deployment of gene-for-gene resistance governs evolution and spread of pathogen populations. *Theor. Ecol.* **2009**, *2*, 229–238.
9. Perkins, T.A.; Phillips, B.L.; Baskett, M.L.; Hastings, A. Evolution of dispersal and life history interact to drive accelerating spread of an invasive species. *Ecol. Lett.* **2013**, *16*, 1079–1087.
10. Burdon, J.J.; Thrall, P.H. Pathogen evolution across the agro-ecological interface: Implications for disease management. *Evolut. Appl.* **2008**, *1*, 57–65.
11. Berngruber, T.W.; Froissart, R.; Choisy, M.; Gandon, S. Evolution of virulence in emerging epidemics. *PLoS Pathog.* **2013**, *9*, e1003209.
12. Cressler, C.E.; McLEOD, D.V.; Rozins, C.; Van Den Hoogen, J.; Day, T. The adaptive evolution of virulence: A review of theoretical predictions and empirical tests. *Parasitology* **2016**, *143*, 915–930.
13. Lannou, C. Variation and selection of quantitative traits in plant pathogens. *Annu. Rev. Phytopathol.* **2012**, *50*, 319–338.
14. Alizon, S.; Michalakis, Y. Adaptative virulence evolution: The good old fitness-based approach. *Trends Ecol. Evol.* **2015**, *30*, 248–254.

15. Dieckmann, U. *Adaptive Dynamics of Pathogen-Host Interactions*; IIASA Interim Report; IIASA: Laxenburg, Austria, 2002.
16. Griette, Q.; Raoul, G.; Gandon, S. Virulence evolution at the front line of spreading epidemics. *Evolution* **2015**, *69*, 2810–2819.
17. Day, T.; Proulx, S.R. A general theory for the evolutionary dynamics of virulence. *Am. Nat.* **2004**, *163*, E40–E63.
18. Day, T.; Gandon, S. Applying population-genetic models in theoretical evolutionary epidemiology. *Ecol. Lett.* **2007**, *10*, 876–888.
19. Bolker, B.M.; Nanda, A.; Shah, D. Transient virulence of emerging pathogens. *J. R. Soc. Interface* **2010**, *7*, 811–822.
20. Bazin, É.; Mathé-Hubert, H.; Facon, B.; Carlier, J.; Ravigné, V. The effect of mating system on invasiveness: Some genetic load may be advantageous when invading new environments. *Biol. Invasions* **2014**, *16*, 875–886.
21. Papaïx, J.; David, O.; Lannou, C.; Monod, H. Dynamics of adaptation in spatially heterogeneous metapopulations. *PLoS ONE* **2013**, *8*, e54697.
22. Fabre, F.; Rousseau, E.; Mailleret, L.; Moury, B. Epidemiological and evolutionary management of plant resistance: Optimizing the deployment of cultivar mixtures in time and space in agricultural landscapes. *Evol. Appl.* **2015**, *8*, 919–932.
23. Bourget, R.; Chaumont, L.; Sapoukhina, N. Timing of pathogen adaptation to a multicomponent treatment. *PLoS ONE* **2013**, *8*, e71926.
24. Chevin, L.M.; Lande, R.; Mace, G.M. Adaptation, plasticity, and extinction in a changing environment: Towards a predictive theory. *PLoS Biol.* **2010**, *8*, e1000357.
25. Segelbacher, G.; Cushman, S.A.; Epperson, B.K.; Fortin, M.J.; Francois, O.; Hardy, O.J.; Manel, S. Applications of landscape genetics in conservation biology: Concepts and challenges. *Conserv. Genet.* **2010**, *11*, 375–385.
26. Lion, S.; Gandon, S. Evolution of spatially structured host–parasite interactions. *J. Evolut. Biol.* **2015**, *28*, 10–28.
27. Borshchev, A.; Filippov, A. From system dynamics and discrete event to practical agent based modeling: Reasons, techniques, tools. In Proceedings of the 22nd International Conference of the System Dynamics Society, Oxford, UK, 25–29 July 2004; Volume 22.
28. Evans, M.R.; Grimm, V.; Johst, K.; Knuuttila, T.; De Langhe, R.; Lessells, C.; Merz, M.; Wilkinson, D.J. Do simple models lead to generality in ecology? *Trends Ecol. Evol.* **2013**, *28*, 578–583.
29. Stillman, R.A.; Railsback, S.F.; Giske, J.; Berger, U.; Grimm, V. Making predictions in a changing world: The benefits of individual-based ecology. *BioScience* **2015**, *65*, 140–150.
30. Le Corre, V.; Machon, N.; Petit, R.J.; Kremer, A. Colonization with long-distance seed dispersal and genetic structure of maternally inherited genes in forest trees: A simulation study. *Genet. Res.* **1997**, *69*, 117–125.
31. Austerlitz, F.; Mariette, S.; Machon, N.; Gouyon, P.H.; Godelle, B. Effects of colonization processes on genetic diversity: Differences between annual plants and tree species. *Genetics* **2000**, *154*, 1309–1321.
32. Yeaman, S.; Guillaume, F. Predicting adaptation under migration load: The role of genetic skew. *Evolution* **2009**, *63*, 2926–2938.
33. Scheiner, S.M.; Barfield, M.; Holt, R.D. The genetics of phenotypic plasticity. XI. Joint evolution of plasticity and dispersal rate. *Ecol. Evolut.* **2012**, *2*, 2027–2039.
34. Soularue, J.P.; Kremer, A. Assortative mating and gene flow generate clinal phenological variation in trees. *BMC Evolut. Biol.* **2012**, *12*, 79.
35. Soularue, J.P.; Kremer, A. Evolutionary responses of tree phenology to the combined effects of assortative mating, gene flow and divergent selection. *Heredity* **2014**, *113*, 485–494.
36. Schiffers, K.; Bourne, E.C.; Lavergne, S.; Thuiller, W.; Travis, J.M. Limited evolutionary rescue of locally adapted populations facing climate change. *Philos. Trans. R. Soc. B* **2013**, *368*, 20120083.
37. Bourne, E.C.; Bocedi, G.; Travis, J.M.; Pakeman, R.J.; Brooker, R.W.; Schiffers, K. Between migration load and evolutionary rescue: Dispersal, adaptation and the response of spatially structured populations to environmental change. *Proc. R. Soc. Lond. B Biol. Sci.* **2014**, *281*, 20132795.
38. Guillaumin, J.J.; Mohammed, C.; Anselmi, N.; Courtecuisse, R.; Gregory, S.C.; Holdenrieder, O.; Rishbeth, J. Geographical distribution and ecology of the Armillaria species in western Europe. *For. Pathol.* **1993**, *23*, 321–341.

39. Morrison, D.; Mallett, K. Silvicultural management of armillaria root disease in western Canadian forests. *Can. J. Plant Pathol.* **1996**, *18*, 194–199.

40. Garbelotto, M.; Gonthier, P. Biology, epidemiology, and control of Heterobasidion species worldwide. *Annu. Rev. Phytopathol.* **2013**, *51*, 39–59.

41. Barrett, L.G.; Thrall, P.H.; Burdon, J.J.; Linde, C.C. Life history determines genetic structure and evolutionary potential of host–parasite interactions. *Trends Ecol. Evolut.* **2008**, *23*, 678–685.

42. Redfern, D.B.; Filip, G.M. Inoculum and infection. *Agric. Handb. USA* **1991**, *691*, 48–61.

43. Labbé, F.; Lung-Escarmant, B.; Fievet, V.; Soularue, J.P.; Laurent, C.; Robin, C.; Dutech, C. Variation in traits associated with parasitism and saprotrophism in a fungal root-rot pathogen invading intensive pine plantations. *Fungal Ecol.* **2017**, *26*, 99–108.

44. Lung-Escarmant, B.; Guyon, D. Temporal and spatial dynamics of primary and secondary infection by Armillaria ostoyae in a Pinus pinaster plantation. *Phytopathology* **2004**, *94*, 125–131.

45. Labbé, F.; Marcais, B.; Dupouey, J.L.; Bélouard, T.; Capdevielle, X.; Piou, D.; Dutech, C. Pre-existing forests as sources of pathogens? The emergence of Armillaria ostoyae in a recently planted pine forest. *For. Ecol. Manag.* **2015**, *357*, 248–258.

46. Bürger, R.; Lynch, M. Adaptation and extinction in changing environments. In *Environmental Stress, Adaptation and Evolution*; Birkhäuser: Basel, Switzerland, 1997; pp. 209–239.

47. Holt, R.D.; Gomulkiewicz, R.; Barfield, M. The phenomenology of niche evolution via quantitative traits in a 'black-hole' sink. *Proc. R. Soc. Lond. B Biol. Sci.* **2008**, *270*, 215–224.

48. Python Software Fundation (US). Available online: https://www.python.org/ (accessed on 8 June 2017).

49. Walt, S.V.D.; Colbert, S.C.; Varoquaux, G. The NumPy array: A structure for efficient numerical computation. *Comput. Sci. Eng.* **2011**, *13*, 22–30.

50. Hunter, J.D. Matplotlib: A 2D graphics environment. *Comput. Sci. Eng.* **2007**, *9*, 90–95.

51. Legrand, P.; Ghahari, S.; Guillaumin, J.J. Occurrence of genets of *Armillaria* spp. in four mountain forests in Central France: The colonization strategy of Armillaria ostoyae. *New Phytol.* **1996**, *133*, 321–332.

52. Prospero, S.; Lung-Escarmant, B.; Dutech, C. Genetic structure of an expanding Armillaria root rot fungus (*Armillaria ostoyae*) population in a managed pine forest in southwestern France. *Mol. Ecol.* **2008**, *17*, 3366–3378.

53. Dutech, C.; Labbé, F.; Capdevielle, X.; Lung-Escarmant, B. Genetic analysis reveals efficient sexual spore dispersal at a fine spatial scale in *Armillaria ostoyae*, the causal agent of root-rot disease in conifers. *Fungal Biol.* **2017**, doi:10.1016/j.funbio.2017.03.001.

54. Bendel, M.; Kienast, F.; Rigling, D. Genetic population structure of three *Armillaria* species at the landscape scale: A case study from Swiss *Pinus mugo* forests. *Mycol. Res.* **2006**, *110*, 705–712.

55. Prospero, S.; Holdenrieder, O.; Rigling, D. Primary resource capture in two sympatric *Armillaria* species in managed Norway spruce forests. *Mycol. Res.* **2003**, *107*, 329–338.

56. Ferguson, B.A.; Dreisbach, T.A.; Parks, C.G.; Filip, G.M.; Schmitt, C.L. Coarse-scale population structure of pathogenic Armillaria species in a mixed-conifer forest in the Blue Mountains of northeast Oregon. *Can. J. For. Res.* **2003**, *33*, 612–623.

57. Kimura, M.; Weiss, G.H. The stepping stone model of population structure and the decrease of genetic correlation with distance. *Genetics* **1964**, *49*, 561.

58. Lynch, M.; Walsh, B. *Genetics and Analysis of Quantitative Traits*; Sinauer: Sunderland, MA, USA, 1998; Volume 1, p. 4.

59. Orr, H.A. Fitness and its role in evolutionary genetics. *Nat. Rev. Genet.* **2009**, *10*, 531–539.

60. Peng, B.; Kimmel, M.; Amos, C.I. *Forward-Time Population Genetics Simulations: Methods, Implementation, and Applications*; John Wiley & Sons, Inc.: Hoboken, NJ, USA, 2012.

61. Rishbeth, J. Armillaria in an ancient broadleaved woodland. *For. Pathol.* **1991**, *21*, 239–249.

62. Hanrahan, J.P.; Eisen, E.J.; Legates, J.E. Effects of population size and selection intensity on short-term response to selection for postweaning gain in mice. *Genetics* **1973**, *73*, 513–530.

63. Hiscox, J.; Savoury, M.; Müller, C.T.; Lindahl, B.D.; Rogers, H.J.; Boddy, L. Priority effects during fungal community establishment in beech wood. *ISME J.* **2015**, *9*, 2246–2260.

64. Lande, R.; Shannon, S. The role of genetic variation in adaptation and population persistence in a changing environment. *Evolution* **1994**, *50*, 434–437.

65. Barrett, R.D.; Schluter, D. Adaptation from standing genetic variation. *Trends Ecol. Evolut.* **2008**, *23*, 38–44.

66. Prentis, P.J.; Wilson, J.R.; Dormontt, E.E.; Richardson, D.M.; Lowe, A.J. Adaptive evolution in invasive species. *Trends Plant Sci.* **2008**, *13*, 288–294.

67. Otto, S.P.; Lenormand, T. Resolving the paradox of sex and recombination. *Nat. Rev. Genet.* **2002**, *3*, 252–261.

68. Austerlitz, F.; Garnier-Géré, P.H. Modelling the impact of colonisation on genetic diversity and differentiation of forest trees: Interaction of life cycle, pollen flow and seed long-distance dispersal. *Heredity* **2003**, *90*, 282–290.

69. Pukkala, T.; Möykkynen, T.; Thor, M.; Rönnberg, J.; Stenlid, J. Modeling infection and spread of Heterobasidion annosum in even-aged Fennoscandian conifer stands. *Can. J. For. Res.* **2005**, *35*, 74–84.

70. Noël, E.; Jarne, P.; Glémin, S.; MacKenzie, A.; Segard, A.; Sarda, V.; David, P. Experimental evidence for the negative effects of selffertilization on the adaptive potential of populations. *Curr. Biol.* **2017**, *27*, 237–242.

![forests logo] *forests*

MDPI

Review

Armillaria Pathogenesis under Climate Changes

Katarzyna Kubiak [1,*], Anna Żółciak [1], Marta Damszel [2], Paweł Lech [1] and Zbigniew Sierota [3]

[1] The Forest Research Institute, Department of Forest Protection, Sękocin Stary, ul. Braci Leśnej 3, 05-090 Raszyn, Poland; a.zolciak@ibles.waw.pl (A.Ż.); p.lech@ibles.waw.pl (P.L.)

[2] Department of Entomology, Phytopathology and Molecular Diagnostics, University of Warmia and Mazury in Olsztyn, Prawocheńskiego 17, 10-721 Olsztyn, Poland; marta.damszel@uwm.edu.pl

[3] Department of Forestry and Forests Ecology, University of Warmia and Mazury in Olsztyn, Plac Łódzki 2, 10-727 Olsztyn, Poland; zbigniew.sierota@uwm.edu.pl

* Correspondence: k.kubiak@ibles.waw.pl; Tel.: +48-22-715-0-635

Academic Editors: Matteo Garbelotto, Paolo Gonthier and Timothy A. Martin
Received: 21 January 2017; Accepted: 23 March 2017; Published: 27 March 2017

Abstract: Climate changes influencing forest ecosystems include increased air temperatures and CO_2 concentrations as well as droughts and decreased water availability. These changes in turn effect changes in species composition of both host plants and pathogens. In the case of *Armillaria*, climate changes cause an increase in the activity of individual species and modify the growth of rhizomorphs, increasing the susceptibility of trees. The relationship between climate changes and the biotic elements of *Armillaria* disease are discussed in overview.

Keywords: *Armillaria*; rhizomorphs; climate changes

---

1. Introduction

The total CO_2 amount together with other greenhouse gases in the atmosphere has led to increasing global temperatures [1,2]. Many authors have suggested that concentrations of CO_2 will increase to as much as 550 ppm by 2050 and 1250 ppm by the end of the twenty-first century [3–5]. Pearson and Dawson [6] reviewed models of climate change impacts on biodiversity and on the distribution of vegetation, and they point out the role of other important factors such as "biotic interactions, evolutionary change and dispersal ability." Global warming will change the special diversity and functional activity of forests—affecting, for example, factors such as photosynthetic rate, tree dieback and wood production, as well as the activity of pests and pathogens—in Central Europe [3,7]. Wargo [8,9] found that defoliation of trees can substantially decrease the starch content in the root wood and the sucrose levels in both bark and cambial tissues of sugar maple roots, which should decrease the attractiveness of these tissues for root pathogens. However, defoliation may also reduce the nutrition available to mycorrhizae and reduce their ability to contribute to water and nutrient uptake for the tree [10].

Taking into consideration that conifers will probably suffer the impacts of global warming in South and Central Europe, including shifting climate zones, one can logically expect changes in the geographic distribution of boreal species. This may include an expansion of deciduous species in the north-east direction, and a decrease in conifer abundance, mainly Norway spruce [11]. "Increased temperatures and subsequent drying due to climate change will increase the presence of "new" hosts on "new" sites." These biotic shifts may favor the initiation of new disease phenomena and increases in invasive species (e.g., those on the European and Mediterranean Plant Protection Organization Alert List), affecting the life of the "new" forest ecosystems [12]. As the coniferous habitats retreat, however, the potential increase in the range of deciduous forests (especially oak, beech, and alder) is unlikely to be fully realized, if only because of the trees dying due to active phytophthorosis and simultaneous *Armillaria* rot in these species.

Atmospheric CO_2 concentrations, warming, and altered precipitation regimes can limit both ecosystem productivity and activity, and change community composition and function [13–15], favoring organisms better adapted to higher temperatures and faster growth rates [16–18]. However, unfavorable relationships between fungi and CO_2 in the forests have also been detected. With the damage to 80- to 100-year-old Scots pine stands by root pathogens reaching 20% of the trees per ha, and up to 80% for Norway spruce, the degradation of roots and stumps could amount to 15.3 m^3/ha of pine and 100.7 m^3/ha of spruce [19]. Assuming that the proportion of the cellulose and lignin together average 70% of wood pulp, one can easily calculate how much CO_2 is released as a result of fungal enzymatic degradation. Sierota [19] estimates that from one hecture of 80-year-old pine stand, 60 tons of CO_2 is released in this way per year, or 3.3 Mt for all Polish forests. However, root pathogens are not the only organisms decaying wood in the forest—one should also take into account saprotrophs colonizing the so-called "dead wood".

Climate change will increase the frequency of extreme climatic conditions, such as droughts, floods, and hurricanes. Forest damage by wind and snow are projected to increase. Some insect species will profit from increasing temperatures, especially at their northern limits. The risk of outbreaks will probably be increased by milder winters that ease the survival of insect and pathogen species during hibernation [20]. In such conditions, there is a high probability that forests will be subject to increased frequency and intensity of fungal diseases. The effects may become more disastrous because of drought and flooding, which are known as factors that predispose trees to several pathogens [21]. Changes in temperature and humidity affect pathogen sporulation and dispersal, and so changes in climatic conditions may favor certain pathogens [22]. Sudden weather changes may increase the incidence of dispersal of pathogens, thereby changing the geographic extent of diseases. Pathogens can meet new hosts and new vectors which may lead to the emergence of new risks due to changes in the species composition of trees or because of invasive species [21].

A. mellea (Vahl. Fr) Kummer and *A. ostoyae* (Romagnesi) Herink are already known to be strong pathogens present in Europe, but they will further benefit from a situation where trees become more stressed due to climatic effects [23] or risks posed to storm-damaged stands [21,24]. *Armillaria* spp. grow at temperatures between 10 and 31 °C, with an optimum between 20 and 22 °C [25,26]. According to Rind and Losel, the mycelial growth of *A. mellea* and *A. gallica* appears to be greater at higher temperatures [21,27]. This may mean that in a warming climate, low soil temperatures probably will not restrict the growth of rhizomorphs during winter as it usually happens in most northern temperate zones [21]. *A. gallica* will probably prefer to attack trees weakened more by changing climatic factors, both because this species of fungus is quite thermophilic (*A. Mellea*), and because it acts like a weak parasite that attacks and sometimes kills weakened trees. *Armillaria gallica* is a likely candidate for an opportunistic pathogen that may become problematic due to climate change, due to increased stress of "host" trees caused by increased temperatures and drought [21,28].

The global rising temperature of the atmosphere and the soils can accelerate the mycelial growth of root pathogens and therefore increase wood decay and CO_2 release. Pastor and Post [29] indicated an important role for temperature, water availability, and nitrogen uptake in the vitality of forest ecosystems. Schwarze et al. [30] found that the temperature range for mycelial growth and wood decay is 5–30 °C, although many fungi can grow at higher temperatures, even +55 °C. However, Witomski [31] found that the optimal temperature range for wood decay is 18–27 °C and below or above this range the decay of wood tissues is decreased.

Pathogens and saprotrophs decomposing timber (mostly polypore basidiomycetes) showed better growth with higher temperatures of growth medium. These fungi benefited from the water formed during enzymatic decomposition, as the humidity of the substrate increases with the degree of decomposition of the timber tissue [32,33]. The development of hyphae of wood decay fungi also occurs in winter, when it is supported by high osmotic pressure and thermal energy emitted during the decomposition of cellulose and lignin. The low concentration of CO_2 in the rhizosphere, on the root surface and inside the wood, has a significant beneficial effect on the development of the spores and mycelia of many different fungi promoting their colonization of the dead wood on the forest floor [34–37]. Wells and Boddy [37] found that increased temperatures had a positive effect on soil phosphorus uptake by the mycelia of some basidiomycetes. Because temperature is the main factor affecting organic matter decomposition, taxa that regulate decomposition, such as fungi and macro-arthropods, may shift their activity or community composition in response to warming [38,39]. According to Austin [38], wood decomposition increases with temperature, but the impact of temperature might vary at sites with different climatic regimes and decomposer communities. The author predicted that warming would have larger impacts on fungal community composition at sites already under heat stress [38].

The direct and indirect impacts of climatic changes affect the health of forests in part by influencing the development of fungal diseases and insect pests [40–42]. Increases in root diseases, particularly those caused by *Armillaria* pathogens [41–43], contribute notably to these impacts. Diseases caused by this fungus can increase under drought conditions and temperature increases [41], which can lead to the reduced tree growth and attacks by secondary pests [44]. Ayres and Lombardero [45] indicated that climate change affects fungal diseases by (i) direct impact on the growth of hosts and pathogens; (ii) changes in tree defense mechanisms; and/or (iii) an indirect effect on the mutualist and competitive organisms and others in the environment, manifesting in changes in abundance and frequency. Klopfenstein et al. [40] proposed a mathematical model to predict the impact of climate change on the pathogen *A. ostoyae* attacking Douglas fir in the northwestern USA. He emphasized that the integration of a variety of tools and data is necessary to improve forecasts for the influence of climate on forest diseases.

The multifunctional interactions between plants and soil communities influenced the selective pressures of pathogens, including *Armillaria* spp., on the functional features of the plants [46,47]. Because the responses of soil microorganisms to *Armillaria* pathogenesis are not well-known, it is still of great importance to investigate how the climatic parameters could influence not only the *Armillaria* pathosystem and host-tree susceptibility, but also soil community composition and microbial activity.

2. The Pathogen

For many decades, root and butt rot caused by *Armillaria mellea* (*sensu lato*) has been a significant threat to the boreal forests of Europe and North America. Currently, it seems to have become the most important phytopathological problem in weakened managed forests [48]. Expansion of the pathogen in colonized tissues is rather fast, depending on the tree species and health status of the host as well as on the vigor of the mycelium (which in turn is affected by the species and clone of *Armillaria* acting as the pathogen) [49,50]. The pathogenesis of *Armillaria* is described in many papers and books; however, the role of the environment in this process is rather neglected [51,52]. The intensity of losses caused by the pathogen is directly connected with changes in the climate and with weather anomalies. Long and frequent drought periods, increases in global and local CO_2, summer reduction of soil moisture, and escape of available water are all relatively strong factors [53]. The deep impact of soil drought on *Armillaria* rot disease has been described by previous authors [40,54–57]. Żółciak et al. [58] found that after drought in Poland in 2000, the disease was noted in an area of 150 thousand hectares in Norway spruce stands as late as 2005. The interactions between air and soil CO_2 and *Armillaria* behavior have been the subject of current and past investigations [19,36,53].

The life cycle of *Armillaria* spp. is very complicated, and is characterized by: different gametogenesis cycles (mono-, di-, eu-karyotic), different phases in ecological status (pathogenic, saprotrophic, orchid-like symbiotic), different methods of infection (basidiospores, mycelium, rhizomorphs), and different host reactions (tissue compartmentalization, resin outflow, host dying, wood decay) [59]. This sophisticated behavior can be additionally modified by site conditions, weather anomalies, and human activity, as it was described in Canadian boreal forests [60]. While *Armillaria ostoyae* favors fertile mountain forests and either pure Norway spruce or mixed spruce-beech stands in Poland [36], other *Armillaria* species can inhabit various forest stands in many sites [61]. There is lack of publications about different species of *Armillaria* in different habitats within the context of climate changes.

The pathogenicity of the known *Armillaria* species depends on the individual virulence, host species, age of the tree, and influence of the environment [62–66]. In Europe, Sicoli et al. [67] found that *A. mellea* (Vahl: Fr.) P. Kummer and *A. gallica* Marxmüller & Romagnesi were the most pathogenic species for some *Quercus* spp. seedlings, whereas other authors showed *A. ostoyae* (Romagnesi) Herink as being the most dangerous for conifer species [21,63]. *Armillaria borealis* Marxmüller & Korhonen, *A. cepistipes* Velenovsky, and *A. gallica* are generally described as weakly pathogenic species or pathogens of weak trees [28,68], while *A. tabescens* (Scop.: Fr.) Emel. is regarded as a typical saprotroph. Nevertheless, some of the opportunistic parasite species such as *A. ectypa* can infect some stressed trees [69]. *Armillaria ectypa* (Fr.) Lamoure plays a rather minor role as a disease perpetrator [70]. The number of *Armillaria* species worldwide is still uncertain; recently some isolated species were identified as *A. nigritula* Orton or renamed *A. altimontana* Brazee, B. Ortiz, Banik & D.L. Lindner (previously described as NABS X) [71–73].

Armillaria rhizomorphs are described as the main source of threat to roots in the pathogenesis of infection. They may also be considered to be an example of a special morphological adaptation of this fungus to different environmental conditions [74,75].

3. The Rhizomorphs

Armillaria commonly occur as rootlike rhizomorphs growing on plant debris or epiphytically attached to the root system of dead, diseased, or healthy host plants [74–79]. Rhizomorphs look like roots or cords that are a dark brown color when old and a reddish brown color when young. Hence, sometimes rhizomorphs have been referred to as mature (black) and immature (red or brown) [80,81] or as maternal (old) and regenerated (young) because of their regenerative abilities [82]. Rhizomorphs grow towards the soil surface, possibly due to the oxygen gradient in soil [83]; however, the seasonal humidity in this layer is what probably adjusts its vertical distribution (e.g., black rhizomorphs were mostly found deeper in dry sites) [25,83]. Rykowski [82] described black rhizomorphs in the soil of rich deciduous stands as persistent organs without infection ability, whereas young, red rhizomorphs were formed mostly in plantations, infecting pines after the removal of stands.

Different species of *Armillaria* produce rhizomorphs with either a monopodial or dichotomous branching pattern [84–86]. Rhizomorphs typically grow in the soil, but they can also occur on dead trees, stumps, and even on the surfaces of living roots. Where the hosts have physical defects, *Armillaria* form apical meristems which can produce a large number of new rhizomorphs [82,87–89]. Rhizomorphs and their growing-tip hyphae are a main source of inoculum, initiating the infection processes and host reactions [82,90].

In the initiation and development of rhizomorphs, environmental factors such as moisture, soil temperature, pH, nutrients, and pollution play an important role [25,51,91–93]. These factors determine the proper functioning of apical meristems and the rhizomorphs' growth [78,82]. Redfern [81] and Kessler and Möser [94] found that low soil moisture and temperature inhibit this growth and branching. A temperature higher than 22 °C is preferred [95]; however, 30 °C limited the growth of the fungus due to enzyme inactivation [96]. Rhizomorphs can grow in different types of forest and farmland soils. Sandy soils, on the other hand, inhibit their production not only because of low nutrients but also due to a high day-night temperature amplitude [25,50,97]. Peat soils are conducive to the development of rhizomorphs, which tend to be concentrated mostly in the humus layer of the soil due to the oxygen concentration gradient. This may be related to increased susceptibility to infections around the root collars of trees [90,97,98]. In addition, *Armillaria* habitats have moist layers of substrate and low levels of oxygen and organic matter, the latter of which is digested by the soil acidic reaction products of decomposers [99]. Mallett and Meynard [100] indicated an increase in the severity of fungus root rot when increasing the content of sand in the mineral substrate layer and decreasing the content of NH^{4+}

in the organic layer of the soil, and Singh [95] found that plants in good condition produce a callus, which initiates an effective barrier against *Armillaria* infection [95].

Lech and Żółciak [101] observed stimulation of *A. ostoyae* rhizomorph production by elevated air CO_2 concentration in a chamber experiment. Hintikka [34] noted the stimulation of rhizomorph growth on a medium under high CO_2 concentrations. Schinner and Concin [102] showed the ability of some wood rotting fungi, including *Armillaria* spp., to assimilate CO_2 from the air. The rate of this assimilation was low, however, reaching only 1.3 nCi/g of dry mass in the case of *Armillaria* spp, which means that 1 g of fungus tissue contained just 0.017 mg of carbon coming from atmospheric CO_2. Unfortunately, studies devoted primarily to the relationship between forest tree species (hosts) and *Armillaria* spp. (pathogens) in a CO_2-enriched atmosphere and/or under increased temperatures are lacking, which makes it extremely difficult to predict the future behavior and functioning of this fungal genus in forest ecosystems under climate change.

4. The Hosts

There is yet another aspect of climate change effects on trees. According to the growth-differentiation balance hypothesis (GDBH) [103], the elevation of air CO_2 may cause a decrease in trees susceptibility towards herbivores as the augmented availability of carbon resources due to intensified photosynthesis is allocated to secondary metabolites rather than to growth. It was found that plants grown in high air CO_2 concentrations were characterized by a higher C:N ratio, and increased lignification and phenolic compound content in the tissues compared to plants from ambient air CO_2 conditions [104,105]. However, Fleischman et al. [106] found atmospheric CO_2 elevation up to ca. 700–800 ppm to cause an increase in beech seedlings' susceptibility to infection by root pathogen *Phytophthora citricola*. Similarly, Tkaczyk et al. [107] reported a decrease of fine root biomass of beech seedlings grown under 800 ppm air CO_2 and exposed to *Phytophthora plurivora* and *Ph. cactorum* artificial inoculation. Oszako et al. [108] had similar results with oak seedlings and *Ph. quercina*.

Pollutants may affect the severity of *Armillaria* root disease on host plants. Impacts associated with SO_2 and other pollutants have been described in the past by many authors [109–113]. Domański [113] found that *Armillaria* root disease was extremely rare in polluted zones but was quite common in plots uninjured by pollution, while Horak and Tesche [114] described an increased mortality of trees both infected by *A. ostoyae* and fumigated with SO_2. According to Wargo et al. [115], lead and other heavy metals present in the soils of spruce-fir sites at high elevations inhibit both mycelial and rhizomorph growth in culture.

Human land-use practices can also influence the host-parasite relationship. Sicoli et al. [116] reported that *A. mellea* attacks on *Cedrus atlantica* were predisposed by the specific soil previously used for pasture causing an iron deficiency (as indicated by reduced growth and chlorotic leaves). Silvicultural treatments such as thinning may result in the spread of *A. ostoyae* via root contacts, mainly in juvenile stands [117]. It should be remembered that the *Armillaria*-decayed roots and rhizomorphs remain as inoculum in the soil for many years. The impact of *Armillaria* inoculum from coniferous stands on the roots of entering deciduous trees (for example, in reconstruction after conifer monocultures) is not fully understood [118]. On the other hand, the higher temperatures and humidity of soil in cleared sites can indirectly protect roots by the mycoparasitism phenomenon, as occurs in the parasitism of *A. gallica* by mycelia of *Trichoderma* species [119]. Note, however, that the reverse relationship can occur: Oomycetes seem to be indicated as the primary pathogens predisposing deciduous trees for *Armillaria* attack [120]. In other biotic interactions of note, Riffle [121] and Cayrol et al. [122] found that nematodes (*Aphelenchus avenae* Bastian, *A. cibolensis* Riffle, and *A. composticola* Franklin) can actively destroy the mycelium of *Armillaria*. The qualitative and quantitative composition of compounds secreted by the roots often determines the development of the antagonistic microorganisms in the soil [123]. For example, *A. mellea* can be inhibited by gastrodianin (an anti-fungal protein from the parasitic plant *Gastrodia* which is affected by *Trichoderma viride* present in the soil [124,125]).

In areas transitioning from coniferous to deciduous trees, seasonal changes in oxygen and CO_2 production due to photosynthesis and respiration are inevitable. Therefore, the additional supply of CO_2. As the result of the decomposition of wood by pathogens and saprotrophs can tip the local and global carbon balance [19].

5. The Microbial Soil Community

Soil microbial communities are responsible for mineralization, decomposition, and nutrient cycling. These communities could be affected by abiotic factors in the climate such as temperature, moisture, and soil nutrient availability, or by biotic factors, particularly interactions with other microorganisms [126]. Certain interactions between antagonists and pathogens are initiated when the antagonistic fungi are under stress, especially nutrient stress, which has a direct impact on the growth, morphogenesis, and organogenesis of the antagonists [127]. The antagonistic fungi can receive nitrogen in the form of ammonium at high doses and they metabolize this form of nitrogen more intensively than do the pathogens. However, the pathogens have a relative advantage over antagonists when ammonium has been used up or the available nitrogen is in another form [127]. It is also known that many of the fungal isolates increase the formation of spores and chlamydospores when the concentration of nitrogen increases [128]. *Armillaria* species probably create rhizomorphs in similar situations.

Many authors report that global warming directly affects the metabolism and respiration of soil communities and the ratio of Gram-positive vs. Gram-negative bacteria [129], because these are sensitive to temperature [17,130–133] both in short-term and long-term scales [134]. The changes in temperature combined with the concomitant changes in soil moisture potentially affect many groups of fungal and bacterial communities [135]. Soil fungal communities shift from one dominant member to another while less-plastic bacterial communities remain more constant [136,137].

Greater or less active protection of plants against pathogens in the soil can result from the lack of ectomycorrhizal fungi as a result of climate changes (mainly drought) [138,139]. The insufficiency of mycorrhizas may stop or slow the adaptation of trees to new sites [140] and affect the ecosystem functioning [137], which could increase the susceptibility of trees to pathogens because trees are stressed. Symbiotic bacteria belonging to the *Rhizobium* genus [141] and mycorrhizal fungi [142] affect plant productivity by providing nutrients to plants. Mycorrhizal fungi can influence free-living bacterial communities to increase the transfer of nitrogen via mycorrhiza to the host plant [143]. However, interactions between mycorrhizal fungi and the host plant are not always mutualistic and can change due to environmental factors or even under plant stress [144]. Rising temperatures will lead to an increase in the allocation of carbon to mycorrhizal hyphae, which, depending on external conditions, can act as symbionts or parasites [145–148].

6. The Interactions with Insect Pests

Pfeffer [149] noted that after 1947, which was a "dry year," over 90% of trees colonized by bark beetle were attacked by *Armillaria* spp. Madziara-Borusiewicz and Strzelecka [150] reported from the Carpathian region that bark beetle first attack trees previously infected by *Armillaria*. James and Goheen [151] found that over 99% of dead or dying trees were affected by root diseases and 80% of them were simultaneously colonized by secondary pests. Capecki [152] has confirmed that stands attacked by *Armillaria* spp. are most strongly threatened by bark beetle in forests of the western Carpathians and that the occurrence of secondary pests is the natural result of greater susceptibility to disease and the poor sanitary state of the stands. In contrast, Christiansen and Husek [153] did not find any significant difference in rot occurrence between dead trees previously attacked by bark beetles and those not attacked by insects. Similarly, Jankovsky et al. [154] found a lack of relationship between the presence of the bark beetle *Ips typographu* and *Armillaria* spp. infection in spruce stands of the Szumawa Mountains (Czech Republic).

Twery et al. [155] reported the distribution of rhizomorphs of *Armillaria* spp. in uninjured mixed oak stands and in stands which were defoliated 1 and 5 years earlier by insects. Trees weakened by biotic stress were infested by *Armillaria*, with an increased abundance of rhizomorphs observed, especially on plots defoliated 5 years before sampling. The authors consider that trees' predisposal to pest invasions was a result of previous attack by *Armillaria* and deep water stress in the whole root system.

Okland et al. [156] supposed that a warmer climate would influence bark beetle populations, which may move north. The north European spruce forest, so far free from bark beetle outbreaks, showed strongly increased susceptibility to *Ips* spp. in climate change models [157–159]. Similarly, Langvall [160] connected the impact of global warming to the conditions of regeneration of Norway spruce towards the northern and higher elevations in Europe, and significant regional differences in the *I. typographus* behavior, such as voltinism. The negative impacts of drought and *I. typographus* populations in the southern range of the Norway spruce have been described by many authors [161–163]. Infection of spruce roots by *Armillaria* spp can cause production and release of specific compounds, for example, limonene, β-phellandrene, camphene, and bornyl acetate by needles and probably the phloem of weakened trees which are secondary pest attractants [164]. These signals can be specific to infection by *Armillaria*, and can encourage *Ips* spp. beetles to respond to these signals by choosing such trees for settlement [165].

7. Conclusions

Armillaria's life cycle, host susceptibility, and interactions with the soil and climate have been summarized in many papers. However, several gaps in research have been identified and further work could help us to predict climate change impacts on the pathogen and the forests. We hypothesize that climatic changes and global warming are not the only factors predisposing the roots of weakened trees to *Armillaria* infections, but that the bacteria and fungi, as well as macro-, meso-, and micro-organisms growing in the soil environment around root systems can also directly or indirectly enhance the proliferation of the pathogen and decrease the immune barriers in roots. The rhizomorphs are probably also colonized by endogenous bacteria and fungi that stimulate the growth of *Armillaria* hyphae and aid in the destruction of cell walls by the secretion of enzymes. This speculation requires further research.

Acknowledgments: Part of the study was financially supported by the Life Plus project HESOFF, Life 11 ENV/PL/459 financed by the European Union and the National Fund for Environmental Protection and Water Management in Warsaw.

Author Contributions: Katarzyna Kubiak and Zbigniew Sierota designed and conducted the text; all authors contributed to the manuscript preparation according to the professional experience.

Conflicts of Interest: There is no conflict of interest.

References

1. Keeling, C.D.; Whorf, T.P.; Wahlen, M.; van der Plicht, J. Inter annual extremes in the rate of rise of atmospheric carbon dioxide since 1980. *Nature* **1995**, *375*, 666–670. [CrossRef]
2. Intergovernmental Panel on Climate Change (IPCC). *IPPC Fourth Assessment Report in Summary for Policymakers*; IPCC: Geneva, Switzerland, 2007.
3. Bazzaz, F.A. The response of natural ecosystems to the rising global CO_2 levels. *Ann. Rev. Ecol. Syst.* **1990**, *21*, 167–196. [CrossRef]
4. Climate Change 2001. *Synthesis Report. A Contribution of Working Groups I, II, and III to the Third Assessment Report of the Integovernmental Panel on Climate Change*; Watson, R.T., Core Writing Team, Eds.; Cambridge University Press: Cambridge, UK; New York, NY, USA, 2001.
5. Long, S.P.; Ainsworth, E.A.; Leakey, A.D.B.; Nösberger, J.; Ort, D.R. Food for thought: Lower-Than expected crop yield stimulation with rising CO_2 concentrations. *Science* **2006**, *312*, 1918–1921. [CrossRef] [PubMed]

6.   Pearson, R.G.; Dawson, T.P. Predicting the impacts of climate change on the distribution of species: Are bioclimate envelope models useful? *Glob. Ecol. Biogeogr.* **2003**, *12*, 361–371. [CrossRef]

7.   Brzeziecki, B.; Kienast, F.; Wildi, O. Modeling potential impact of climate change on the spatial distribution of zonal forest communities in Switzerland. *J. Veg. Sci.* **1995**, *6*, 257–268. [CrossRef]

8.   Wargo, P.M.; Parker, J.; Houston, D.R. Starch content in roots of defoliated sugar maple. *For. Sci.* **1972**, *18*, 203–204.

9.   Jung, T. Beech decline in Central Europe driven by the interaction between *Phytophthora* infections and climatic extremes. *For. Pathol.* **2009**, *39*, 77–94.

10.   Trocha, L.K.; Weiser, E.; Robakowski, P. Interactive effects of juvenile defoliation, light conditions, and interspecific competition on growth and ectomycorrhizal colonization of *Fagus sylvatica* and *Pinus sylvestris* seedlings. *Mycorrhiza* **2016**, *26*, 47–56. [CrossRef] [PubMed]

11.   Bonan, G.B. Forests and climate change: Forcings, feedbacks, and the climate benefits of forests. *Science* **2008**, *320*, 1444–1449. [CrossRef] [PubMed]

12.   Santini, A.; Ghelardini, L.; De Pace, C.; Desprez-Loustau, M.L.; Capretti, P.; Chandelier, A.; Cech, T.; Chira, D.; Diamandis, S.; Gaitniekis, T.; et al. Biogeographical patterns and determinants of invasion by forest pathogens in Europe. *New Phytol.* **2013**, *197*, 238–250. [CrossRef] [PubMed]

13.   Williams, W.M.A. Response of microbial communities to water stress in irrigated and drought-prone tallgrass prairie soils. *Soil Biol. Biochem.* **2007**, *39*, 2750–2757. [CrossRef]

14.   Blankinship, J.C.; Niklaus, P.A.; Hungate, B.A. A meta-analysis of responses of soil biota to global change. *Oecologia* **2011**, *165*, 553–565. [CrossRef] [PubMed]

15.   De Angelis, K.M.; Pold, G.; Topcuoglu, B.D.; van Diepen, L.T.A.; Varney, R.M.; Blanchard, J.L.; Melillo, J.; Frey, S.D. Long-Term forest soil warming alters microbial communities in temperate forest soils. *Front. Microbiol.* **2015**, *6*, 104.

16.   Pettersson, M.; Baath, E. Temperature-Dependent changes in the soil bacterial community in limed and unlimed soil. *FEMS Microbiol. Ecol.* **2003**, *45*, 13–21. [CrossRef]

17.   Bradford, M.A.; Davies, C.A.; Frey, S.D.; Maddox, T.R.; Melillo, J.M.; Mohan, J.E.; Reynolds, J.F.; Treseder, K.K.; Wallenstein, M.D. Thermal adaptation of soil microbial respiration to elevated temperature. *Ecol. Lett.* **2008**, *11*, 1316–1327. [CrossRef] [PubMed]

18.   Hagerty, S.B.; van Groenigen, K.J.; Allison, S.D.; Hungate, B.A.; Schwartz, E.; Koch, G.W.; Kolka, R.K.; Dijkstra, P. Accelerated microbial turnover but constant growth efficiency with warming in soil. *Nat. Clim. Chang.* **2014**, *4*, 903–906. [CrossRef]

19.   Sierota, Z. Wpływ grzybów rozkładających korzenie drzew leśnych na uwalnianie CO_2 - próba waloryzacji [Effect of fungi decomposing roots of forest trees on CO_2 release—An attempt of evaluation]. *Sylwan* **2012**, *156*, 128–136. (In Polish)

20.   Lindner, M.; Garcia-Gonzalo, J.; Kolström, M.; Green, T.; Reguera, R. *Impacts of Climate Change on European Forests and Options for Adaptation No. AGRI-2007-G4–06*; Report to the European Commission Directorate-General for Agriculture and Rural Development: Brussels, Belgium.

21.   La Porta, N.; Capretti, P.; Thomsen, I.M.; Kasanen, R.; Hietala, A.M.; Von Weissenberg, K. Forest pathogens with higher damage potential due to climate change in Europe. *Can. J. Plant Pathol.* **2008**, *30*, 177–195. [CrossRef]

22.   Tubby, K.V.; Webber, J.F. Pests and diseases threatening urban trees under a changing climate. *Forestry* **2010**, *83*, 451–459. [CrossRef]

23.   Szynkiewicz, A.; Kwasna, H. The susceptibility of forest trees to *Armillaria* root rot. *Sylwan* **2008**, *148*, 25–33. (In Polish)

24.   Brockerhoff, E.G.; Jactel, H.; Goldarazena, A.; Berndt, L.; Bain, J. *Risk assessment of European pests of Pinus radiata*; Client Report No. 12216; ENSIS: Rotorua, New Zealand, 2006.

25.   Rishbeth, J. Effects of soil temperature and atmosphere on growth of *Armillaria* rhizomorphs. *Trans. Brit. Mycol. Soc.* **1978**, *70*, 213–220. [CrossRef]

26.   Keca, N. Characteristics of *Armillaria* species development and their growth at different temperatures. *Bull. Fac. Forstry* **2005**, *91*, 149–162. (In Serbian) [CrossRef]

27.   Rind, B.; Losel, D.M. Effect of nutrients and temperature on the growth of *Armillaria mellea* and other fungi. *Indus J. Biol. Sci.* **2005**, *2*, 326–331.

28. Guillaumin, J.J.; Legrand, P. *Armillaria* root rot. In *Infectious Forest Diseases*; Gonthier, P., Nicolotti, G., Eds.; CABI: Oxfordshire, UK, 2005; Chapter 8; pp. 159–177.

29. Pastor, J.; Post, W.M. Influence of Climate, Soil Moisture, and Succession on Forest Carbon and Nitrogen Cycles. *Biogeochemistry* **1986**, *2*, 3–27. [CrossRef]

30. Schwarze, F.W.M.R.; Engels, J.; Mattheck, C. *Fungal Strategies of Wood Decay in Trees*; Springer-Verlag: Berlin/Heidelberg, Germany, 2000.

31. Witomski, P. *Zmiany Wybranych właściwości fizycznych i Chemicznych Drewna Sosny Zwyczajnej (Pinus sylvestris L.) pod Wpływem Rozkładu Białego i Brunatnego*; SGGW: Warsaw, Poland, 2008. (In Polish)

32. Rayner, A.D.M.; Boddy, L. *Fungal Decomposition of Wood. Its Biology and Ecology*; J. Wiley & Sons Ltd.: Chichester, Sussex, UK, 1988.

33. Sierota, Z. Dry weight loss of wood after the inoculation of Scots pine stumps with *Phlebiopsis gigamtea*. *Eur. J. For. Path.* **1997**, *27*, 179–185. [CrossRef]

34. Hintikka, V. Notes on the ecology of *Armillaria mellea* in Finland. *Karstenia* **1974**, *14*, 12–31.

35. Redfern, D.B. Infection of *Picea sitchensis* and *Pinus contorta* stumps by basidiospores of *Heterobasidion annosum*. *Forest Pathol.* **1982**, *12*, 11–25. [CrossRef]

36. Lech, P.; Żółciak, A. Wzrost sadzonek sosny zwyczajnej i rozwój ryzomorf opienki ciemnej w warunkach podwyższonej koncentracji CO_2 w powietrzu. (Growth of Scots pine seedlings and *Armillaria ostoyae* rhizomorphs under elevated air CO_2, concentration conditions). *Leś. Pr. Bad.* **2006**, *4*, 17–34. (In Polish)

37. Wells, J.M.; Boddy, L. Effect of temperature on wood decay and translocation of soil-derived phosphorus in mycelial cord systems. *New Phytol.* **1995**, *129*, 289–297. [CrossRef]

38. Austin, E. Wood Decomposition in a Warmer World. Ph.D. Thesis, University of Tennessee, Knoxville, TN, USA, 2013.

39. Dang, C.K.; Schindler, M.; Chauvet, E.; Gessner, M.O. Temperature oscillation coupled with fungal community shifts can modulate warming effects on litter decomposition. *Ecology* **2009**, *90*, 122–131. [CrossRef] [PubMed]

40. Klopfenstein, N.B.; Kim, M.; Hanna, J.; Richardson, B.A.; Smith, A.L.; Maffei, H. Predicting potential impacts of climate change on *Armillaria* root disease in the inland northwestern USA. *Phytopathology* **2009**, *99*, S65.

41. Kliejunas, J.T.; Geils, B.; Glaeser, J.M.; Goheen, E.M.; Hennon, P.; Kim, M.-S.; Kope, H.; Stone, J.; Sturrock, R.; Frankel, S. *Climate and Forest Diseases of Western North America: A Literature Review*; U.S. Department of Agriculture, Forest Service, Pacific Southwest Research Station: Albany, CA, USA, 2008; p. 36.

42. Ramsfield, T.D.; Bentz, B.J.; Faccoli, M.; Jactel, H.; Brockerhoff, E.G. Forest health in a changing world: Effects of globalization and climate change on forest insect and pathogen impacts. *Forestry* **2016**, *89*, 245–252. [CrossRef]

43. Worrall, J. *Armillaria* root disease. In *The Plant Health Instructor*; The American Phytopathological Society (APS): St. Paul, MN, USA, 2004. [CrossRef]

44. Battles, J.J.; Robards, T.; Das, A.; Waring, K.; Gilless, J.K.; Biging, G.; Schurr, F. Climate change impacts on forest growth and tree mortality: A data-driven modeling study in a mixed-conifer forest of the Sierra Nevada. *Clim. Chang.* **2008**, *87*, S193–S213. [CrossRef]

45. Ayres, M.P.; Lombardero, M.J. Assessing the consequences of global change for forest disturbance from herbivores and pathogens. *Sci. Total Environ.* **2000**, *262*, 263–286. [CrossRef]

46. Lau, J.A.; Lennon, J.T. Evolutionary ecology of plant-microbe interactions: Soil microbial structure alters selection on plant traits. *New Phytol.* **2011**, *192*, 215–224. [CrossRef] [PubMed]

47. Horst, C.P.; Zee, P.C. Eco-Evolutionary dynamics in plant–soil feedbacks. *Funct. Ecol.* **2016**, *30*, 1062–1072.

48. Hood, I.A.; Redfern, D.B.; Kile, G.A. Armillaria in planted host. In *Armillaria Root Diseases*; Agricultural Handbook No. 691; Show, C.G., III, Kile, G.A., Eds.; U.S.D.A. Forest Service: Washington, DC, USA, 1991; pp. 122–149.

49. Worrall, J.J.; Sullivan, K.F.; Harrington, T.C.; Steimel, J.P. Incidence, host relations and population structure of *Armillaria ostoyae* in Colorado campgrounds. *For. Ecol. Manag.* **2004**, *192*, 191–206. [CrossRef]

50. Narayanasamy, P. Detection of Fungal Pathogens in Plants. In *Microbial Plant Pathogens-Detection and Disease Diagnosis: Fungal Pathogens*; Springer Science+Business Media: Dordrecht, The Netherlands, 2011; Volume 1.

51. McDonald, G.I.; Martin, N.E.; Harvey, A.E. *Armillaria in the Northern Rockies: Pathogenicity and Host Susceptibility on Pristine and Disturbed Areas*; USDA, FS Intermountain Research Station: Ogden, UT, USA, 1987.

52. Van der Putten, W.H.; Klironomos, J.N.; Wardle, D.A. Microbial ecology of biological invasions. *ISME J.* **2007**, *1*, 28–37. [CrossRef] [PubMed]

53. Manabe, S.; Spelman, M.J.; Stouffer, R.J. Transient responses of a coupled ocean atmosphere model to gradual changes of atmospheric CO_2. Part II: Seasonal response. *J. Clim.* **1992**, *5*, 105–126. [CrossRef]

54. Loretto, F.; Burdsall, H.; Tirro, A. *Armillaria* infection and water stress influence gas-exchange properties of mediterranean trees. *Hort Sci.* **1993**, *28*, 222–224.

55. Hadfield, J.S.; Goheen, D.J.; Filip, G.M.; Schmitt, C.L.; Harvey, R.D. *Root Diseases in Oregon and Washington Conifers - R6-FPM-250-86*; U.S.D.A. Forest Service: Portland, OR, USA, 1986.

56. Mullen, J.; Hagan, A. Alabama Cooperative Extension System. Available online: http://www.aces.edu/pubs/docs/A/ANR-0907/ANR-0907.pdf (accessed 15 October 2004).

57. Szewczyk, W.; Kwaśna, H.; Behnke-Borowczyk, J. *Armillaria* population in flood-plain forest of natural pedunculate oak showing oak decline. *Pol. J. Environ. Stud.* **2016**, *25*, 1253–1262. [CrossRef]

58. Żółciak, A.; Lech, P.; Małecka, M.; Sierota, Z. Opieńkowa zgnilizna korzeni a stan zdrowotny drzewostanów świerkowych w Beskidach. *PAU* **2009**, *11*, 61–72. (In Polish)

59. Lamoure, D.; Guillaumin, J.J. The life cycle of the *Armillaria* mellea complex. *Euro. J. For. Pathol.* **1985**, *15*, 288–293. [CrossRef]

60. Maynard, D.G.; Paré, D.; Thiffault, E.; Lafleur, B.; Hogg, K.E.; Kishchuk, B. How do natural disturbances and human activities affect soils and tree nutrition and growth in the Canadian boreal forest? *Environ. Rev.* **2014**, *22*, 161–178. [CrossRef]

61. Żółciak, A. Występowanie grzybów z rodzaju *Armillaria* (Fr.: Fr.) Staude w kompleksach leśnych w Polsce (The occurence of *Armillaria* (Fr.: Fr.) Staude in forests stands in Poland). *Pr. Inst. Bad. Leśn. Ser. A* **1999**, *888*, 21–40. (In Polish)

62. Gregory, S.C.; Rishbeth, J.; Shaw, C.G., III. Pathogenicity and Virulence. In *Armillaria Root Disease*; Agricultural Handbook No. 691; Show, C.G., III, Kile, G.A., Eds.; U.S.D.A. Forest Service: Washington, DC, USA, 1991; pp. 76–87.

63. Guillaumin, J.J.; Mohammed, C.; Anselmi, N.; Courtecuisse, R.; Gregory, S.C.; Holdenrieder, O.; Intini, M.; Lung, B.; Marxmüller, H.; Morrison, D.; et al. Geographical distribution and ecology of the *Armillaria* species in western Europe. *Eur. J. For. Path.* **1993**, *23*, 321–341. [CrossRef]

64. Fox, R.T.V. Pathogenicity. In *Armillaria Root Rot: Biology, and Control of Honey Fungus Section 3: Pathology*; Fox, R.T.V., Ed.; Intercept Ltd.: Andover, UK, 2000; pp. 113–136.

65. Cleary, M.; van der Kamp, B.J.; Morrison, D.J. Pathogenicity and virulence of *Armillaria sinapina* and host response to infection in Douglas-fir, western hemlock and western redcedar in the southern interior of British Columbia. *For. Patolh.* **2012**, *42*, 481–491.

66. Ross-Davis, A.L.; Stewart, J.E.; Hanna, J.W.; Kim, M.S.; Knaus, B.J.; Cronn, R.; Rai, H.; Richardson, B.A.; McDonald, G.I.; Klopfenstein, N.B. Transcriptome of an *Armillaria* root disease pathogen reveals candidate genes involved in host substrate utilization at the host-pathogen interface. *For. Patolh.* **2013**, *43*, 468–477. [CrossRef]

67. Sicoli, G.; Annese, V.; de Gioia, T.; Luisi, N. *Armillaria* pathogenicity tests on oaks in southern Italy. *J. Plant Path.* **2002**, *84*, 107–111.

68. Lygis, V.; Vasiliauskas, R.; Larsson, K.; Stenlid, J. Wood-Inhabiting fungi in stems of *Fraxinus excelsior* in declining ash stands of northern Lithuania, with particular reference to *Armillaria cepistipes*. *Scand. J. For. Res.* **2005**, *20*, 337–346. [CrossRef]

69. Moricca, S.; Ginetti, B.T.B.; Scanu, B.; Franceschini, A.; Ragazzi, A. Endemic and emerging pathogens threatening cork oak trees: Management options for conserving a unique forest ecosystem. *Plant Dis.* **2016**, *100*, 2184–2193. [CrossRef]

70. Żółciak, A.; Bouteville, R.-J.; Tourvieille, J.; Roeckel-Drevet, P.; Nicolas, P.; Guillaumin, J.-J. Occurrence of *Armillaria ectypa* (Fr.) Lamoure in peat bogs of the Auvergne—the reproduction system of the species. *Cryptogam. Mycol.* **1997**, *18*, 299–313.

71. Brazee, N.J.; Ortiz-Santana, B.; Banik, M.T.; Lindner, D.L. *Armillaria altimontana*, a new species from the western interior of North America. *Mycologia* **2012**, *104*, 1200–1205. [CrossRef] [PubMed]

72. Pegler, D.N. Taxonomy, Nomenclature and Description of *Armillaria*. In *Armillaria Root Rot: Biology and Control of Honey Fungus, Section 2: Diversity*; Fox, R.T.V., Ed.; Intercept Ltd.: Andover, UK, 2000; pp. 81–93.

73. Kim, M.-S.; Klopfenstein, N.B.; Hanna, J.W.; McDonald, G.I. Characterization of North American *Armillaria* species: Genetic relationships determined by ribosomal DNA sequences and AFLP markers. *For. Pathol.* **2006**, *36*, 145–164. [CrossRef]

74. Garett, S.D. Rhizomorph behavior in *Armillaria mellea* (Fr.) Quel., III Saprophytic colonization of woody substrates in soil. *Ann. Bot.* **1960**, *24*, 275–285.

75. Morrison, D.J. Rhizomorph growth, habit, saprophytic ability and virulence of 15 *Armillaria* species. *For. Pathol.* **2004**, *34*, 15–26. [CrossRef]

76. Leach, R. Biological control and ecology of *Armillaria mellea* (Vahl) Fr. *T. Brit. Mycol. Soc.* **1939**, *23*, 320–329. [CrossRef]

77. Raabe, R.D.; Trujillo, E.E. *Armillaria mellea* in Hawaii. *Plant Dis. Rep.* **1963**, *47*, 776.

78. Redfern, D.B.; Filip, G.M. Inoculum and Infection. In *Armillaria Root Disease*; Agricultural Handbook No. 691; Show, C.G., III, Kile, G.A., Eds.; U.S.D.A. Forest Service: Washington, DC, USA, 1991; pp. 48–61.

79. Kile, G.A. Behaviour of *Armillaria* in some *Eucalyptus obliqua* – *Eucalyptus regnans* forests in Tasmania and its role in their decline. *Eur. J. For. Pathol.* **1980**, *10*, 278–296. [CrossRef]

80. Morrison, D.J. Studies on the Biology of *Armillaria mellea*. Ph.D. Thesis, University of Cambridge, Cambridge, UK, 1972.

81. Redfern, D.B. Growth and behaviour of *Armillaria mellea* rhizomorphs in soil. *Trans. Br. Mycol. Soc.* **1973**, *61*, 569–581. [CrossRef]

82. Rykowski, K. Niektóre troficzne uwarunkowania patogeniczności *Armillaria mellea* (Vahl) Quèl. w uprawach sosnowych. *Prace Inst. Bad. Leśn.* **1985**, *640*, 1–140. (In Polish)

83. Morrison, D.J. Vertical distribution of *Armillaria mellea* rhizomorphs in soil. *Trans. Brit. Mycol. Soc.* **1976**, *66*, 393–399. [CrossRef]

84. Morrison, D.J.; Thomson, A.J.; Chu, D.; Peet, F.G.; Sahota, T.S. Variation in isozyme patterns of esterase and polyphenol oxidase among isolates of *Armillaria ostoyae* from British Columbia. *Can. J. Plant Pathol.* **1989**, *11*, 229–234. [CrossRef]

85. Mihail, J.D.; Bruhn, J.N. Foraging behaviour of *Armillaria* rhizomorph systems. *Mycol. Res.* **1995**, *109*, 1195–1207. [CrossRef]

86. Mihail, J.D.; Obert, M.; Bruhn, J.N. Fractal geometry of diffuse mycelia and rhizomorphs of *Armillaria* species. *Mycol. Res.* **1995**, *99*, 81–88. [CrossRef]

87. Żółciak, A.; Sierota, Z. Zabiegi hodowlane a zagrożenie drzewostanów przez patogeny korzeni. *Prace Inst. Bad. Leśn. seria B* **1997**, *31*, 71–84. (In Polish)

88. Żółciak, A. Przydatność herbicydu Roundup do ograniczania rozwoju ryzomorf opieniek w uprawach leśnych.). *Prace Inst. Bad. Leśn. Seria A* **2001**, *910*, 65–83. (In Polish)

89. Żółciak, A. Refraining the regeneration of *Armillaria* rhizomorphs using *Trichoderma*. Bulletin of the Polish Academy of Sciences. *Biol. Sci.* **2001**, *49*, 265–273.

90. Baumgartner, K.; Coetzee, M.P.A.; Hoffmeister, D. Secrets of the subterranean pathosystem of *Armillaria*. *Mol. Plant Path.* **2011**, *12*, 515–534. [CrossRef] [PubMed]

91. Morrison, D.J. Effects of soil organic matter on rhizomorph growth by *A. mellea*. *Trans. Brit. Mycol. Soc.* **1982**, *78*, 201–207. [CrossRef]

92. Singh, P. *Armillaria* root rot: Influence of soil nutrients and pH on the susceptibility of conifer species to the disease. *Eur. J. For. Pathol.* **1983**, *13*, 92–101. [CrossRef]

93. Mihail, J.D.; Bruhn, J.N.; Leininger, T.D. The effects of moisture and oxygen availability on rhizomorphs generation by *Armillaria tabescens* in comparison with *A. gallica* and *A. mellea*. *Mycol. Res.* **2002**, *106*, 697–704. [CrossRef]

94. Kessler, W.; Moser, S. Moglichkeiten der Vorbeugung gegen Schaden durch Hallimasch in Kiefernkulturen. *Beitr. fur die Forstwirtsch.* **1974**, *8*, 86–89.

95. Rishbeth, J. The growth rate of *Armillaria mellea*. *Trans. Brit. Mycol. Soc.* **1986**, *51*, 575–586. [CrossRef]

96. Pearce, M.H.; Malajczuk, N. Factors affecting the growth of *Armillaria luteobubalina* rhizomorphs in soil. *Mycol. Res.* **1990**, *94*, 38–48. [CrossRef]

97. Blenis, P.V.; Mugala, M.S.; Hiratsuka, Y. Soil affects *Armillaria* root rot of lodgepole pine. *Can. J. For. Res.* **1989**, *19*, 1638–1641. [CrossRef]

98. Singh, P. *Armillaria* root rot: Influence of soil nutrients and pH on the susceptibility of conifer species to the disease. *For. Pathol.* **2007**, *13*, 92–101. [CrossRef]

99. Prescott, L.M.; Harley, J.P.; Klein, D.A. Human Diseases Caused by Bacteria, Part X Microbial Diseases and Their Control. In *Microbiology*, 5th ed.; the McGraw-Hill Companies: Columbus, OH, USA, 2002; pp. 900–940.

100. Mallett, K.I.; Meynard, D.G. *Armillaria* root disease, stand characteristics, and soil properties in young lodgepole pine. *For. Ecol. Manag.* **1998**, *105*, 37–44. [CrossRef]

101. Lech, P.; Żółciak, A. Uwarunkowania występowania opieńkowej zgnilizny korzeni w lasach Beskidu Żywieckiego. *Leś. Pr. Bad.* **2006**, *2*, 33–49. (In Polish).

102. Schinner, F.; Concin, R. Carbon dioxide fixation by wood rotting fungi. *Eur. J. For. Pathol.* **1981**, *11*, 120–123. [CrossRef]

103. Herms, D.A.; Mattson, W.J. The dilemma of plants: to grow or defend. *Q. Rev. Biol.* **1992**, *67*, 283–335. [CrossRef]

104. Drigo, B.; Kowalchuk, C.A.; van Veen, J.A. Climate change goes underground: Effects of elevated atmospheric CO_2 on microbal community structure and activities in the rhizosphere. *Biol. Fert. Soils* **2008**, *44*, 667–679. [CrossRef]

105. Sallas, L.; Kainulainen, P.; Utrainen, J.; Holopainen, T.; Holopainen, J.K. The influence of elevated O_3 and CO_2 concentrations on secondary metabolites of Scots pine (*Pinus sylvestris* L.) seedlings. *Glob. Chang. Biol.* **2001**, *7*, 303–311. [CrossRef]

106. Fleischmann, F.; Raidl, W.F.; Oβwald, W.F. Changes in susceptibility of beech (*Fagus sylvatica*) seedlings towards *Phytophthora citricola* under the influence of elevated atmospheric CO_2 and nitrogen fertilization. *Environ. Pollut.* **2010**, *158*, 1051–1060. [CrossRef] [PubMed]

107. Tkaczyk, M.; Sikora, K.; Nowakowska, J.A.; Kubiak, K.; Oszako, T. Effects of CO_2 enhancement on beech (*Fagus sylvatica* L.) seedling root rot due to *Phytophthora plurivora* and *Phytophthora cactorum*. *Folia For. Pol. Ser. A* **2014**, *56*, 149–156. [CrossRef]

108. Oszako, T.; Sikora, K.; Borys, M.; Kubiak, K.; Tkaczyk, M. *Phytophthora quercina* infections in elevated CO_2 concentrations. *Folia For. Pol. Ser. A* **2016**, *58*, 131–141. [CrossRef]

109. Grzywacz, A.; Ważny, J. The impact of industrial air pollutants on the occurrence of several important pathogenic fungi of forest trees in Poland. *Eur. J. For. Pathol.* **1973**, *3*, 129–141. [CrossRef]

110. Jancarik, U. Uyskyt drevokaznych hub u Kourem poskozovane oblasti Krusnych hor. *Lesnictvi* **1962**, *1*, 677–692.

111. Kudela, M.; Novakova, E. Lesni skudci a skody sveri v lesich poskozovanych Kourem. *Lesnictvi* **1962**, *6*, 493–502.

112. Schaeffer, T.C.; Hedgecock, G.G. *Injury to Northwestern Forest Trees by Sulfur Dioxide from Smelters*; U.S. Forest Service Tech. Bull. No. 1117; U.S.D.A.: Washington, DC, USA, 1955; pp. 1–49.

113. Domanski, S.; Kowalski, S.; Kowalski, T. Fungi; occurring in forests injured by industrial air pollutants in the upper Silesia and Krakow industrial regions Poland: IV Higher fungi causing root diseases within forest stands not rebuilt in the years 1971–1975. *Acta Agrar. Silv. Ser. Silv.* **1976**, *16*, 61–74.

114. Horak, M.; Tesche, M. Einfluss von SO_2 auf die Infektione von Fichtensamlingen durch *Armillaria ostoyae*. *Forstwiss. Cbl.* **1993**, *112*, 93–97. [CrossRef]

115. Wargo, P.M.; Carey, A.C. Effects of metals and pH on in vitro growth of *Armillaria ostoyae* and other root and butt rot fungi of red spruce. *For. Pathol.* **2001**, *41*, 5–24. [CrossRef]

116. Sicoli, G.; Luisi, N.; Manicone, R.P. *Armillaria* species occurring in southern Italy. In Proceedings of the Eighth International Conference on Root and Butt Rots, Wik, Sweden and Haikko, Finland, 9–16 August 1993; Johansson, M., Stenlid, J., Eds.; Swedish University of Agriculture Sciences: Uppsala, Sweden, 1994; pp. 383–387.

117. Cruickshank, M.G.; Morrison, D.J.; Punja, Z.K. Incidence of *Armillaria* species in precommercial thinning stumps and spread of *Armillaria ostoyae* to adjacent Douglas-fir-trees. *Can. J. For. Res.* **1997**, *27*, 481–490. [CrossRef]

118. Lygis, V. Root Rot in North-Temperate Forest Stands: Biology, Management and Communities of Associated Fungi. Ph.D. Thesis, Swedish University of Agricultural Sciences, Forestry Faculty, Uppsala, Sweden, 2005.

119. Dumas, M.T.; Boyonoski, N.W. Scanning electron microscopy of mycoparasitism of *Armillaria* rhizomorphs by species *Trichoderma*. *Eur. J. For. Pathol.* **1992**, *22*, 379–383. [CrossRef]

120. De Wit, P.J.G.M. Plant Pathogenic Fungi and Oomycetes. In *Principles of Plant-Microbe Interactions*; Lugtenberg, B., Ed.; Springer Int. Publisher: Cham, Switzerland, 2015; pp. 79–90.

121. Riffle, J.W. Effect of two mycophagus nematodes on *Armillaria mellea* root rot of *Pinus ponderosa* seedlings. *Plant Dis. Rep.* **1973**, *57*, 355–357.

122. Cayrol, J.C.; Dubos, B.; Guillaumin, J.-J. Etude preliminare in vitro de l'agressivite de quelque nematodes mycophages vis-à-vis de *Trichoderma viride* Pers., *T. polysporum* (Link. Ex Pers.) Rifai et *Armillaria mellea* (Vahl) Karst. *Ann. Phytopathol.* **1978**, *10*, 177–185.

123. Weller, D.M. Biological control of soilborne plant pathogens in the rhizosphere. *Ann. Rev. Phytopathol.* **1988**, *26*, 379–407. [CrossRef]

124. Sa, Q.; Wang, Y.; Li, W.; Zhang, L.; Sun, Y. The promoter of an antifungal protein gene from *Gastrodia elata* confers tissue-specific and fungus-inducible expression patterns and responds to both salicylic acid and jasmonic acid. *Plant Cell Rep.* **2003**, *22*, 79–84. [CrossRef] [PubMed]

125. Wang, H.X.; Yang, T.; Zeng, Y.; Hu, Z. Expression analysis of the gastrodianin gene *ga4B* in an achlorophyllous plant *Gastrodia elata* Bl. *Plant Cell Rep.* **2007**, *26*, 253–259. [CrossRef] [PubMed]

126. Singh, B.K.; Dawson, L.A.; Macdonald, C.A.; Buckland, S.M. Impact of biotic and abiotic interaction on soil microbial communities and functions: A field study. *Appl. Soil Ecol.* **2009**, *41*, 239–248. [CrossRef]

127. Celar, F. Competition for amonium and nitrate forms of nitrogen between some phytopathogenic and antagonistic soil fungi. *Biol. Control* **2003**, *28*, 19–24. [CrossRef]

128. Watanabe, N.; Lewis, J.A.; Papavizas, G.C. Influence of nitrogen fertilizers on growth, spore production and germination, and biological potential of *Trichoderma* and *Gliocladium*. *J. Phtopathol.* **1987**, *120*, 337–346. [CrossRef]

129. Zogg, G.P.; Zak, D.R.; Ringelberg, D.B.; Macdonald, N.W.; Pregitzer, K.S.; White, D.C. Compositional and functional shifts in microbial communities because of soil warming. *Soil Sci. Soc. Am. J.* **1997**, *61*, 475–481. [CrossRef]

130. Karhu, K.; Auffret, M.D.; Dungait, J.A.J.; Hopkins, D.W.; Prosser, J.I.; Singh, B.K.; Subke, J.A.; Wookey, P.A.; Ågren, G.I.; Sebastià, M.T.; et al. Temperature sensitivity of soil respiration rates enhanced by microbial community response. *Nature* **2014**, *513*, 81–84. [CrossRef] [PubMed]

131. Anderson, O.R. Soil respiration, climate change and the role of microbial communities. *Protist* **2011**, *162*, 679–690. [CrossRef] [PubMed]

132. Giardina, C.P.; Litton, C.M.; Crow, S.E.; Asner, G.P. Warming-Related increases in soil CO$_2$ efflux are explained by increased below-ground carbon flux. *Nat. Clim. Chang.* **2014**, *4*, 822–827. [CrossRef]

133. Carey, J.C.; Tang, J.; Templer, P.H.; Kroeger, K.D.; Crowther, T.W.; Burton, A.J.; Dukes, J.S.; Emmett, B.; Frey, S.D.; Heskel, M.A.; et al. Temperature response of soil respiration largely unaltered with experimental warming. *Proc. Natl. Acad. Sci. USA* **2016**, *113*, 13797–13802. [CrossRef] [PubMed]

134. Frey, S.D.; Lee, J.; Melillo, J.M.; Six, J. The temperature response of soil microbial efficiency and its feedback to climate. *Nat. Clim. Chang.* **2013**, *3*, 395–398. [CrossRef]

135. Briones, M.J.I.; McNamara, N.P.; Poskitt, J.; Crow, S.E.; Ostle, N.J. Interactive biotic and abiotic regulators of soil carbon cycling: Evidence from controlled climate experiments on peatland and boreal soils. *Glob. Chang. Biol.* **2014**, *20*, 2971–2982. [CrossRef] [PubMed]

136. Kaisermann, A.; Maron, P.A.; Beaumelle, L.; Lata, J.C. Fungal communities are more sensitive indicators to non-extreme soil moisture variations than bacterial communities. *Appl. Soil Ecol.* **2015**, *86*, 158–164. [CrossRef]

137. Classen, A.T.; Sundqvist, M.; Henning, J.A.; Newman, G.S.; Moore, J.A.M.; Cregger, M.; Moorhead, L.C.; Patterson, C.M. ESA Centennial Paper: Direct and indirect effects of climate change on soil microbial and soil microbial-plant interactions: What lies ahead? *Ecosphere* **2015**, *6*, 130. [CrossRef]

138. Campbell, A.H.; Harder, T.; Nielsen, S.; Kjelleberg, S.; Steinberg, P.D. Climate change and disease: Bleaching of a chemically defended seaweed. *Glob. Chang. Biol.* **2011**, *17*, 2958–2970. [CrossRef]

139. Morriën, E.; Engelkes, T.; van der Putten, W.H. Additive effects of aboveground polyphagous herbivores and soil feedback in native and range-expanding exotic plants. *Ecology* **2011**, *92*, 1344–1352. [CrossRef] [PubMed]

140. Nuñez, M.A.; Horton, T.R.; Simberloff, D. Lack of belowground mutualisms hinders *Pinaceae* invasions. *Ecology* **2009**, *90*, 2352–2359. [CrossRef] [PubMed]

141. De Bello, F.; Lavorel, S.; Díaz, S.; Harrington, R.; Cornelissen, J.H.C.; Bardgett, R.D.; Berg, M.P.; Cipriotti, P.; Feld, C.K.; Hering, D.; et al. Towards an assessment of multiple ecosystem processes and services via functional traits. *Biodivers. Conserv.* **2010**, *19*, 2873–2893. [CrossRef]

142. Yang, G.; Yang, X.; Zhang, W.; Wei, Y.; Ge, G.; Lu, W.; Sun, J.; Liu, N.; Kan, H.; Shen, Y.; et al. Arbuscular mycorrhizal fungi affect plant community structure under various nutrient conditions and stabilize the community productivity. *OIKOS* **2016**, *125*, 576–585. [CrossRef]

143. Van der Heijden, M.G.A.; de Bruin, S.; Luckerhoff, L.; van Logtestijn, R.S.P.; Schlaeppi, K. A widespread plant-fungal-bacterial symbiosis promotes plant biodiversity, plant nutrition and seedling recruitment. *ISME J.* **2016**, *10*, 389–399. [CrossRef] [PubMed]

144. Streitwolf-Engel, R.; van der Heijden, M.G.A.; Wiemken, A.; Sanders, I.R. The ecological significance of arbuscular mycorrhizal fungal effects on clonal reproduction in plants. *Ecology* **2001**, *82*, 2846–2859. [CrossRef]

145. Grimoldi, A.A.; Kavanová, M.; Lattanzi, F.A.; Schnyder, H. Phosphorus nutrition-mediated effects of arbuscular mycorrhiza on leaf morphology and carbon allocation in perennial ryegrass. *New Phytol.* **2005**, *168*, 435–444. [CrossRef] [PubMed]

146. Elmore, W.C. Population and Identification of Mycorrhizal Fungi in St. Augustinegrass in Florida and Their Effect on Soilborne Pathogens. Ph.D. Thesis, University of Florida, Gainesville, FL, USA, 2006.

147. Olsson, P.A.; Hansson, M.C.; Burleigh, S.H. Effect of P availability on temporal dynamics of carbon allocation and *Glomus intraradices* high-affinity P transporter gene induction in arbuscular mycorrhiza. *Appl. Environ. Microbiol.* **2006**, *72*, 4115–4120. [CrossRef] [PubMed]

148. Hawkes, C.V.; Hartley, I.P.; Ineson, P.; Fitter, A.H. Soil temperature affects carbon allocation within arbuscular mycorrhizal networks and carbon transport from plant to fungus. *Glob. Chang. Biol.* **2008**, *14*, 1181–1190. [CrossRef]

149. Pfeffer, A. Sucha 1947 a kurovci na smrku v r. 1949. *CsL. Les.* **1950**, *30*, 176–179. (In Czech)

150. Madziara-Borusiewicz, K.; Strzelecka, H. Conditions of spruce (*Picea excelsa*) infestations by the engraver beetle (*Ips typographus* L.) in mountains in Poland. I. Chemical composition of volatile oils from healthy trees and those infested with the honey fungus (*Armillaria mellea* (Vahl) Quel. *J. Appl. Entomol.* **1977**, *83*, 409–415.

151. James, R.L.; Goheen, D.J. Conifer mortality associated with root disease and insects in Colorado. *Plant Dis.* **1981**, *65*, 506–507. [CrossRef]

152. Capecki, Z. Rejony zdrowotności zachodniej części Karpat. *Prace IBL* **1994**, *781*, 61–125. (In Polish)

153. Christiansen, E.; Husek, K.J. Infestation ability of *Ips typographus* in Norway spruce, in relation to butt rot, tree vitality and increment. *Medd. Nor. Inst. Skogforsk.* **1980**, *35*, 468–482.

154. Jankovsky, L.; Cudlin, P.; Moravec, I. Root decays as a potential predisposition factor of a bark beetle disaster in the Sumava Mts. *J. For. Sci.* **2003**, *49*, 125–132.

155. Twery, M.J.; Mason, G.N.; Wargo, P.M.; Gottschalk, K.W. Abundance and distribution of rhizomorphs of *Armillaria* spp. in defoliated mixed oak stands in western Maryland. *Can. J. For. Res.* **1990**, *20*, 674–678. [CrossRef]

156. Okland, B.; Krokene, P.; Lange, H. Science Nordic. The Effect of Climate Change on the Spruce Bark Beetle. Available online: http:www.cicero.uio.no/fulltex/index (accessed on 21 April 2015).

157. Seidl, R.; Rammer, W.; Lexer, M.J. Schätzung von Bodenmerkmalen und Modellparametern für die Waldökosystemsimulation auf Basis einer Großrauminventur. *Allg. Forst-Jagdztg* **2009**, *180*, 35–44. (In German)

158. Seidl, R.; Fernandes, P.M.; Fonseca, T.F.; Gillet, F.; Jönsson, A.M.; Merganicova, K.; Netherer, S.; Arpaci, A.; Bontemps, J.D.; Bugmann, H.; et al. Modelling natural disturbances in forest ecosystems: A review. *Ecol. Model.* **2011**, *222*, 903–924. [CrossRef]

159. Seidl, R.; Rammer, W.; Lexer, M.J. Climate change vulnerability of sustainable forest management in the Eastern Alps. *Clim. Chang.* **2011**, *106*, 225–254. [CrossRef]

160. Langvall, O. Impact of climate change, seedling type and provenance on the risk of damage to Norway spruce (*Picea abies* (L.) Karst.) seedling in Sweden due to early summer frosts. *Scand. J. For. Res.* **2011**, *26*, 56–63. [CrossRef]

161. Faccoli, M. Effect of weather on *Ips typographus* (Coleoptera Curculionidae) phenology, voltinism, and associated spruce mortality in the southeastern. *Alps. Environ. Entomol.* **2009**, *38*, 307–316. [CrossRef] [PubMed]

162. Marini, L.; Ayres, M.P.; Battisti, A.; Faccoli, M. Climate affects severity and altitudinal distribution of outbreaks in an eruptive bark beetle. *Clim. Chang.* **2012**, *115*, 327–341. [CrossRef]

163. Marini, L.; Liedelow, A.; Jönsson, A.M.; Wulff, S.; Schroeder, L.M. Population dynamics of the spruce bark beetle: A long term study. *Oikos* **2013**, *112*, 1768–1776. [CrossRef]
164. Pham, T.; Chen, H.; Yu, J.; Dai, L.; Zhang, R.; Vu, T.Q.T. The Differential Effects of the Blue-Stain Fungus *Leptographium qinlingensis* on Monoterpenes and Sesquiterpenes in the Stem of Chinese White Pine (*Pinus armandi*) Saplings. *Forests* **2014**, *5*, 2730–2749. [CrossRef]
165. Schiebe, C.; Hammerbacher, A.; Birgersson, G.; Witzell, J.; Brodelius, P.; Gershenzon, J.; Hansson, B.S.; Krokene, P.; Schyler, F. Inducibility of chemical defences in Norway spruce bark is correlated with unsuccessful mass attacks by the spruce bark beetle. *Oecologia* **2012**, *170*, 183–198. [CrossRef] [PubMed]

forests

MDPI

Article

Soil and Stocking Effects on Caliciopsis Canker of *Pinus strobus* L.

Isabel A Munck [1,*], Thomas Luther [1], Stephen Wyka [1], Donald Keirstead [2],
Kimberly McCracken [2], William Ostrofsky [3], Wayne Searles [3], Kyle Lombard [4],
Jennifer Weimer [4] and Bruce Allen [4]

[1] Northeastern Area State & Private Forestry, U.S. Department of Agriculture (USDA) Forest Service (USFS), Durham, NH 03861, USA; imunck@fs.fed.us (T.L.); stephenwyka@gmail.com (S.W.)
[2] USDA Natural Resources Conservation Service (NRCS), Dover, NH 03861, USA; donald.keirstead@nh.usda.gov (D.K.); kimberly.mccracken@nh.usda.gov (K.M.)
[3] Maine Forest Service, Maine Department of Agriculture, Conservation and Forestry, Augusta, MA 04692, USA; bill.ostrofsky@maine.gov (W.O.); Wayne.Searles@maine.gov (W.S.)
[4] New Hampshire Department of Resources and Economic Development, New Hampshire Division of Forests and Lands, Concord, NH 03861, USA; kyle.lombard@dred.state.nh.us (K.L.); Jennifer.Weimer@dred.nh.gov (J.W.); Bruce.Allen@dred.nh.gov (B.A.)
* Correspondence: imunck@fs.fed.us; Tel.: +1-603-868-7636

Academic Editors: Matteo Garbelotto and Paolo Gonthier
Received: 6 September 2016; Accepted: 4 November 2016; Published: 11 November 2016

Abstract: Soil and stand density were found to be promising predictive variables associated with damage by the emerging disease of eastern white pine, Caliciopsis canker, in a 2014 survey with randomly selected eastern white pine (*Pinus strobus* L.) stands. The objective of this study was to further investigate the relationship between soil and stocking in eastern white pine forests of New England by stratifying sampling across soils and measuring stand density more systematically. A total of 62 eastern white pine stands were sampled during 2015–2016. Stands were stratified across soil groups and several prism plots were established at each site to measure stand density and determine stocking. Caliciopsis canker incidence in mature trees was greater in sites with drier or shallow soils compared to sites with loamy soils and in adequately stocked stands compared to understocked stands ($p < 0.0001$). Caliciopsis canker signs and symptoms were observed in all size classes. Live crown ratio, a measure of forest health, decreased with increasing Caliciopsis canker symptom severity. The fungal pathogen, *Caliciopsis pinea* Peck, was successfully isolated from cankers on trees growing in each soil group. Forest managers will need to consider damage caused by Caliciopsis canker related to stand factors such as soil and stocking when regenerating white pine stands.

Keywords: forest health monitoring; eastern white pine; tree density; tree disease

---

1. Introduction

In New England and New York, where the forest cover surpasses 60% of the land area and the annual value of forest products industry exceeds $18.8 billion, forests are vital to the region's economy [1]. Eastern white pine (*Pinus strobus* L.) is an important component of the region's forests. For example, in Massachusetts white pine forest types comprise 25% of the forest [2]. White pine often grows in pure stands, but it can also grow well in association with other conifers and hardwoods. The geographic range of eastern white pine extends from Canada to South Carolina and west to Iowa. Eastern white pine grows in all the soils throughout this range, but does best in sandy, well-drained soils where it is not outcompeted by hardwoods [3,4].

Caliciopsis pinea Peck is a fungal pathogen that causes cankers to pine and other conifer hosts in North America and Europe [5]. In eastern North America, *C. pinea* frequently causes damage to eastern white pine [6,7]. Reports of Caliciopsis canker were frequent during the 1930s [8,9], but since then, the disease has been mostly overlooked until more recent reports of unexpected damage by this disease [7,10,11]. Concern over loss of value to eastern white pine led to a 2014 survey of randomly selected sites throughout Maine, Massachusetts, and New Hampshire—states with the greatest concentration of white pine in New England [2,6]. Results from that initial survey indicated that soil group and stand density were promising, predictive factors of the probability of a site having symptoms associated with Caliciopsis canker [12]. In that study, Caliciopsis canker symptoms were more frequently observed on trees growing in excessively drained (86%) or poorly drained soils (78%) than in well drained more fertile soils (59%) ($p = 0.1$). Stand density was greater for stands with Caliciopsis than for stands without Caliciopsis ($p = 0.1$). The relationship between presence of Caliciopsis canker symptoms and soil or stand density, however, was not statistically significant at $p = 0.05$. In addition, stand density measurements in that study were limited. The objective of this study, therefore, was to further explore the relationship between Caliciopsis canker symptoms and soil or tree density in eastern white pine stands. We hypothesize that eastern white pine growing in more productive soils with adequate soil moisture are less likely to have Caliciopsis canker symptoms than trees growing in nutrient poor, excessively drained, and poorly drained soils. We also hypothesize that incidence of Caliciopsis canker symptoms will increase with increasing stand density.

2. Materials and Methods

2.1. Site Selection

A modified version of the New Hampshire Important Forest Soil groups was developed specifically for conditions potentially related to white pine growth and productivity [13]. Soils from Belknap, Merrimack, and Hillsborough counties in New Hampshire (NH) and Oxford, Androscoggin, Sagadahoc, and York counties in Maine (ME), where eastern white pine is abundant, were classified into four soil groups (Table 1). Layers of these modified soil groups for use in Geographic Information System (GIS) were generated for New Hampshire and Maine counties with the greatest concentration of eastern white pine basal area. Modified methods from Munck et al. (2015) were used to sample eastern white pine sites within each soil group during two consecutive years. In 2015, at least 12 sites per soil group with more than 2 contiguous hectares with more than 75% basal area of eastern white pine as predicted by the National Insect and Disease Risk Map (NIDRM) eastern white pine host layer for use in GIS were randomly selected for sampling [14]. The sites visited during 2015 were sawtimber (mean stand diameter at breast height, 1.3 m from the ground: DBH > 23 cm) stands because most of the eastern white pine resource in New England is mature and these site were randomly selected. Previously, Caliciopsis canker symptoms were more frequently observed on pole-size trees (DBH = 11.5–22.9 cm) [6]. In addition, during 2015, Caliciopsis cankers and fruiting bodies were frequently observed on white pine regeneration (Figure 1). In 2016, therefore, poletimber stands and eastern white pine regeneration within these were surveyed to better quantify probability of Caliciopsis canker symptoms in these stands at greater risk of damage by *C. pinea*. Inventories of New Hampshire State Lands were used to locate three sites with poletimber (mean stand DBH = 11.5 to 23 cm) white pine stands within each soil group. Latitude and longitude coordinates for the center point of each stand were generated and imported into a Global Positioning System (GPS) receiver (GPSmap 64, Garmin International Inc., Olathe, KS, USA).

Table 1. New Hampshire important forest soil groups (NHFSG) modified for Belknap, Merrimack, and Hillsborough counties in New Hampshire (NH) and Oxford, Androscoggin, Sagadahoc, and York counties in Maine (ME). Groups are generally rated from 1–4 along a continuum of water holding capacity, drainage class, and depth. Soils in Group 1 are very dry (low water holding capacity and excessively drained) to Groups 4 where water is more plentiful due to finer soil texture, landscape position, or root restrictive layers.

Soil Group	Soil Properties and Interpretations	Brief Description
Very dry	NHFSG = 1C Drainage Class = excessively drained Flooding Frequency = none or very rare	Soils are coarse textured and have a significant amount (>35%) of gravel and are somewhat excessively drained or excessively drained. Fertility and soil moisture is limiting for hardwoods but may produce high quality softwoods. Pine should dominate with *Pinus rigida* P. Mill. and *P. resinosa* Aiton.
Dry	NHFSG = 1C Drainage class = somewhat excessively drained (NH and ME), and well-drained (ME only) Flooding frequency = none, very rare, rare	Soils are coarse textured with only a small amount (10%) of gravel and are somewhat excessively drained. Fertility and soil moisture is limiting for hardwoods but may produce high quality softwoods. No apparent water table. Pine should dominate with oak and beech.
Loamy	NHFSG = 1B Drainage class = well drained Flooding frequency = none, rare, occasional	Loamy sand to loamy textures (with or without gravel/cobble). Not moderately well drained or wetter. Productive soils with adequate, but not excessive, soil moisture.
Shallow	NHFSG = 1A, 1B, 1C, or 2B Shallow soils and soils with layer restricting water movement and root growth (ME only)	Wetter soils. Soils are considered poor for most plant growth due to excess moisture or shallow depth. These soils have features (bedrock or dense sub-surface layer) restricting movement of water and root growth—no additional grouping by drainage class or flooding frequency.

Figure 1. Eastern white pine (*Pinus strobus* L.) seedling in Rhode Island, USA, photographed on April of 2013 with Caliciopsis canker symptoms and signs (black, hair-like fruiting structures, 2–3 mm).

2.2. Field Sampling

In the field, a GPS receiver and compass were used to locate the center point of each site generated by the intersection of GIS soil groups and white pine layers. A 10 basal-area-factor prism was used at plot center to select sample trees (DBH > 11.5 cm). Three more prism plots at 120° (0, 120, and 240) and 17 m from the initial plot center were installed for a total of four prism plots per site). Trees per prism plot were converted to trees per hectare by calculating an expansion factor unique to each tree's DBH. In addition, the following data were collected for each eastern white pine: live crown ratio, presence of *C. pinea* fruiting bodies, incidence and severity of Caliciopsis canker symptoms. Caliciopsis canker symptom severity was visually assessed by dividing the bole of the tree into thirds (bottom, middle, and upper stem) and counting the number of resin streaks in each section up to ten streaks per section. This assessment was conducted on two opposing faces of each tree and added for a maximum score of 60 resin streaks per tree. Resinosis associated with white pine blister rust symptoms, insect boring, decayed branch stubs, or mechanical damage was not considered in Caliciopsis canker disease severity assessments. In 2015, the presence of *C. pinea* in understory regeneration was recorded for each prism plot, but not for individual seedlings. In 2016, consequently, the presence of Caliciopsis fruiting bodies from the five closest eastern white pine seedlings (DBH < 2.54 cm, height > 30 cm) to each prism plot center as well as maximum distance to prism plot center were recorded. Caliciopsis fruiting bodies and infected plant tissue were collected from symptomatic eastern white pines trees or seedlings during 2015 for diagnoses.

2.3. Isolation and Identification of Caliciopsis pinea

Modified methods described by Munck et al (2015) were used to isolate and identify *C. pinea* isolates. Briefly, a piece of the bark containing the fruiting structure was placed onto the lid of an inverted petri dish, so that the lid was on the bottom and potato dextrose agar (PDA) media on top, and then sealed with Parafilm to induce sporulation. Small colonies that were formed on the PDA were transferred onto fresh PDA plates after two to three days. Once pure cultures were established, isolates were transferred onto 2% PDA plates overlaid with a cellophane membrane and grown for one to two weeks. Using direct colony polymerase chain reaction (PCR) [15] on 21 of these strains, at least three from each soil group, the internal transcribed spacer (ITS1-5.8s-ITS2) region of the rDNA using the primers ITS1 and ITS4 was amplified and sequenced in the forward direction (GENEWIZ, South Plainfield, NJ, USA).

2.4. Statistical Analyses

To explore the effect of soil and stocking on the incidence of eastern white pine with symptoms associated with Caliciopsis canker, single binary regressions were executed with the GLIMMIX procedure (Statistical Analyses Software v. 9.2, SAS Institute Inc., Cary, NC, USA) in SAS by specifying binomial distribution and the logit link function for the response variables "presence or incidence of Caliciopsis canker symptoms". Before data analysis, the mean counts of trees or seedlings with Caliciopsis damage per site (y) were calculated by averaging prism plot counts for each site and transformed (y + 0.0001). Stocking guides developed for eastern white pine which took into account tree density and basal area were used to determine stocking at each site [16]. Sites with stocking levels below the "unmanaged B-line" were considered to be under-stocked whereas sites with stocking levels between the "unmanaged B-line" and the "A-line" were considered to be adequately stocked. One-way analyses of variances (ANOVA) were performed using the GLIMMIX procedure to explore the main effects of soil group on tree density. The response variable was either "basal area per hectare" of "trees per hectare". Similarly, to determine the main effect of soil group on seedling density linear mixed model (PROC GLIMMIX) was used with "soil group" as a fixed effect, "site" was a random factor, and the response variable was "seedlings per hectare". To quantify the relationship between symptoms on the main stem and tree crown health, one-way analyses of variances (ANOVA) were performed using the GLIMMIX procedure to investigate main effects of Caliciopsis canker severity on live crown ratios. Caliciopsis canker disease severity categories were defined as: "None" = Caliciopsis canker symptoms absent; "Low" = 1 to 4 resin streaks per tree, "Medium" = 5 to 19 resin streaks per tree, and "High" = 20 to 60 resin streaks per tree. The response variable was the live crown ratio of eastern white pine. For all analyses, when the main effects were significant ($\alpha = 0.05$), a Tukey–Kramer test was used to identify differences between categories.

3. Results

A total of 62 white pine stands or sites were sampled in Maine and New Hampshire during 2015–2016 (Figure 2). Fifty of these (81%) had symptoms associated with Caliciopsis canker. During 2015, 50 sawtimber stands, at least 11 in each soil group (Table 1) were sampled in New Hampshire and Maine, 13 of which were privately owned. For 2015, mean stand diameter was 34 cm, mean tree density per stand was 336 trees per hectare, on average eastern white pine comprised 80% of the stand basal area, and 20% of eastern white pines exhibited symptoms associated with Caliciopsis canker. During 2016, 12 poletimber stands, at least three in each soil group were sampled in New Hampshire all on lands owned and managed by the State. For 2016, mean stand diameter was 23 cm, mean tree density per stand was 459 trees per hectare, on average eastern white pine comprised 86% of the stand basal area, and 66% of eastern white pines exhibited symptoms associated with Caliciopsis canker.

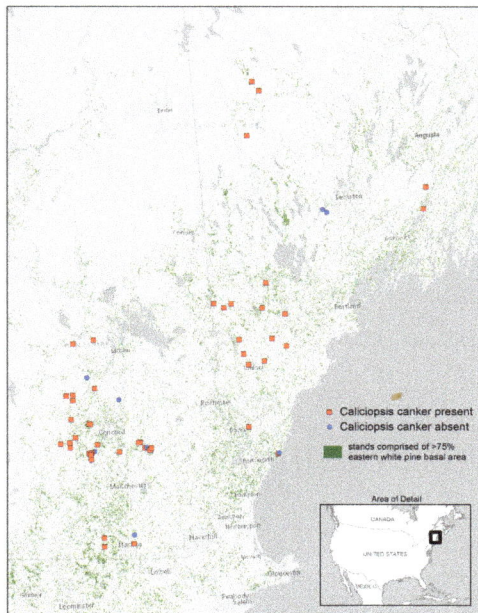

Figure 2. Location of 62 *Pinus strobus* stands in Maine and New Hampshire sampled during 2015 and 2016 and the presence of symptoms associated with Caliciopsis canker in those sites.

The proportion of trees (DBH > 11.5 cm) or seedlings (DBH < 2.54 cm, height > 30 cm) with symptoms associated with Caliciopsis canker differed significantly across soil groups ($p < 0.0001$) and stocking categories (Table 2, Figure 3). Trees in poletimber stands were more likely to have Caliciopsis canker symptoms than trees in sawtimber stands. For example, the probability (ranging from 0 to 1) of having Caliciopsis canker symptoms was 0.7 for trees in poletimber stands in the dry soil group compared to 0.3 for trees in poletimber stands in the dry soil group.

Table 2. Type III test for fixed-effect model results for the probability of symptoms associated with Caliciopsis canker.

Sample	Parameter	F-test (Degrees of Freedom = *df*)	*p*-Value
Live trees (>11.5 cm DBH) in sawtimber (>23 cm mean stand diameter) stands sampled in 2015	Soil	48.81 (3, 46)	<0.0001
	Stocking	59.56 (1, 48)	<0.0001
Live trees in poletimber (11.5 to 23 cm mean stand diameter) stands sampled in 2016	Soil	42.79 (3, 7)	<0.0001
	Stocking	45.79 (1, 9)	<0.0001
Seedlings (>30.5 cm height, and <2.54 cm DBH) in poletimber stands sampled in 2016	Soil	859.01 (3, 3)	<0.0001
	Stocking	14,248.3 (1, 5)	<0.0001

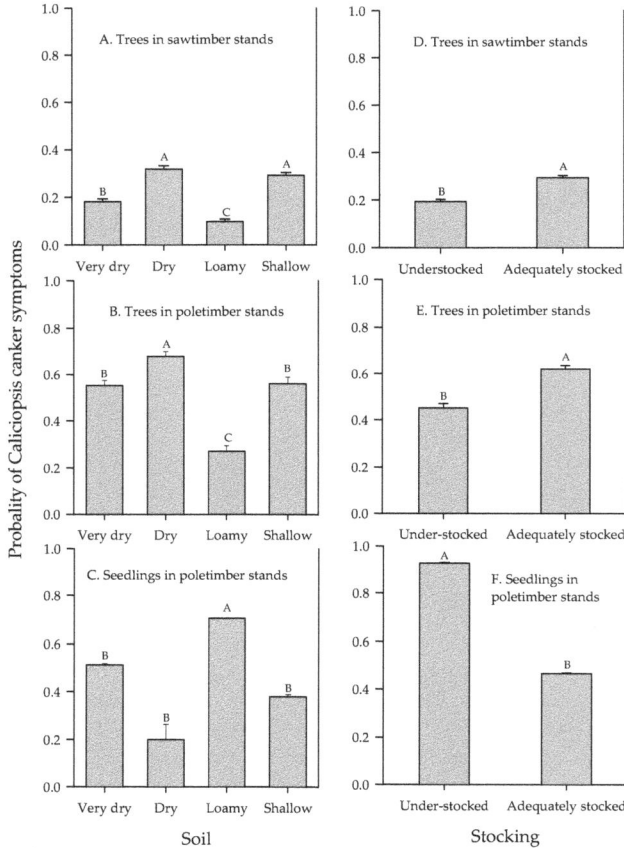

Figure 3. Incidence of symptoms associated with Caliciopsis canker in relation to soil groups for (**A** and **D**) *Pinus strobus* trees (DBH > 11.5 cm) in sawtimber stands (mean stand DBH > 23 cm) sampled in New Hampshire and Maine during 2015; (**B** and **E**) *P. strobus* trees in poletimber (mean stand DBH = 11.5 to 23 cm) stands sampled in New Hampshire during 2016; and (**C** and **F**) *P. strobus* seedlings (height > 30.5 cm and DBH < 2.54 cm) sampled in New Hampshire during 2016. Values with the same letter within each graph are not statistically different (α = 0.05).

Trees in loamy soils and understocked stands were less likely to be damaged by Caliciopsis canker than trees in other soil groups or adequately stocked stands. The proportion of live trees with symptoms associated with Caliciopsis canker was less for loamy soils than any other soil group, and greatest for dry soils (Figure 3A,B). The proportion of live trees with symptoms associated with Caliciopsis canker was less in understocked stands than in adequately stocked stands (Figure 3D,E). Pinus strobus seedlings in poletimber stads were more likely to be damaged by *C. pinea* in sites with loamy soils (Figure 3C) and understocked over story (Figure 3F). Stand density was not related to soil groups (for basal area per hectare: $p = 0.47$, F-value $= 0.86$, $df = 3, 58$; or for trees per hectare: $p = 0.9$, F-value $= 2.3$, $df = 3, 58$) (Figure 4A,B).

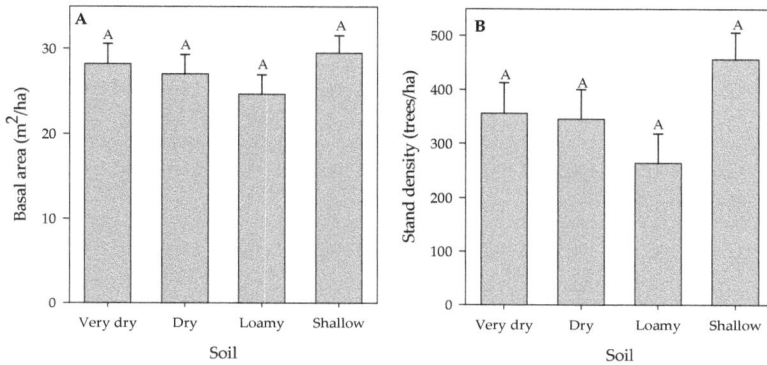

Figure 4. Relationship between soil group and stand density expressed as: (**A**) basal area per hectare; or (**B**) trees per hectare, for *P. strobus* trees sampled in New Hampshire and Maine during 2015–2016. Values with the same letter within each graph are not statistically different ($\alpha = 0.05$).

In contrast to results from the tree density analyses, seedling density of seedling in poletimber stands was related to soil type ($p = 0.0239$, F-value $= 3.49$, $df = 3, 42$). Seedling density was greater in poletimber stands on loamy soils than those on shallow soils (Figure 5).

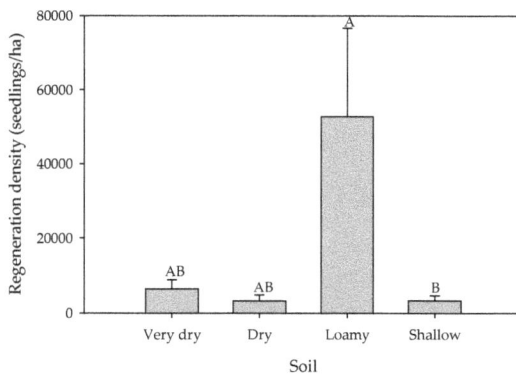

Figure 5. Relationship between soil group and *P. strobus* seedling density in poletimber stands sampled in New Hampshire during 2016. Values with the same letter are not statistically different ($\alpha = 0.05$).

Live crown ratio of eastern white pines decreased with increasing Caliciopsis canker symptom severity (Figure 6), ranging from 28% for trees without symptoms to 20% for trees with high disease severity ($p = 0.002$). BLASTn analyses of ITS locus of *C. pinea* isolates displayed 100% homology to the ITS locus of *C. pinea* in the GenBank CBS 139.64 (Accession # KP881691.1).

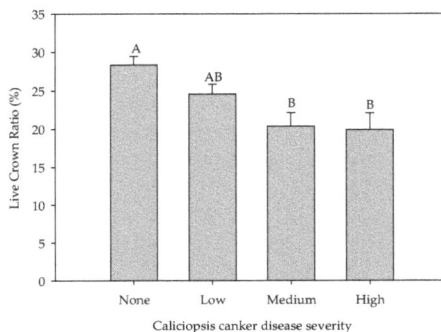

Figure 6. Live crown ratio of eastern white pines in relation to Caliciopsis canker symptom severity for trees sampled in Maine and New Hampshire in 2015–2016. Values with the same letter are not statistically different ($\alpha = 0.05$).

4. Discussion

In New England and New York, *Pinus strobus* grows from sea level to 460 m [3] and competes better on drier soils with low to medium site quality that do not favor growth of hardwood species. In this study, we sampled 62 eastern white pine stands across a variety of soils. We did not have prior knowledge of Caliciopsis canker incidence in these sites prior to sampling. The 62 sites had an average slope of 3.4% ranging from 0% to 25%, therefore, soil was the most salient topographic feature differing among sites. Incidence of symptoms associated with Caliciopsis canker in mature white pines was greater in drier and shallow soils compared to loamy soils. Conceivably, white pine trees growing in soils with more nutrients, and adequate but not excessive water-holding capacity, likely have more resources to grow and allocate towards defense than trees growing in more nutrient poor and excessively or poorly drained sites. Plants have to grow to compete for resources such as light and water and also to synthesize plant defense compounds, thus facing a continuous tradeoff between growth and defense [17].

The water holding capacity of soils affects the amount of water available for tree growth. Other studies have associated canker diseases of pine, such as Diplodia canker of red pine (*Pinus resinosa* Aiton), to soil type [18]. Drier soils are more conducive to drought stress which might render trees more susceptible to damage by Caliciopsis canker because these trees might have less water to allocate towards growth and defense. Many published studies have found a positive association between drought and canker diseases [18]. For example, red pines suffering from drought stress exhibited greater symptom severity when inoculated with *Diplodia sapinea* (Fr.) [19]. Pathogens and drought can interact and effect the carbon economy of trees resulting in tree death [20].

In contrast to mature trees, white pine regeneration in loamy soils exhibited greater incidence of Caliciopsis canker symptoms than seedlings growing in other soil groups. Seedling density was also greater in loamy soils compared to other soils. Stem density increases the probability of damage by Caliciopsis canker. For example, mature trees in stands that were adequately or fully stocked were more likely to have Caliciopsis canker symptoms than trees in understocked stands.

Stem density could affect disease development because trees in denser stands or seedlings growing at high densities have less resources to allocate towards defense due to competition. Stem density could also influence disease development by creating an environment more favorable to pathogen reproduction and dissemination. Thickets of white pine regeneration are probably more likely to retain moisture and create high humidity conditions that could favor reproduction and dissemination of *C. pinea*. Furthermore, stem density could also affect tree physiology. Trees growing in denser stands typically have thinner bark, which is associated with greater Caliciopsis canker severity, allegedly because the pathogen is better able to penetrate thinner bark. Currently, all these hypotheses are highly speculative because the epidemiology of *C. pinea* has not been investigated. For example, we do not know what temperature and humidity favor growth and reproduction of *C. pinea*. The time of the year at which inoculum is most abundant and how inoculum is disseminated are both also unknown.

Caliciopsis pinea appears to be the primary pathogen associated with white pine damage in the Northeast, however, other insects (*Matsucoccus macrocicatrices* Richards) and pathogens (*Diplodia scrobiculata* J. de Wet, Slippers & M.J. Wingf.) have been associated with white pine cankers [6,7]. The pathogenicity of *C. pinea* to eastern white pine was demonstrated by Ray in 1936 [8]. In this study, isolates of *C. pinea* were obtained from fruiting bodies protruding from cankers, mostly of seedlings. Caliciopsis canker has been successfully isolated from wood in the cankers of mature trees and inoculation trials are underway to clarify pathogenicity of *C. pinea* and other opportunistic fungi [21].

Forest health management of white pine in past decades has focused on reducing damage by white pine blister rust (*Cronartium ribicola* J. C. Fisch.) and white pine weevil (*Pissoides strobi* Peck) [22]. In this study, trees of all size classes were affected by *Caliciopsis pinea*, did not have white pine blister rust symptoms, and were rarely damaged by the white pine weevil. Management recommendations to reduce damage from white pine blister rust, such as promoting high density of young trees to promote pruning of branches that are susceptible to white pine blister rust, may have improved conditions for the development of *C. pinea*. It is difficult to test this hypotheses due to lack of past baseline data. Most of the eastern white pine resource is mature, and thus, white pine regeneration is important to the future of the resource [2].

5. Conclusions

The objective of this study was to further investigate the relationship between soil and stocking in eastern white pine forests of New England. Eastern white pines growing in excessively drained, poorly drained, and nutrient poor soils are at greater risk of being damaged by Caliciopsis canker. Stem density was also positive correlated with Caliciopsis canker damage for mature trees and regeneration. Poor soils and high stem density could predispose trees to Caliciopsis canker or could affect environmental conditions that favor disease development. The epidemiology of Caliciopsis canker is not understood at this time. Given the prevalence of Caliciopsis canker in important white pine growing regions, this topic deserves further investigation. Foresters will have to take *C. pinea* into account in eastern white pine management.

Acknowledgments: This work was partially funded by the U.S. Department of Agriculture (USDA) Forest Service (USFS) grants # 14-DG-11420004-155, 14-DG-11420004-161, and 14-DG-1142004-229. We are grateful to two anonymous reviewers and editors for the thoughtful suggestions that have improved this work. We would like to thank JacobMavrogeorge from the USFSPathway Program for field assistance, Michelle Cram, Elizabeth Burril, Michael Bohne, and Kevin Dodds from USFS, Lyndsay Hodgman from the USDA Natural Resources Conservation Service (NRCS), and Kara Constanza, William Livingston, and Shawn Fraver from the University of Maine for technical assistance and support. Lastly, we appreciate wisdom and guidance provided by William Leak, USFS. USDA is an equal opportunity provider and employer.

Author Contributions: I.M., K.L., W.O., B.A., J.W., and W.S. developed field sampling methods and collected field data. I.M. analyzed data and wrote manuscript. S.W. cultured *C. pinea* and extracted DNA. T.L. conducted GIS analyses and created maps. D.K. and K.M. provided technical expertise and created GIS layers for soils.

Conflicts of Interest: The authors declare no conflict of interest.

References

1.   North East State Foresters Association. The Economic Importance of Forest-Based Economies of Maine, New Hampshire, New York and Vermont. Available online: http://www.nefainfo.org/uploads/2/7/4/5/27453461/nefa13_econ_importance_summary_aw_feb05.pdf (accessed on 9 March 2016).

2.   Widmann, R.; McWilliams, W. Managing white pine in a new millennium. In *An Overview of the White Pine Resource in New England Using Forest Inventory and Analyses Data*; Bennett, K., Desmarais, K., Eds.; University of New Hampshire Cooperative Extention: Hillsborough, NH, USA; pp. 1–8.

3. Wendel, G.W.; Smith, H.C. Eastern white pine (*Pinus strobus* L.). In *Silvics of North America, Agriculture Handbook 654*; Burns, R.M., Honkala, B.H., Eds.; USDA Forest Service: Washington, DC, USA, 1990; Volume 1, pp. 476–488.

4. Joyce, D.G.; Rehfeldt, G.E. Climatic niche, ecological genetics, and impact of climate change on eastern white pine (*Pinus strobus* L.): Guidelines for land managers. *For. Ecol. Manag.* **2013**, *295*, 173–192. [CrossRef]

5. Funk, A. Studies in the genus Caliciopsis. *Can. J. Bot.* **1963**, *41*, 530–543. [CrossRef]

6. Munck, I.; Livingston, W.; Lombard, K.; Luther, T.; Ostrofsky, W.; Weimer, J.; Wyka, S.; Broders, K. Extent and severity of caliciopsis canker in New England, USA: An emerging disease of eastern white pine (*Pinus strobus* L.). *Forests* **2015**, *6*, 4360–4373. [CrossRef]

7. Mech, A.M.; Asaro, C.; Cram, M.M.; Coyle, D.R.; Gullan, P.J.; Cook, L.G.; Gandhi, K.J.K. Matsucoccus macrocicatrices (hemiptera: Matsucoccidae): First report, distribution, and association with symptomatic eastern white pine in the southeastern United States. *J. Econ. Entomol.* **2013**, *106*, 2391–2398. [CrossRef]

8. Ray, W.W. Pathogenicity and cultural experiments with *Caliciopsis pinea. Mycologia* **1936**, *28*, 201–208. [CrossRef]

9. McCormack, H.W. The morphology and development of *Caliciopsis pinea. Mycologia* **1936**, *28*, 188–196. [CrossRef]

10. Asaro, C. What is killing white pine in the highlands of west Virginia? In *Forest Health Review May 2011*; Virginia Department of Forestry: Charlottesville, VA, USA, 2011.

11. Lombard, K. Caliciopsis canker (pine canker) *Caliciopsis pinea.* In *UNH Cooperative Extension Publication*; UNH: Durham, NH, USA, 2003.

12. Munck, I.A.; Livingston, W.; Lombard, K.; Luther, T.; Ostrofsky, W.D.; Weimer, J.; Wyka, S.; Broders, K. Extent and severity of caliciopsis canker in New England, USA: An emerging disease of eastern white pine (*Pinus strobus* L.). *Forests* **2015**, *6*, 4360–4373. [CrossRef]

13. Extension, U.O.N.H.C. Important Forest Soil Groups. Available online: https://extension.unh.edu/goodforestry/html/app-soils.htm (accessed on 16 January 2014).

14. Krist, F.J.; Ellenwood, J.R.; Woods, M.E.; McMahan, A.J.; Cowardin, J.P.; Ryerson, D.E.; Sapio, F.; Zweifler, M.O.; Romero, S.A. 2013–2014 National Insect and Disease Forest Risk Assessment. Available online: http://www.fs.fed.us/foresthealth/technology/nidrm.shtml#NIDRMReport (accessed on 16 June 2015).

15. Broders, K.D.; Wallhead, M.W.; Austin, G.D.; Lipps, P.E.; Paul, P.A.; Mullen, R.W.; Dorrance, A.E. Association of soil chemical and physical properties with pythium species diversity, community composition, and disease incidence. *Phytopathology* **2009**, *99*, 957–967. [CrossRef] [PubMed]

16. Leak, W.B.; Lamson, N.I. *Revised White Pine Stocking Guide for Managed Stands*; U.S. NA-TP-01-99 Department of Agriculture, Forest Service, Northeastern Area State and Private Forestry: Newtown Square, PA, USA, 1999; p. 2.

17. Stamp, N. Out of the quagmire of plant defense hypotheses. *Q. Rev. Biol.* **2003**, *78*, 23–55. [CrossRef] [PubMed]

18. Marie-Laure, D.-L.; Benoit, M.; Louis-Michel, N.; Dominique, P.; Andrea, V. Interactive effects of drought and pathogens in forest trees. *Ann. For. Sci.* **2006**, *63*, 597–612.

19. Blodgett, J.T.; Kruger, E.L.; Stanosz, G.R. Sphaeropsis sapinea and water stress in a red pine plantation in central Wisconsin. *Phytopathology* **1997**, *87*, 429–434. [CrossRef] [PubMed]

20. Oliva, J.; Stenlid, J.; Martinez-Vilalta, J. The effect of fungal pathogens on the water and carbon economy of trees: Implications for drought-induced mortality. *New Phytol.* **2014**, *203*, 1028–1035. [CrossRef] [PubMed]

21. Costanza, K.K.L.; W.H. Livingston, S.; Fraver, I.A.; Munck, K.; Lombard, W.; Ostrofsky, R.W. Impact of Caliciopsis Pinea on White Pine Biology, Wood Quality, and Lumber Yield, San Francisco, CA, USA, 29 March–3 April 2016; American Association of Geographers: Washington, DC, USA.

22. Ostry, M.E.; Laflamme, G.; Katovich, S.A. Silvicultural approaches for management of eastern white pine to minimize impacts of damaging agents. *For. Pathol.* **2010**, *40*, 332–346. [CrossRef]

Article

Environmental Factors Driving the Recovery of Bay Laurels from *Phytophthora ramorum* Infections: An Application of Numerical Ecology to Citizen Science

Guglielmo Lione [1,2], Paolo Gonthier [1] and Matteo Garbelotto [2,*]

[1]   Department of Agricultural, Forest and Food Sciences, University of Torino, Largo P. Braccini 2,
      I-10095 Grugliasco, Italy; guglielmo.lione@unito.it (G.L.); paolo.gonthier@unito.it (P.G.)
[2]   Department of Environmental Science, Policy and Management, University of California at Berkeley,
      151 Hilgard Hall, Berkeley, CA 94720, USA
*   Correspondence: matteog@berkeley.edu; Tel.: +1-510-643-4282

Received: 24 July 2017; Accepted: 10 August 2017; Published: 13 August 2017

Abstract: *Phytophthora ramorum* is an alien and invasive plant pathogen threatening forest ecosystems in Western North America, where it can cause both lethal and non-lethal diseases. While the mechanisms underlying the establishment and spread of *P. ramorum* have been elucidated, this is the first attempt to investigate the environmental factors driving the recovery of bay laurel, the main transmissive host of the pathogen. Based on a large dataset gathered from a citizen science program, an algorithm was designed, tested, and run to detect and geolocate recovered trees. Approximately 32% of infected bay laurels recovered in the time period between 2005 and 2015. Monte Carlo simulations pointed out the robustness of such estimates, and the algorithm achieved an 85% average rate of correct classification. The association between recovery and climatic, topographic, and ecological factors was assessed through a numerical ecology approach mostly based on binary logistic regressions. Significant ($p < 0.05$) coefficients and the information criteria of the models showed that the probability of bay laurel recovery increases in association with high temperatures and low precipitation levels, mostly in flat areas. Results suggest that aridity might be a key driver boosting the recovery of bay laurels from *P. ramorum* infections.

Keywords: biological invasions; climate; disease triangle; epidemiology; forest; geographic information system; modelling; Oomycetes; plant disease; sudden oak death

---

1. Introduction

The "disease triangle" model [1,2] frames pathogenesis as a process relying on the trophic interconnection between susceptible hosts and pathogens, provided that the environmental conditions are conducive to infection by the pathogen and to disease progression. The interest of forest pathologists in unraveling environmental factors driving plant diseases has been amplified in the last decades by the onset of relevant epidemics caused by emerging pathogens such as *Phytophthora ramorum* Werres, De Cock and Man in't Veld in Western North America, *Heterobasidion irregulare* Garbelotto and Otrosina, *Hymenoscyphus fraxineus* (T. Kowalski) Baral, Queloz and Hosoya, and *Gnomoniopsis castaneae* G. Tamietti in Europe, just to cite a few relevant examples [3–12]. The main environmental drivers underlying the success of such novel epidemics have often been identified through a numerical ecology approach, based on computational and multivariate statistical techniques suitable to deal with complex ecological datasets [10,11,13–22].

Although environmental factors play a key role in boosting plant diseases, they may also unbalance the interaction between pathogen and host, favoring the latter and promoting recovery of the host. Recovery is not only a theoretical possibility, since it has been extensively documented for a broad range of plant diseases [23–28]. However, few studies have proposed a reverse approach to the "disease triangle" paradigm in forest ecosystems, investigating the association between environmental factors and the actual or potential recovery from infectious diseases [29–31].

As mentioned above, the invasion by the oomycete *P. ramorum* has led to one of the most disruptive epidemics in forest ecosystems of North America, especially in California and Oregon [32]. Depending on the host, different diseases displaying a wide range of symptoms have been reported for the same pathogen [33–38]. The most relevant among such diseases is "sudden oak death" (SOD), a rapid and severe decline associated with stem bleeding cankers and high mortality rates of *Quercus* spp. (section Lobatae) and tanoak (*Notholithocarpus densiflorus* (Hook. & Arn.) Manos, Cannon and S.H. Oh). "Ramorum leaf blight" is instead a foliar disease caused by *P. ramorum* on bay laurels (*Umbellularia californica* (Hook. and Arn.) Nutt.), *Camellia* spp., *Kalmia* spp. and *Rhododendron* spp., whose main symptoms are leaves spots, stains and diffuse necroses, as well as browning or blackening, followed by premature leaf fall. Although bay laurels are highly susceptible, they generally display only moderate or mild symptoms, playing a key role as reservoirs of *P. ramorum* inoculum [39]. However, bay laurels support maximum sporulation levels, while the contribution of tanoaks and oaks in this sense is notably less relevant [40]. Interestingly, *P. ramorum* is characterized by unusual biological traits within the genus *Phytophthora*, being a pathogen on aerial parts of the infected hosts, where it differentiates deciduous sporangia acting as the main infective inoculum [38,41]. Once mobilized and aerially spread during rainfalls, sporangia release the zoospores, whose motility in water or wet surfaces facilitates the infection of new hosts [40,41]. Because of its high pathogenicity, spread potential and invasiveness, *P. ramorum* is considered a serious threat for forests ecosystems in Western North America. Moreover, the pathogen has been sporadically reported in Europe, where it was included by the European and Mediterranean Plant Protection Organization (EPPO) in the A2 list of alien organisms recommended for regulation [35].

Most studies investigating the ecology of *P. ramorum* agree on the overall conclusion that the pathogen spread is boosted by increasing host densities and high rainfalls levels, while it is hindered by aridity [32,42–47]. An innovative approach based on citizen science was recently designed, tested and applied to predict the infection risk by *P. ramorum* in California and southern Oregon [17,48]. From 2005 to 2015, during the so called "SOD blitzes", volunteer citizens were engaged to survey forests and urban parks, where they collected soil, water and tree samples geolocated with amatorial GPS devices. Samples coordinates were stored in a geodatabase along with the results provided by laboratory analyses confirming, or not, the presence of *P. ramorum*. As a result, maps of the current distribution of *P. ramorum*, as well as estimates of infection risk as a function of latitude and longitude were obtained and integrated in user friendly applications for computers and mobile devices (i.e., "SODmap" and "SODmap Mobile") [17,48].

Although citizen science has disclosed an interesting potential to support applied environmental research [49–51], the appealing possibility of a cost-efficient realization of huge datasets is not risk-free. In fact, the poor scientific background of most citizens along with constraints related to equipment availability may lead to procedural errors, biases an inaccuracy potentially affecting data quality [17,49,50]. For instance, during the "SOD blitzes" [17,48], citizens were not expected to own professional GPS devices. Amatorial GPS are widespread, since they are often integrated by default in popular smartphones, but their error in collecting the right location may vary from a tenth to some hundred meters [52]. Hence, the possibility that some trees were inadvertently resampled in different "SOD blitzes" is likely. Despite being an unsought bias, resampling could be turned into an interesting opportunity allowing the follow-up of the infection by *P. ramorum* in the same trees, along with the detection of trees whose infection status switched from positive to negative (i.e., recovered trees).

Under the above premise, the main goals of this work were (I) to design and test a general method for the identification and geolocation of the recovered trees, based on a citizen science dataset; (II) to apply the above method to the "SODmap" database; and (III) to test the association between the probability of recovery from *P. ramorum* infections in bay laurel and some climatic, topographic and ecological factors.

2. Materials and Methods

2.1. Samplings and Laboratory Analyses

    Samplings and laboratory analyses leading to the realization of the "SODmap" database were performed within the "SOD Blitz" program, whose details were previously published [17]. From 2005 to 2015 volunteer citizens were recruited to collect water, soil or plant tissue samples. In this study, only the latter were considered to investigate host recovery. Specific training was offered to all volunteers in order to properly detect disease symptoms and standardize sampling procedures. The distribution area of *P. ramorum* in California and Southern Oregon was systematically monitored and random samplings were carried out in forests sites, urban parks and private gardens. Storage packets were provided to the volunteers for sampling preservation prior to laboratory analyses. The detection of *P. ramorum* in host-plant tissues was carried out using a species-specific molecular assay previously described [53]. Such assay is based on a highly efficient polymerase chain reaction (PCR) using a $5'$ fluorogenic exonuclease (TaqMan) chemistry that combines two species specific primers (Pram5 and Pram6) to an internal probe (Pram6), allowing the detection of target DNA up to a minimum threshold of 15 fg [53]. When the assay detected the presence of *P. ramorum* DNA, the infection status of the corresponding sample was scored as 1, otherwise as 0. Sampling time, trees species, coordinates and infection status were included in the "SODmap" database.

2.2. Design of the Algoritm to Detect and Geolocate the Recovered Trees

2.2.1. Rationale

    Let $T = \{T_1, T_2, T_3 \ldots T_K\}$ be a set of K points (i.e., sampled trees locations) whose Cartesian coordinates x_T and y_T (m) are exact, but unknown. T is composed by the sets $V = \{V_1, V_2, V_3 \ldots V_N\}$ and $W = \{W_1, W_2, W_3 \ldots W_M\}$ including N and M trees, respectively, with $N + M = K$, $V \cup W = T$ and $V \cap W = \varnothing$. Trees coordinates for sets V (x_V, y_V) and W (x_W, y_W) are estimated with a GPS device whose median value ε of the radial error r (m) is known [54]. Coordinates collection is performed once for V set (i.e., single-sampled trees), and more than once for the W set (i.e., resampled trees). Hence, the set $V' = \{V'_1, V'_2, V'_3 \ldots V'_N\}$, derived from V, includes N trees with estimated coordinates $x_{V'}$ and $y_{V'}$. Similarly, the set $W' = \{W'_1, W'_2, W'_3 \ldots W'_{M'}\}$ is gathered from W and includes $M' > M$ trees with coordinates $x_{W'}$ and $y_{W'}$. Let $T' = V' \cup W' = \{T'_1, T'_2, T'_3 \ldots T'_{K'}\}$ be a set of K' points with coordinates $x_{T'}$ and $y_{T'}$. All elements of T' are in bijective association with K' elements of the set $I = \{I_1, I_2, I_3 \ldots I_{K'}\}$, expressing the infection status (i.e., if positive to the pathogen 1, else 0), and with K' elements of the set $S = \{S_1, S_2, S_3 \ldots S_{K'}\}$, indicating the sampling time. Under this premise, a citizen science dataset includes T', I, S with the corresponding GPS coordinates $x_{T'}$ and $y_{T'}$, while the partitioning of trees locations between the sets V' and W' is unknown.

    Let Φ_Λ be an algorithm operating as a binary classifier based on partitioning [55] and able to split T' into the two mutually exclusive sets $\widetilde{W}'$ and $\widetilde{V}'$, estimates of W' and V', respectively. The partitioning mechanism is based on the application of a virtual squared grid over the point features representing the T' trees. A grid-system cannot discriminate distinct points locations if their distance is below its spatial resolution, which is determined by the length of the grid-cells edge (i.e., pixel size). Such points are merged by the grid-system, as if they were not spatially distinct [56]. However, the separation observed among some point features (i.e., W') can be an artifact deriving from the iterated estimation of the same locations (i.e., W) with a GPS device affected by some spatial location error. Hence, Φ_Λ can classify such points as deriving from accidental resampling (i.e., $\widetilde{W}'$) through a grid-based system Λ_ϑ with adequate spatial resolution ϑ. Φ_Λ assigns to the set $\widetilde{W}'$ the neighboring points features falling into the same pixel, since they cannot be resolved by the spatial resolution of Λ_ϑ (Figure 1). The set $\widetilde{W}$

is derived from $\widetilde{W}'$ by averaging its points coordinates at pixel level, while the other points are then classified as $\widetilde{V}'$ elements. Finally, the set of recovered trees $R = \{R_1, R_2, R_3 \dots\}$ is identified by Φ_Λ as the subset of $\widetilde{W}$ whose infection status switched from 1 to 0, from the first to the last sampling.

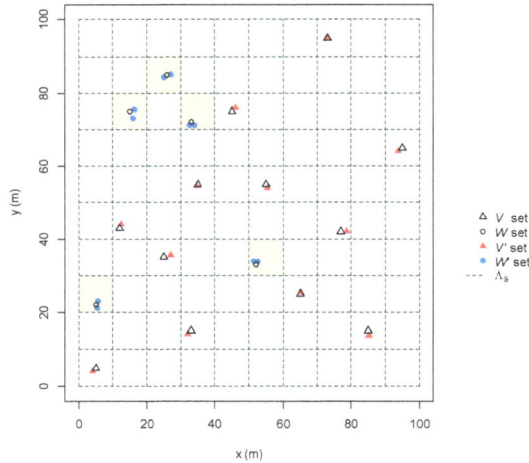

Figure 1. Schematic representation of exact, but unknown, single-sampled and resampled trees locations (sets V and W) with the associated estimates (sets V' and W'), whose coordinates are collected with a GPS device. As a consequence of its spatial resolution, the grid-based system Λ_ϑ cannot resolve trees locations estimates included within the borders of the same pixel (highlighted in yellow if the case occurs). For this reason, the algorithm Φ_Λ can assess which trees coordinates are likely to have been collected more than once. As shown in this example, all the elements of W', which are by definition replicated estimates of W elements, were detected.

2.2.2. Modelling GPS Error and Spatial Resolution of the Grid-System

The Φ_Λ algorithm is based on the selection of an adequate spatial resolution ϑ which is dependent upon the GPS error scattering the resampled trees features (i.e., W') around their true location (i.e., W). The mathematical structure of the GPS error was modeled prior to its relation with the spatial resolution of the grid-system.

The Rayleigh distribution $f_{(r|\sigma)} = \frac{r}{\sigma^2} e^{-\frac{r^2}{2\sigma^2}}$ parametrized by σ and with median value $E = \sigma\sqrt{2\ln2}$ was assumed as the probability distribution function (PDF) of r [54,57]. The goodness-of-fit of the Rayleigh PDF to r was experimentally tested with an in-field trial. In January 2017, a point location with known coordinates (389,095.5 m, 4,990,997.5 m, UTM WGS 1984 zone 32N, EPSG 32632) was identified in an open space flat area of Western North Italy. The point coordinates were collected 200 times in different days and daylight hours with a professional GPS device (Magellan MobileMapper 6), resulting in 200 point location estimates (Dataset S1). The GPS radial error r was calculated for each point location estimate as the Euclidean distance separating the true coordinates from the assessed ones [54]. The adequacy of the Rayleigh PDF to model r was tested with the Kolmogorov-Smirnov goodness-of-fit test [58], parametrizing the Rayleigh curve with the maximum likelihood estimate of σ [57].

A functional relation $\vartheta = f_{(\varepsilon)}$ was assessed in silico with Monte Carlo (MC) methods to allow the selection of ϑ from ε, which is a parameter associated with the GPS device [52]. A set of 10^3 GPS coordinates collection of a known virtual tree location was simulated 10^3 times for each integer value of ε included between 1 and 3×10^3. The GPS location error was randomly sampled from a Rayleigh PDF. The maximum pairwise distance obtained for each set of GPS coordinates was set as ϑ value.

After a visual examination of the scatterplot displaying the ε and the associated ϑ values, a linear regression model with no intercept was fit. Its performances were assessed in terms of ε coefficient (β_ε) significance, minimum Akaike information criterion (AIC) and maximum AIC weight (AIC$_w$), setting as reference baseline the null model [58–60]. Details about the algorithm performing the MC simulations are reported in Note S1.

Once ϑ is defined based on ε, Λ_ϑ can be integrated as a sliding grid-system into Φ_Λ to provide different alternative scenarios estimating R. Such alternatives account for different possible positions of the grid-system.

2.2.3. Design of the Algorithm

The algorithm was designed in R language as a two-step process consisting in distinct blocks of code embedded into specific R-functions [58]. The first block, starting from the trees coordinates $x_{T'}$ and $y_{T'}$ and from the median value ε^* of the GPS radial error, builds up the virtual grid-system Λ_ϑ by fixing its spatial extent and spatial resolution ϑ. The grid layer is located over the map of the study area including all trees locations and randomly fixing the coordinates of its left-bottom corner. Trees whose location cannot be resolved based on ϑ are sent to the second block of code (Φ_Λ). The latter processes the sampling year and the infection status for each cell, detecting trees sampled in different years and selecting the ones whose infection status switched from 1 to 0 (i.e., recovered trees R). Since the classifier Φ_Λ includes Λ_ϑ as a stochastic component (i.e., the left-bottom vertex coordinates of the grid), alternative scenarios estimating R are possible. Their assessment can be achieved by embedding Λ_ϑ as a sliding grid-system into Φ_Λ through iterate runs of the two algorithm blocks (i.e., loop [58]). Technical details about the algorithm design are included in Note S2.

2.2.4. Algorithm Performances Assessment

Algorithm performances were assessed in silico with a series of MC simulations [61]. In a virtual squared area extended for 10^4 km^2, the sets V' with N = 200 and W with M = 100 were simulated by randomly extracting the associated coordinates $x_{V'}$, $y_{V'}$, x_W and y_W from a uniform distribution [57] limited between 0 and 100 km. A constant infection status and sampling year was assigned to the elements of V'. A virtual GPS with median radial error $\varepsilon^* = 100$ m was used to mimic the process of coordinates collection on W, setting two samplings for each element in W, and resulting in M' = 200 estimated trees locations included in the set W'. The radial error r of every virtual GPS coordinates collection was randomly extracted from a Rayleigh distribution whose parameter σ was calculated by solving the equation $E = \varepsilon^*$ [57]. Each element of W was then associated with the couple of related elements in W'. Random consecutive sampling years were assigned to the above couples. One of the four permutations of the infection status was randomly assigned to each fourth of the couples.

Φ_Λ was applied to the set of trees locations based on 10^3 different random positions of the sliding grid-system Λ_ϑ. The ability of Φ_Λ in partitioning the trees between W' and V' was assessed with the Cohen's k statistics and overall rate of correct classification (%) [62–64], which were averaged within all 10^3 positions of Λ_ϑ. The above MC simulation was iterated 10^3 times.

2.3. Detection and Geolocation of the Trees Recovered from P. ramorum Infections Based on the "SODmap" Database

Only the input needed by the Φ_Λ algorithm was retained from the original "SODmap" database, which was filtered to remove all unnecessary information [17,48]. Sampled trees coordinates were reprojected into UTM NAD83 zone 11N (EPSG 4326) and included within a polygon shapefile in a GIS environment [65].

Assuming that trees coordinates had been collected by volunteers endowed with GPS sensors embedded in smartphones, iPhones or comparable devices [17,48], two scenarios were analyzed depending on the GPS median radial error ε^*. Based on thresholds reported in previous studies [52], the minimum and maximum ε^* were set to 10 m and 500 m, respectively, defining two alternative scenarios (i.e., scenario-10 m and scenario-500 m). For both scenarios, the detection and geolocation of trees recovered from *P. ramorum* infections were performed by applying Φ_Λ, with 10^3 iterations, accounting for different locations of the grid-system. The algorithm was run separately for bay laurel, tanoak and oaks, allowing reproducibility by fixing a unique seed number to each run (provided in Dataset S2). The output datasets were ranked within host species and scenarios depending on the number of elements included in the set $\widetilde{W}$ at each iteration. Datasets displaying the maximum number of elements were selected for further analyses. If several of the above datasets resulted from the same tree species and scenario (i.e., alternative datasets), they were tested for equivalence based on the comparison of their associated two-dimensional Gaussian kernel density estimate (KDE) matrices [58,66,67]. KDE matrices and their associated rasters were obtained by setting a common bandwidth [67] and the cells edge to 10 km, with extension and resolution allowing for a perfect cell-wise alignment. Comparisons among KDE matrices were performed with the Kruskal-Wallis or Mann-Whitney tests for the mean and with the Fligner-Killeen test on rank-transformed data for the variance [58,68]. Since some alternative datasets were equivalent (see Results), a random selection of the final datasets to include in the further analyses was performed. The number of resampled trees included in $\widetilde{W}$ for each scenario and host species was calculated and compared to the corresponding number of trees from the screened "SODmap" dataset. Similarly, the incidence of trees recovered from *P. ramorum* infections was calculated as absolute count and in percent. The latter, including 95% confidence intervals, derived from ratio between the recovered trees and the trees which have been infected at least once [69].

2.4. Analysis of the Association between Environmental Factors and Recovery from P. ramorum Infections

Because of the low number of oaks and tanoaks recovered from *P. ramorum* infections (see Results), the association between recovery and environmental factors was tested only for bay laurels. The averages of the minimum, maximum and mean temperatures (T_{min}, T_{max}, T_{mean}, °C) and cumulate precipitations (P, mm) were calculated at pixel level based on the yearly values of the rasters derived from parameter-elevation regressions on independent slopes models (PRISM-4 km) in the period 2005–2015 [70,71]. Elevation (El, m) rasters were obtained from digital elevation models (DEM) provided by the NASA Shuttle Radar Topographic Mission in the 30 arc-sec (~900 m) version (SRTM30) [56,72]. Estimates of bay laurel density (Bld, trees/ha) were gathered from an available source raster [73]. All rasters were reprojected into UTM NAD83 zone 11N (EPSG 4326) and cropped within the borders of the study area. Aspect (As, as azimuth, °), slope (Sl, %), terrain ruggedness (as Terrain Ruggedness Index—TRI [74]), topographic position (as Topographic Position Index—TPI [75]) were derived from GIS geomorphological analyses carried out on the DEM [76]. All rasters were converted into matrices [58], whose basic descriptive statistics (range, average, coefficient of variation (CV)) were calculated, with the exception of As. The Bld raster was trimmed discarding pixels below 150 and above 1500 to improve analyses robustness [77].

The associations between recovery from *P. ramorum* infections and T_{min}, T_{max}, T_{mean}, P (i.e., climatic factors) El, Sl, TRI, TPI (i.e., topographic factors), and Bld (i.e., ecological factor) were tested through a numerical ecology approach based on correlation analyses and binary logistic regressions [58,63]. For both scenarios, the binary dependent variable was coded as 1 for recovered trees, as 0 for the other trees whose infection status to *P. ramorum* had been positive at least once. The continuous predictors were obtained by extracting from the rasters of the environmental factors the cell values corresponding to each tree location. In addition, trees latitude (Lat) and longitude (Long) in UTM NAD83 zone 11N (EPSG 4326) were included as predictors. Colinearity among predictors, except As and Bld (see below), was tested by calculating all possible pairwise Spearman's ρ coefficients and results were visualized through circular correlograms [58]. Binary logistic regressions were fitted with single and multiple predictors, including the intercept (β_0). Combinations of multiple predictors were included only in the absence of significant pairwise correlations [61,63]. All models within the same scenario were compared in terms of sign and significance of the predictor coefficient (β), range of its associated 95% CI, minimum AIC and maximum AIC_w, setting the null model as reference [58–60,63].

The robustness of the results obtained from binary logistic regression models was assessed through a MC simulation trial [61] accounting for a putative overall misclassification rate associated with Φ_Λ. Since the Φ_Λ algorithm displayed an average overall rate of correct classification theoretically attaining 85.3% (see Results), the putative misclassification rate was set to 15%. Hence, all binary logistic regression models were run on a dependent variable whose 15% of values was randomly selected and reverted at each iteration, until 10^5 iterations were conducted. From the resulting β coefficients distributions, the associated 95% CI were calculated (i.e., MC 95% CI) [78] and the bound signs were compared with the corresponding 95% CI obtained from the fit on the original dataset.

Bay laurel density models were fitted only as single predictor models and excluded from the comparative assessment with the others because of missing data. As was not included as predictor because of data circularity [79]. Hence, azimuth distributions were compared between recovered and not recovered bay laurels with the Mardia-Watson-Wheeler test and the Rao's homogeneity test [80,81]. A complementary correlation and partial correlation analysis was carried out between KDE matrices of recovered trees and matrices of T_{min}, T_{max}, T_{mean}, P, El, Sl, TRI, TPI, and Bld (Note S3).

2.5. Software and Libraries Used for Statistical, GIS and Numerical Analyses

Statistical and numerical analyses were run in R version 3.2.3. with libraries *binGroup, circular, extraDistr, Hmisc, igraph, maptools, MASS, MuMIn, ppcor, psych, raster, rasterVis, REdaS, spaa, spatstat.* GIS data manipulation and related analyses were performed both in R and in QGIS 2.8.5-Wien. All statistical tests were carried out with a cut-off significance threshold set to 0.05.

3. Results

3.1. Design of the Algorithm to Detect and Geolocate the Recovered Trees

3.1.1. GPS Error and Spatial Resolution of the Grid-System

The 200 GPS radial errors r ranged between 0.07 m and 4.52 m, with an average of 1.64 m. The maximum likelihood estimate of σ attained 1.32 m and the resulting Kolmogorov-Smirnov test was non-significant ($p = 0.553$) confirming the adequacy of the Rayleigh PDF to model r.

The MC experiment produced a total of 3×10^9 GPS estimates of the single tree location, split in 10^3 assessments of 10^3 independent sets of points for each of the 3×10^3 values of the GPS median radial error ε (Figure S1). The linear regression model displayed a significant coefficient $\beta_\varepsilon = 6.033$ ($p < 0.05$) and outperformed the null model for all metrics considered (AIC = 47457202, vs. AIC = 59928055 and $AIC_w = 1$ vs. $AIC_w = 0$). Hence, the functional relation $\vartheta = f_{(\varepsilon)}$ was assessed as $\vartheta = 6.033\varepsilon$.

3.1.2. Design of the Algorithm

The first block of code resulted in a 32 lines script in R language embedded in the function *LAMBDA(x, y, theta)*, which defines the grid-based system Λ_ϑ. *LAMBDA* inputs are the vectors of coordinates $x = x_{T'}$ and $y = y_{T'}$, associated with trees in T', and the *theta* $= \vartheta$ value gathered from the equation $\vartheta = 6.033\varepsilon$ (see 3.1.1). *LAMBDA* output is a vector γ assigned to each element of T', NA included, identifying with a unique cell-code the locations that cannot be resolved based on ϑ. A map of T' locations along with the associated grid with spatial resolution ϑ is provided as additional outcome.

The second block of code, defining the classifier Φ_Λ, resulted in a 35 lines R-script incorporated into the function *PHI(x, y, I, S, gamma)*. In addition to the coordinates of trees in T', the function requires input vectors indicating the infection status (*I*), the sampling time (*S*) and the γ (*gamma*) values provided by *LAMBDA*. *PHI* output consists of a matrix object with one row per each element of the set $\widetilde{W}$ and five columns indicating the associated "switch-code", x and y coordinates, the first and the last sampling time, respectively. Column names are returned by the function as switch, xmed, ymed, fy and ly, respectively. "Switch-code" values refer to the transition of infection status from the first to the last sampling, indicated as: "1" from 0 to 0, "2" from 0 to 1, "3" from 1 to 0 (i.e., set *R*, recovered trees), and "4" from 1 to 1. Functions *LAMBDA* and *PHI* are provided as R-scripts in Algorithm S1 and S2 [58].

3.1.3. Algorithm Performances Assessment

A total of 10^6 runs of Φ_Λ resulted in 10^3 average values of the Cohen's k statistic and of the overall rate of correct classification, each one based on 10^3 different positions of the Λ_θ sliding grid-system. The average Cohen's k ranged from 0.654 to 0.755, with a global average of 0.706, while the overall rate of correct classification was included between 82.7% and 87.8%, with an overall mean attaining 85.3% (Figure 2). Based on the threshold reported [64] for Cohen's k, the classification performed by Φ_Λ ranks at the 2nd place out of 6 classes, displaying "substantial agreement" between observations (i.e., simulated sets) and predictions (i.e., Φ_Λ classifications).

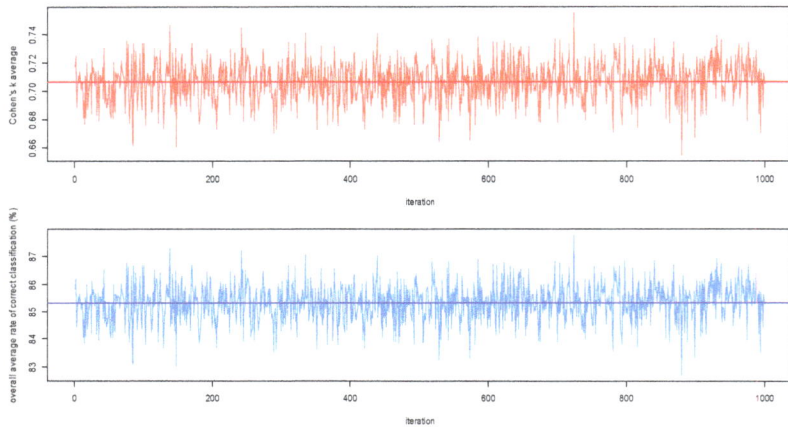

Figure 2. Results of the Monte Carlo validation assessing the performances of the classifier Φ_Λ. The graphs show the averages of the Cohen's k and of the overall rate of correct classification achieved by Φ_Λ at each iteration of the 10^3 applications of Λ_θ. Horizontal thick lines indicate the overall averages.

3.2. Detection and Geolocation of Trees Recovered from P. ramorum Infections Based on the "SODmap" Database

The filtering of the original "SODmap" database resulted in a matrix with 19,211 records, one per sampled tree, and 8 fields providing the input for the classification algorithm (Dataset S3). A topographic surface of over 36,700 km^2 was delimited within the borders of the study area, with an extension of approximately 920 km along the major axis N-NW/S-SE oriented. Within this area, Φ_Λ detected a single dataset for bay laurel and some alternative datasets for oaks and tanoak in both scenarios. Such alternative datasets were equivalent based on the outcomes of the KDE matrices comparisons (see Note S4). Host species final datasets are provided as Datasets S4 and S5 for scenario-10 m and scenario-500 m, respectively.

In the filtered "SODmap" dataset, 14,666 out of 19,211 trees were bay laurels (76.3%), 2468 tanoaks (12.8%) and 157 oaks (0.82%), while the remaining plants (10%) were classified as other host species. In scenario-10 m the algorithm detected 1044 resampled trees ($\widetilde{W}$) including 943 bay laurels (90.3%), 88 tanoaks (8.4%) and 13 oaks (1.2%). Similarly, 284 bay laurels (75.3%), 79 tanoaks (21.0%) and 14 oaks (3.7%) out of 377 resampled trees were detected in scenario-500 m. In percent, resampled trees accounted for 1.96% and 5.43% of the sampled trees in scenario-10 m and scenario-500 m, respectively. The percentage of trees recovered from *P. ramorum* infections (*R*) ranged from 9.3 to 32.9% depending on host species and scenario, while the overall percentage was substantially homogeneous between scenarios (29.7% in scenario-10 m and 28.6% in scenario-500 m). Bay laurel attained the largest incidence of recovered trees (approximately 32–33%), followed by oaks and tanoak (Table 1). The distribution of resampled and recovered trees in the study area is shown in Figure 3 for both scenarios.

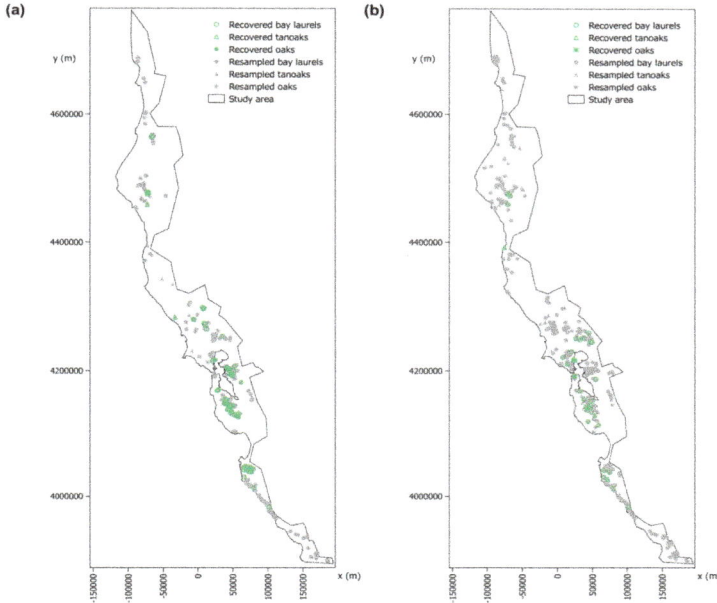

Figure 3. Distribution of resampled and recovered bay laurels, tanoaks and oaks in scenario-10 m (**a**) and scenario-500 m (**b**). Coordinates are in UTM NAD83 zone 11N (EPSG 4326).

Table 1. Trees recovered from *P. ramorum* infections.

Species	Scenario	Trees Infected at Least Once by *P. ramorum*	Recovered Trees	Recovered Trees (%)	95% CI Lower Bound (%)	95% CI Upper Bound (%)
Bay laurel	10 m	359	116	32.3	27.5	37.3
	500 m	82	27	32.9	23.4	43.8
Tanoak	10 m	43	4	9.3	3.2	21.5
	500 m	36	7	19.4	8.8	35.8
Oaks	10 m	6	1	16.7	0.9	59.4
	500 m	8	2	25.0	4.6	64.1
Overall species	10 m	408	121	29.7	25.3	34.3
	500 m	126	36	28.6	21.2	37.2

For each scenario, data are provided separately for bay laurel, tanoak, oak, and jointly for all host species (i.e., overall species). The lower and upper bounds of the 95% confidence interval associated with the percent of recovered trees are reported.

3.3. Analysis of the Association between Environmental Factors and Recovery from P. ramorum Infections

3.3.1. Environmental Factors

A series of 10 derived rasters were obtained for the environmental factors T_{min}, T_{max}, T_{mean}, P, El, As, Sl, TRI, TPI, and Bld (Figure 4). All rasters were continuously distributed within the study area with the exception of Bdl, whose patchy covering was due to missing data from the source raster. On average, temperatures during the timeframe 2005–2015 attained 7.62 °C, 19.68 °C, 13.65 °C for T_{min},

T_{max} and T_{mean}, respectively, with the highest variability observed for T_{min} (CV = 18.69), followed by T_{max} (CV = 10.84) and T_{mean} (CV = 10.40). In the same period, precipitations across the study area were more variable than temperatures (CV = 56.07), with a yearly average of 1236.43 mm. The study area was also characterized by an average El of 375.26 m (CV = 84.36) and a mean Sl value of 8.19% (CV = 74.81), while TRI and TPI attained averages of 80.83 (CV = 61.91) and 0.5 (CV = 12 × 10^3). Finally, the average value achieved by Bld was 297 trees/ha (CV = 49.92).

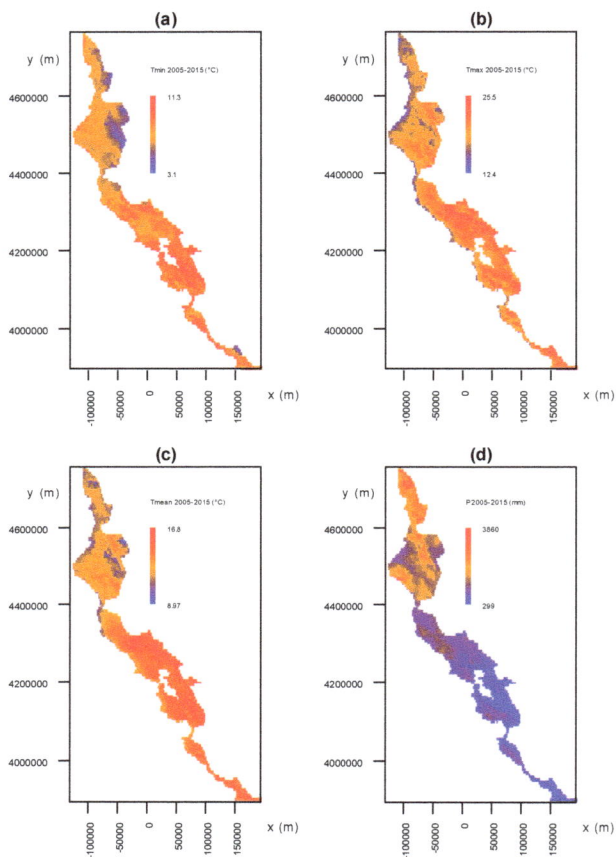

Figure 4. *Cont.*

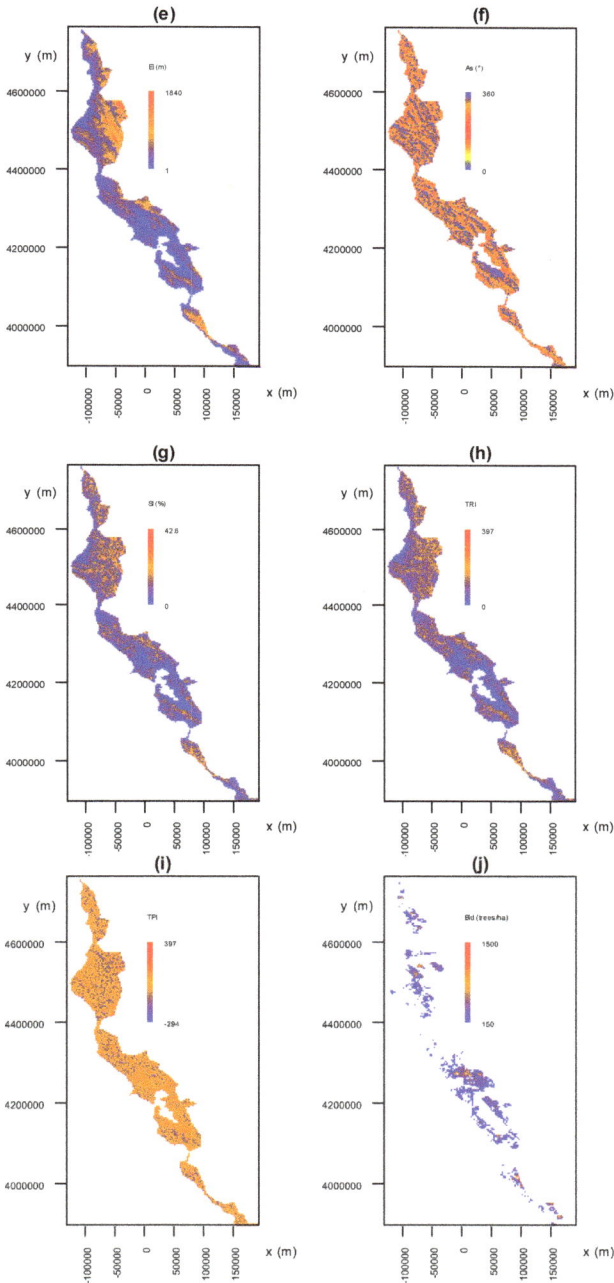

Figure 4. Raster maps of the environmental factors across the study area. Each panel refers to a specific factor: (**a**) T_{min}; (**b**) T_{max}; (**c**) T_{mean}; (**d**) P; (**e**) El; (**f**) As; (**g**) Sl; (**h**) TRI; (**i**) TPI and (**j**) Bld. Along with coordinates, the unit of measurement and the range are reported for each factor. A color gradient from blue to red shows the transition between low and high values of the factor, with the exception of As, where colors represent transitions between contiguous cardinal points (N—blue, E—yellow, S—red, W—orange). Coordinates are in UTM NAD83 zone 11N (EPSG 4326). For factors acronyms, see the main text.

3.3.2. Binary Logistic Regression Models

Correlograms showed that 36 and 32 out of 45 pairwise Spearman's ρ correlation coefficients between predictors were significant ($p < 0.05$) in scenario-10 m and scenario-500 m, respectively (Figure 5). Hence, the 9 couples of uncorrelated predictors were included in multiple binary logistic regression models in scenario-10 m. Similarly, multiple binary logistic regressions were fitted on the 13 couples and one triad of uncorrelated predictors in scenario-500 m.

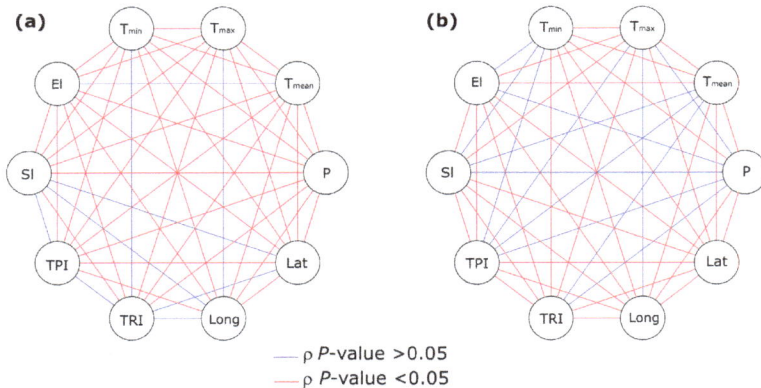

Figure 5. Circular correlograms showing the significance of all possible pairwise Spearman's ρ coefficients calculated between environmental predictors, including latitude and longitude, in scenario-10 m (**a**) and scenario-500 m (**b**). For factors acronyms, see the main text.

A total of 14 out of 21 and 7 out of 26 binary logistic regression models displayed significant β coefficients ($p < 0.05$) in scenario-10 m and scenario-500 m, respectively (Table S1). P was significantly ($p < 0.05$) and negatively ($\beta < 0$) associated with the probability of bay laurel recovery in all models where it was included as predictor, namely, one model in scenario-10 m and 7 models in scenario-500m. Such models ranked first based on AIC and AIC_w, the latter attaining 87.1% in scenario-10m and a cumulative value of 94.4% in scenario-500 m. In all cases, the bounds of the 95% CI associated with P displayed negative lower and upper bounds, hence confirming the results gathered from *p*-values. The same outcome was observed in MC 95% CI, supporting the robustness of the negative association between P and the probability of bay laurel recovery when assuming a 15% putative misclassification rate in the Φ_Λ algorithm. In scenario-500 m, no other climatic and topographic factors displayed either significant β coefficients or 95% CI with lower and upper bounds of the same sign. Accordingly, the signs of the bounds were discordant also in the associated MC 95% CI. However, in comparison to the model with P as single predictor ($AIC_w = 6.8\%$), the environmental factors T_{max} and TPI (with $\beta > 0$), TRI and Sl (with $\beta < 0$) improved model performances when included in multiple regressions along with P (AIC_w from 10.4 to 37.5%). In the same scenario, Bld was characterized by a negative β coefficient with a 95% CI excluding 0, but with a *p*-value > 0.05 and a MC 95% CI with a negative lower bound and a positive upper one. In the other scenario, in addition to P, significance of the β coefficient ($p < 0.05$) was achieved by T_{max}, T_{mean} (with $\beta > 0$), TRI and Sl (with $\beta < 0$), indicating that the probability of bay laurel recovery is positively associated with increasing mean and maximum temperatures, but negatively associated with increasing terrain ruggedness and slope. Such associations were confirmed by the concordant signs of the lower and upper bounds displayed by both 95% CI and MC 95% CI. The graphs of the logistic equations modelling the probability of bay laurel recovery based on the single significant predictors detected in the two scenarios are shown in Figure 6. Although in some models Long was positively and significantly associated with the probability of bay laurel recovery ($\beta > 0$ and $p < 0.05$), such association was never confirmed by MC 95% CI, including

0 between the bounds. The remaining predictors were non-significant ($p > 0.05$) and showed both 95% CI and MC 95% CI with discordant signs, pointing out the lack of association with the probability of bay laurel recovery.

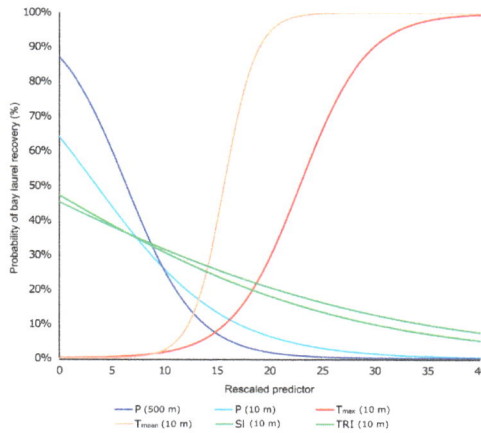

Figure 6. Graphs of the logistic equations modelling the probability of bay laurel recovery based on the single significant predictors detected in scenario-500 m (500 m) and scenario-10 m (10 m). The abscissa (rescaled predictor) represents each factor eventually rescaled so that one unit equals: 100 mm for precipitations (P), 1 °C for temperatures (T), 1% for slope (Sl) and 10 points of terrain ruggedness index (TRI). For more details about factors acronyms, see the main text.

3.3.3. Aspect Analysis

Regardless of the scenario, the azimuth distributions extracted from the As raster (Figure S2) did not display significant differences ($p > 0.05$) between recovered and not recovered bay laurels (Table 2). The outcomes of Mardia-Watson-Wheeler and Rao's homogeneity tests were consistent within each scenario.

Table 2. Comparisons of the azimuth distributions between recovered and not recovered bay laurels.

Scenario	W	W p-Value	R_1	R_1 p-Value	R_2	R_2 p-Value
10 m	1.974	0.373	1.026	0.311	0.178	0.673
500 m	0.236	0.888	0.428	0.513	0.006	0.938

W refers to the Mardia-Watson-Wheeler statistic, while R_1 and R_2 indicate the Rao statistics testing for the equality of polar vectors and for the equality of dispersions, respectively. *p*-values are reported in association with each statistic. Test outcomes are divided per scenario.

4. Discussion

Citizen science is an unsurpassable opportunity for basic and applied research, allowing the collection of huge amount of data with a limited financial investment, yet data quality may be questionable as the result of recruiting non-specialized volunteers [49–51]. Nonetheless, as shown in our study, some of the errors intrinsic to citizen science datasets might be turned into valuable information, providing new insights into biological and ecological processes, such as plant disease dynamics.

A massive amount of epidemiological data had been gathered from the long-term monitoring (2005–2015) and mapping of the alien invasive plant pathogen *Phytophthora ramorum* in Western North America, during one of the largest citizen science experiments ever conducted in forest research [17,48].

The resulting "SODmap" database included thousands of records reporting the infection status of each sampled host-tree, along with its estimated coordinates [17,48]. However, such coordinates had not been collected with professional GPS, and trees had not been marked or labeled in the field. Hence, the risk that volunteers had accidentally resampled some of the trees was rather likely. Despite being an unsought side-effect of the data collection process [17,48], resampling turned into an interesting occasion to follow up the infection status in the same trees and detect the ones which had recovered. By reverting the traditional "disease triangle" paradigm, our study was focused on testing the association between recovery from *P. ramorum* infections in bay laurel and the underlying climatic, topographic and ecological (i.e., environmental) factors through an innovative approach combining numerical ecology to citizen science.

A classification algorithm based on the notion of spatial resolution was designed to extract from the "SODmap" database the "hidden" information about the location of putatively resampled trees [56]. Whenever tree coordinates are repeatedly collected with a GPS device, the resulting location estimates are scattered around the true location as a consequence of the error associated with the GPS [52,54]. If a squared window (i.e., pixel) with suitable spatial resolution is overlapped to such estimates, it cannot spatially resolve them as separate, thus revealing the presence of a single resampled tree location. Basically, the classification algorithm detects resampled trees by incorporating this window as cell of a regular grid, whose spatial resolution is gathered from the GPS median radial location error.

The trials performed in the field with a GPS device, and in silico with Monte Carlo (MC) simulations, confirmed the plausibility of the assumptions underlying the classification algorithm, namely I) that the unknown GPS radial error can be modeled by a Rayleigh probability distribution, and, II) that the spatial resolution of the grid and the median of the above error can be functionally related. Although assuming a spatial homogeneous function to model the GPS radial error might lead to an excessive simplification, especially in mountain areas where the geomorphology could interfere with the GPS signal, the adequacy of the Rayleigh distribution was supported by the outcomes of the Kolmogorov-Smirnov test [57,58]. Moreover, collecting GPS coordinates in different hours of the day, for approximately one month, should have successfully prevented satellite configuration biases potentially influencing the above results. Despite coordinates used for the Kolmogorov-Smirnov test having been collected in Italy, the displacement from the Californian sampling sites seems to be an unlikely source of error in such a long-term trial, especially because the coverage offered by GPS satellites is comparable between Europe and USA. Finally, the functional relation between median GPS radial error and spatial resolution of the grid was clearly shown by the adequacy of the linear regression model fitted on the MC outcomes.

The algorithm performances were satisfactory, as demonstrated by the large overall rate of correct classification (on average over 85%) and by the substantial agreement between the simulated values and the classification outcomes (average Cohen's k over 0.70) [62–64]. However, such performances were not optimal, probably due to the fact that the algorithm classification process is based on a grid with rigid shape and fixed orientation. Once the grid is virtually overlapped on the map of trees locations, it may erroneously split fake locations deriving from the same tree among different pixels. Conversely, the grid might also incorrectly include within the same pixel both resampled and not resampled trees. It is worth noting that, while stem diameters of trees do not reach the meter, the GPS median error and the associated spatial resolution of the grid are in the order of several meters, hence the potential effect of tree size on the classification process should be negligible. Additional sources of misclassification might derive from false-positive or false-negative rates in laboratory assay, yet their occurrence is also unlikely, based on evidence previously published [82]. Finally, misclassification associated with the co-occurrence of localized infections by *P. ramorum* and sampling of uninfected tissues within the same tree seems quite improbable, since an accurate training about symptoms detection and samples collection was offered to all volunteers [17,48]. Considering that a non-null misclassification rate was somehow predictable, the grid-system was incorporated into the algorithm as a sliding layer, allowing stochastic iterate applications on the same dataset, leading to alternative outcomes [61]. While in simulation trials the best position of the grid can be determined, this is not possible during in-field applications, hence the equivalence/divergence of the outcomes should be assessed based on statistical and geostatistical approaches, depending on the overall aim of the analysis [65]. The same issues would have also arisen if the classification algorithm had been based on continuous probability, rather than on a discrete grid-system. In fact, even assuming that the probability of being a resampled/not resampled tree could be calculated for each GPS location, the assignment to one of the two classes would depend on the selection of an arbitrary probability threshold, ultimately leading to multiple outcomes, as documented for other classification processes [63,83]. Although we recognize that different computational approaches might have been proposed, this study was aimed neither at identifying an optimal theoretical solution, nor at contrasting and ranking alternative classification algorithms. Rather, our major goal was to design and test a single method, whose performances could be adequate for practical applications. Moreover, outcomes of the algorithm are fully reproducible, since the code is released in R, a popular open-source programming language for scientific data analysis [58].

The classification algorithm was run under two alternative scenarios, each one accounting for different levels of GPS precision. The median radial error was set to 10 m and 500 m, respectively, based on the assumption that the most practical way to collect coordinates for citizen scientists was to use the GPS embedded in their smartphone. Such thresholds are consistent with data reported in the literature [52]. Although larger values cannot be excluded [52], the most likely precision of sampling efforts should be around 10 m. In fact, non-specialized volunteers would have easily detected sampling positioning errors in the range of hundreds of meters with any mobile mapping application (e.g., Google Maps, Google Earth, OruxMaps or others) even at a coarse spatial scale. In such cases, coordinates would have been discarded by the volunteers based on the quality control phase of the "SOD blitz" program, a phase performed by the volunteers themselves after data are published on the web [17,48].

Results gathered from the classification algorithm identified only 2–5% accidental resampling rates. Such low percentages lead to a negligible error when modelling the distribution of *P. ramorum* at statewide scale, further confirming the reliability of the citizen science approach underlying the "SODmap" [17,48]. However, and as an unexpected side benefit, the number of resampled trees provided an adequate sampling size to investigate the probability of recovery, at least for bay laurels. In fact, host abundance detected by the algorithm mirrored the species composition of the original "SODmap" database, with a noticeable prevalence of bay laurels (75–90%), followed by tanoaks and oaks. While the absolute frequencies of resampled trees differ between scenarios, reflecting the different levels of GPS precision, the homogeneity of the relative frequencies across scenarios suggests that the classification algorithm can provide consistent results regardless of the scale.

Interestingly, based on our findings it can be concluded that recovery from *P. ramorum* infection is neither a rare, nor a negligible phenomenon, since it can be observed in a relevant number of infected trees (approximately 30%). In particular, recovery is more likely in bay laurel (32%) than in tanoak and oaks (ranging between 9% and 25%). Although the number of recovered trees was assessed by an algorithm, it is worth noting that recovery from *P. ramorum* is not simply a theoretical possibility, but a fact that has already been reported, based on repeated sampling of the same set of trees [46]. The large recovery rate detected among bay laurels is in agreement with the notion that this species is a transmissive host of *P. ramorum*, acting as a reservoir of inoculum and supporting sporulation of the pathogen, but with infection limited to leaves without ever spreading even to the twigs [3,22,34,38,40]. *P. ramorum* is not associated with severe symptoms, relevant physiological disequilibrium, or mortality on bay laurels [39,84,85]. Hence, it seems plausible that recovery might be more likely for a host that reportedly is tolerant to the pathogen, displaying only infection of leaves that can be dropped during the dry season. Conversely, and in agreement with our observations, a lower recovery rate might be expected for oaks and tanoaks, the dead-end-hosts of *P. ramorum*. In fact, on such hosts, the pathogen can induce severe bole cankers whose rapid progression leads to lethal outcomes in the majority of cases [86]. Although the association between disease severity and recovery rate has still not been tested for *P. ramorum* and its transmissive or dead-end-hosts, experiments conducted in other model systems have reported that the two aspects are negatively correlated [87,88].

The role of environmental factors in boosting the spread of *P. ramorum* has been investigated in several previous studies [17,22,32,42–46,48,85], yet this is the first attempt to elucidate their possible effects on host recovery. Correlation analyses showed high colinearity among environmental factors. Hence, in order to prevent unstable coefficient estimates and inaccurate variances, only single or uncorrelated predictors were included in binary logistic regressions [58,63,89]. Some differences in correlation patterns were observed between the correlograms of the two alternative scenarios (10 m vs. 500 m GPS error), with a slight decrease in the number of significant Spearman's coefficients with increasing GPS median radial error. Such differences might be ascribed to the different sampling size of the two scenarios, a factor notoriously affecting *p*-values [90]. However, all environmental factors were included at least once in binary logistic regressions, and the substantial consistency of the outcomes gathered from the two scenarios suggested that such differences should not have resulted in relevant effects on the analyses.

Distribution maps of bay laurels showed that, while resampled trees were present across the whole study area, recovered ones were mostly confined to the central-southern regions. Color gradients of temperatures and precipitations maps suggest that areas with a warmer and driest climate were more prone to harbor a large number of bay laurels recovered from *P. ramorum*. Such qualitative inferences were confirmed by the outcomes of the binary logistic regression models. In both scenarios, regressions including precipitation values as predictors were the most informative, and outcompeted all alternative models, as indicated by the weights of the Akaike index. In all cases, the negative association between precipitation and probability of recovery in bay laurel was significant. Logistic equations determined that the probability of recovery is less than 10% when yearly average cumulated rainfall exceeds 2000 mm. The expected probability of bay laurel recovery smoothly approaches 25% as precipitation values decrease from 2000 to 1000 mm. When rainfall values are less than 1000 mm, the likelihood of recovery increases steeply, exceeding 50% when precipitations drop below 500 mm. Although *P. ramorum* is an atypical *Phytophthora* species, being a pathogen on aerial parts of its hosts, its sporulation and spread at local scale is largely dependent on the occurrence of abundant rainfalls and on the presence of running water and wet surfaces, as largely documented [17,22,32,38,40–46,48]. By reducing loads of infectious inoculum, aridity could hinder *P. ramorum* and promote the recovery of bay laurels through the interruption of reinfection cycles and leaf drop. In addition, published studies have suggested that aridity is likely to inhibit mycelial growth *in planta*, weakening the pathogen and reducing its survival chances due to loss in nutrients' absorption [32,85,91]. A prolonged aridity has been associated with the absence of novel *P. ramorum* symptoms in bay laurels [85], but at the same time aridity is likely to be well tolerated by bay laurels, trees whose ecological plasticity and adaptability to a wide range of habitats, climate and soil types have been largely documented [92]. Interestingly, a recent study [22] disclosed the causal relation between rainfalls and the occurrence of events in which *P. ramorum* sporulation was high enough to reach the threshold necessary to infect oaks. Such events tend to be rare with low precipitation levels, thus we cannot exclude that this phenomenon could be associated with an increased recovery among bay laurels. If so, aridity could be indirectly favorable to oaks and tanoaks, since remission in bay laurels would translate in a reduced possibility of pathogen transmission from the transmissive to the dead-end hosts. If the drought trend observed in California during the latest decade [22,93] persisted, according to our models the frequency of recovered bay laurels should substantially grow, hindering the spread and the incidence of *P. ramorum* in the next future.

Binary logistic regression models including temperatures detected an increased probability of bay laurel recovery with raising maximum and mean temperatures. High temperatures not only boost aridity when co-occurring with low precipitations, but they are also climatic factors documented to depress growth, sporulation potential and vitality of *P. ramorum* both *in vitro* and in nature [85,94].

Nonetheless, the effect of temperature was less pronounced than that of precipitation, as indicated by the coefficients p-values associated with this variable, as well as by AIC_w of temperature-related models. Yearly average maximum and mean temperature values of 23 °C and 16 °C, respectively, could be regarded as reference thresholds for bay laurel recovery, since the probability of recovery increases to over 50% with temperatures higher than those. Considering that on sunny days temperatures on foliar surfaces can be 20 °C higher than air temperature [95,96], the above thresholds seem to be consistent with critical values reported for *P. ramorum* sporulation and survival in culture and in plant tissues. In fact, regardless of the growing substrate, temperatures over 30 °C have been previously shown to be detrimental to *P. ramorum* [85,97–99], but not to bay laurels [100]. In the study area, days approaching or exceeding the above values are rather frequent in interior regions. Moreover, it has been hypothesized that climate change is going to result in higher frequencies of extreme heat days in countries with a Mediterranean climate, such as California [85,101,102]. Consequently, and according to our results, recovery from infection should increase in time, if current climatic predictions were correct. Although southern exposures were expected to harbor a larger incidence of recovered bay laurels, due to increasing temperatures in such settings, circular statistics failed to confirm this hypothesis. Nonetheless, it should be noted that this negative result might have been caused by the coarse approximation of local aspect values related to the DEM spatial resolution.

The regression analyses performed on topographic factors showed that only slope and terrain ruggedness can be associated with the probability of recovery in bay laurels. The fragmented pattern of coefficients significance along with the relatively low values of the weights attained by the Akaike information criterion suggest that topographic predictors play a minor role in explaining recovery. This is not surprising considering that landscape geomorphology is strongly spatial scale dependent, and averaging topographic variables over large areas may lead to information losses [103]. Logistic equations pointed out that bay laurel recovery is less likely with increasing slope and terrain ruggedness. When slopes are over 20%, the associated probability of recovery is less than a half of the probability estimated for plain areas, and the same conclusion can be drawn for TRI values over 150–200. Conversely, elevation and TPI were probably not detected as significant factors, due to the limited altitudinal range characterizing most of the study area. Moreover, it is worth noting that the TPI raster showed a high level of heterogeneity among neighboring pixels, probably associated with the spatial resolution of the underlying DEM. Since other DEM could have been used to derive topographic rasters, the possibility that our results were influenced by the available spatial resolution cannot be excluded. Nonetheless, the positive association between plain areas and recovery from *P. ramorum* infections is consistent with previous findings [32]. Interestingly, plain areas might not only hinder the pathogen, but also trigger the recovery by providing more favorable ecological conditions for the hosts. Soils evolving in planar regions are generally characterized by greater thicknesses, jointly with improved physical and chemical properties, whose suitability for the growth of both cultivated crops and natural vegetation has been largely documented [104,105]. Hydraulic and hydrologic properties of plain soils might help host trees to successfully overcome drought periods, since root systems can reach deep water supplies. Not surprisingly, bay laurel is characterized by roots whose extension can easily explore deep soil horizons [100]. However, in-field trials would be necessary to test the above hypothesis.

Although bay laurel density has been acknowledged as a key factor for the short-range transmission of *P. ramorum* [22,40,41,44,106], no evidence of its association with recovery was detected by binary logistic regressions. This result was somehow surprising, since thinnings and localized eradication of bay laurels were proved to be an effective control strategy against *P. ramorum*, resulting in a reduction of infectious inoculum [22]. Hence, low densities were expected to increase the probability of recovery through the inhibition of infection cycles among neighboring hosts, especially considering that the strains of *P. ramorum* harbored by bay laurels often display high virulence levels [107]. However, binary logistic regression models might have failed to point out the role of bay laurel density as a result of missing data, suggesting the need of future investigations to elucidate this point.

The possibility that our results might have been influenced by the mere location of recovered bay laurels, rather than by the underlying variation of environmental factors, was assessed by testing the association between the probability of recovery, latitude and longitude values. While latitude was never significant, longitude displayed significant *p*-values, but only in two out of six models where it was included as predictor. These findings seem to exclude a relevant role of the mere geographic location of trees on the probability of their recovery. In addition, the complementary correlation analyses based on KDE matrices confirmed the outcomes of the binary logistic regressions. Although *p*-values might have been biased by the KDE technique, potentially inflating the number of significant environmental factors, it is worth noting that most signs of correlation coefficients and of the associated logistic coefficients were concordant. More remarkably, the large majority of such signs did not change in partial correlations controlling for sampling density, hence excluding that the latter could have altered the results. Our findings proved to be robust also against a 15% putative misclassification rate, as demonstrated by the outcomes of the Monte Carlo simulations, providing confidence intervals consistent with the confidence intervals calculated for the regression coefficients. Only significant longitudes were not confirmed, hence suggesting that this geographic variable may not be relevant.

Models predicting spread rates of emerging pathogens in forest ecosystems are on the rise, as documented in the last few years for *Heterobasidion irregulare* and *Gnomoniopsis castaneae*, just to cite two relevant examples other than *P. ramorum* [11,16,18,19,108]. Notwithstanding this increased effort, unraveling factors favorable to recovery may help to improve the modelling of disease dynamics and the prediction of related economic losses, as well as to plan more effective mitigation strategies under different climatic or geographic scenarios.

5. Conclusions

In this study, we investigated the role of some key climatic, topographic and ecological factors as drivers of recovery of bay laurels from infection by the alien and invasive plant pathogen *Phytophthora ramorum* in Western North America. By combining the results gathered from a large citizen science disease survey program with an innovative numerical ecology approach based on a newly designed classification algorithm, we were able to determine the location of trees resampled over time, detecting the ones which had recovered from *P. ramorum*. The information contained within the huge "SODmap" database generated by citizen scientists was extracted using the newly designed algorithm, estimating that approximately 32% of the infected bay laurels recovered between 2005 and 2015. This large recovery rate is epidemiologically relevant given that bay laurels, when present, are the main transmissive hosts of *P. ramorum*. Recovery of bay laurels is also plausible, given that infection by *P. ramorum* on this host is exclusively limited to the foliage, and bay laurel foliage has been reported to be prematurely dropped when infected, especially in drier conditions. Multivariate analyses carried out over the entire current range of *P. ramorum* indicated that the probability of recovery increased in association with lower precipitation levels and higher temperatures, especially in flat regions. These findings are also consistent with the known key ecological traits of *P. ramorum*, a pathogen whose survival is hampered by aridity and enhanced by rainfall. It is worth noting that while the combination of prolonged drought and heat is notably detrimental to *P. ramorum*, it is well tolerated by bay laurels, which are key components of those Western North American landscapes that are characterized by a Mediterranean climate. In addition, the physical properties of soils in flat areas may facilitate the survival of bay laurels during droughts, since plausibly root systems may have better access to deep water sources on flat rather than on rugged terrains.

Reverting the traditional "disease triangle paradigm" by seeking conditions leading to plant recovery rather than to infection may substantially improve our understanding and modelling accuracy of plant disease dynamics in forest ecosystems. Accurate models are especially pivotal when dealing with biological invasions such as that by *P. ramorum*, and when attempting to predict future spread and to plan effective management strategies. Finally, this study showcases how pitfalls and limitations of crowdsourced data may be turned into valuable sources of information. In order to take full advantage of this opportunity, though, specific computational tools must be designed and made available to the scientific community, in order to best utilize the large potential hidden within citizen science datasets.

Supplementary Materials: The following are available online at www.mdpi.com/1999-4907/8/8/293/s1, Figure S1: Scatterplot of the values of ϑ and ε resulting from the Monte Carlo experiment assessing $\vartheta = f_{(\varepsilon)}$, Figure S2: Azimuth distributions extracted from the associated raster, Table S1: Binary logistic regressions modelling the probability of recovery among trees whose infection status to *P. ramorum* was positive at least once, Algorithm S1: R-script for function *LAMBDA*, Algorithm S2: R-script for function *PHI*, Dataset S1: Point location estimates collected with a GPS device to assess of the goodness-of-fit of the Rayleigh distribution to the GPS radial error, Dataset S2: Seed values associated with the R-functions *LAMBDA* and *PHI* to allow the reproducibility of the process leading to the identification of trees recovered from *P. ramorum* infections, Dataset S3: Database with 8 field and 19,212 records resulting from the screening of the original "SODmap" database, Dataset S4: Overall dataset merging individual host species final datasets for scenario-10m, resulting from the application of the classifier Φ_Λ integrated in the R-functions *LAMBDA* and *PHI*, Dataset S5: Overall dataset merging individual host species final datasets for scenario-500 m, Note S1: Assessment of the relation between GPS median error and spatial resolution of the grid-system, Note S2: Technical details about the classification algorithm, Note S3: Complementary correlation and partial correlation analysis between recovery and environmental factors, Note S4: Assessment of equivalence among alternative datasets detected for oaks and tanoaks.

Acknowledgments: The research was made possible thanks to the following funding sources to M.G.: the US Forest Service, Region 5, State and Private Forestry; the PG & E Foundation, and the National Science Foundation, Ecology of Infectious Diseases panel. The Authors wish to thank Doug Schmidt and John Vogler for their support in data management.

Author Contributions: G.L. and M.G. conceived and designed the experiments; M.G. provided the original "SODmap" dataset, G.L. designed the algorithms, performed the computational and statistical analyses, interpreted the results and wrote the manuscript; M.G. and P.G. integrated results interpretations and critically revised the manuscript.

Conflicts of Interest: The authors declare no conflict of interest. The founding sponsors had no role in the design of the study; in the collection, analyses, or interpretation of data; in the writing of the manuscript, and in the decision to publish the results.

References

1. McNew, G.L. The nature, origin and evolution of parasitism. In *Plant Pathology: An Advanced Treatise*; Horsfall, J.G., Dimond, A.E., Eds.; Academic Press: New York, NY, USA, 1960; pp. 19–69.
2. Scholthof, K.G. The disease triangle: Pathogens, the environment and society. *Nat. Rev. Microbiol.* **2007**, *5*, 152–156. [CrossRef] [PubMed]
3. Rizzo, D.M.; Garbelotto, M.; Hansen, E.M. *Phytophthora ramorum*: Integrative research and management of an emerging pathogen in California and Oregon forests. *Annu. Rev. Phytopathol.* **2005**, *43*, 309–335. [CrossRef] [PubMed]
4. Gonthier, P.; Nicolotti, G.; Linzer, R.; Guglielmo, F.; Garbelotto, M. Invasion of European pine stands by a North American forest pathogen and its hybridization with a native interfertile taxon. *Mol. Ecol.* **2007**, *16*, 1389–1400. [CrossRef] [PubMed]
5. Kowalski, T.; Holdenrieder, O. Pathogenicity of *Chalara fraxinea*. *For. Pathol.* **2009**, *39*, 1–7. [CrossRef]
6. Visentin, I.; Gentile, S.; Valentino, D.; Gonthier, P.; Tamietti, G.; Cardinale, F. *Gnomoniopsis castanea* sp. nov. (Gnomoniaceae, Diaporthales) as the causal agent of nut rot in sweet chestnut. *J. Plant Pathol.* **2012**, *94*, 411–419.
7. Hauptman, T.; Piškur, B.; Groot, M.D.; Ogris, N.; Ferlan, M.; Jurc, D. Temperature effect on *Chalara fraxinea*: Heat treatment of saplings as a possible disease control method. *For. Pathol.* **2013**, *43*, 360–370.
8. Santini, A.; Ghelardini, L.; Pace, C.D.; Desprez-Loustau, M.L.; Capretti, P.; Chandelier, A.; Cech, T.; Chira, D.; Diamandis, S.; Gaitniekis, T.; et al. Biogeographical patterns and determinants of invasion by forest pathogens in Europe. *New Phytol.* **2012**, *197*, 238–250. [CrossRef] [PubMed]
9. Gonthier, P.; Anselmi, N.; Capretti, P.; Bussotti, F.; Feducci, M.; Giordano, L.; Honorati, T.; Lione, G.; Luchi, N.; Michelozzi, M.; et al. An integrated approach to control the introduced forest pathogen *Heterobasidion irregulare* in Europe. *Forestry* **2014**, *87*, 471–481. [CrossRef]

10. Garbelotto, M.; Guglielmo, F.; Mascheretti, S.; Croucher, P.J.P.; Gonthier, P. Population genetic analyses provide insights on the introduction pathway and spread patterns of the North American forest pathogen *Heterobasidion irregulare* in Italy. *Mol. Ecol.* **2013**, *22*, 4855–4869. [CrossRef] [PubMed]

11. Lione, G.; Giordano, L.; Sillo, F.; Gonthier, P. Testing and modelling the effects of climate on the incidence of the emergent nut rot agent of chestnut *Gnomoniopsis castanea*. *Plant Pathol.* **2015**, *64*, 852–863. [CrossRef]

12. Sillo, F.; Giordano, L.; Zampieri, E.; Lione, G.; De Cesare, S.; Gonthier, P. HRM analysis provides insights on the reproduction mode and the population structure of *Gnomoniopsis castaneae* in Europe. *Plant Pathol.* **2016**, *66*, 293–303. [CrossRef]

13. Legendre, P.; Legendre, L. *Numerical Ecology*; Elsevier: Amsterdam, The Netherlands, 2012.

14. Kéry, M. *Introduction to WinBUGS for Ecologists: Bayesian Approach to Regression, ANOVA, Mixed Models and Related Analyses*; Academic Press: London, UK, 2010.

15. Gonthier, P.; Brun, F.; Lione, G.; Nicolotti, G. Modelling the incidence of *Heterobasidion annosum* butt rots and related economic losses in alpine mixed naturally regenerated forests of northern Italy. *For. Pathol.* **2012**, *42*, 57–68. [CrossRef]

16. Gonthier, P.; Lione, G.; Giordano, L.; Garbelotto, M. The American forest pathogen *Heterobasidion irregulare* colonizes unexpected habitats after its introduction in Italy. *Ecol. Appl.* **2012**, *22*, 2135–2143. [CrossRef] [PubMed]

17. Meentemeyer, R.K.; Dorning, M.A.; Vogler, J.B.; Schmidt, D.; Garbelotto, M. Citizen science helps predict risk of emerging infectious disease. *Front. Ecol. Environ.* **2015**, *13*, 189–194. [CrossRef]

18. Lione, G.; Gonthier, P. A permutation-randomization approach to test the spatial distribution of plant diseases. *Phytopathology* **2016**, *106*, 19–28. [CrossRef] [PubMed]

19. Lione, G.; Giordano, L.; Ferracini, C.; Alma, A.; Gonthier, P. Testing ecological interactions between *Gnomoniopsis castaneae* and *Dryocosmus kuriphilus*. *Acta Oecol.* **2016**, *77*, 10–17. [CrossRef]

20. Paparella, F.; Ferracini, C.; Portaluri, A.; Manzo, A.; Alma, A. Biological control of the chestnut gall wasp with *T. sinensis*: A mathematical model. *Ecol. Model.* **2016**, *338*, 17–36. [CrossRef]

21. Zampieri, E.; Giordano, L.; Lione, G.; Vizzini, A.; Sillo, F.; Balestrini, R.; Gonthier, P. A nonnative and a native fungal plant pathogen similarly stimulate ectomycorrhizal development but are perceived differently by a fungal symbiont. *New Phytol.* **2017**, *213*, 1836–1849. [CrossRef] [PubMed]

22. Garbelotto, M.; Schmidt, D.; Swain, S.; Hayden, K.; Lione, G. The ecology of infection between a transmissive and a dead-end host provides clues for the treatment of a plant disease. *Ecosphere* **2017**, *8*, e01815. [CrossRef]

23. Benda, G.T.A.; Naylor, A.W. On the tobacco ringspot disease. III. Heat and recovery. *Am. J. Bot.* **1958**, *45*, 33–37. [CrossRef]

24. Ghoshal, B.; Sanfaçon, H. Temperature-dependent symptom recovery in *Nicotiana benthamiana* plants infected with tomato ringspot virus is associated with reduced translation of viral RNA2 and requires ARGONAUTE 1. *Virology* **2014**, *456*, 188–197. [CrossRef] [PubMed]

25. Chinnaraja, C.; Viswanathan, R. Variability in yellow leaf symptom expression caused by the sugarcane yellow leaf virus and its seasonal influence in sugarcane. *Phytoparasitica* **2015**, *43*, 339–353. [CrossRef]

26. Morone, C.; Boveri, M.; Giosuè, S.; Gotta, P.; Rossi, V.; Scapin, I.; Marzachì, C. Epidemiology of flavescence dorée in vineyards in northwestern Italy. *Phytopathology* **2007**, *97*, 1422–1427. [CrossRef] [PubMed]

27. Galetto, L.; Miliordos, D.; Roggia, C.; Rashidi, M.; Sacco, D.; Marzachì, C.; Bosco, D. Acquisition capability of the grapevine Flavescence dorée by the leafhopper vector *Scaphoideus titanus* Ball correlates with phytoplasma titre in the source plant. *J. Pest Sci.* **2014**, *87*, 671–679. [CrossRef]

28. Karajeh, M.R.; Al-Momany, A.R. Effect of postplanting soil solarization and solar chamber on Verticillium wilt of olive. *Jordan J. Agric. Sci.* **2008**, *4*, 335–342.

29. Aguayo, J.; Elegbede, F.; Husson, C.; Saintonge, F.X.; Marçais, B. Modeling climate impact on an emerging disease, the *Phytophthora alni*-induced alder decline. *Glob. Chang. Biol.* **2014**, *20*, 3209–3221. [CrossRef] [PubMed]

30. Rosenvald, R.; Drenkhan, R.; Riit, T.; Lõhmusc, A. Towards silvicultural mitigation of the European ash (*Fraxinus excelsior*) dieback: The importance of acclimated trees in retention forestry. *Can. J. For. Res.* **2015**, *45*, 1206–1214. [CrossRef]

31. Brown, N.; Jeger, M.; Kirk, S.; Xu, X.; Denman, S. Spatial and temporal patterns in symptom expression within eight woodlands affected by Acute Oak Decline. *For. Ecol. Manag.* **2016**, *360*, 97–109. [CrossRef]

32. Venette, R.C.; Cohen, S.D. Potential climatic suitability for establishment of *Phytophthora ramorum* within the contiguous United States. *For. Ecol. Manag.* **2006**, *231*, 18–26. [CrossRef]
33. Rizzo, D.M.; Garbelotto, M.; Davidson, J.M.; Slaughter, G.W.; Koike, S.T. *Phytophthora ramorum* as the cause of extensive mortality of *Quercus* spp. and *Lithocarpus densiflorus* in California. *Plant Dis.* **2002**, *86*, 205–214. [CrossRef]
34. Rizzo, D.M.; Garbelotto, M.; Davidson, J.M.; Slaughter, G.W.; Koike, S.T. *Phytophthora ramorum* and Sudden Oak Death in California: I. Host Relationships. In *General Technical Report PSW-GTR-184, Proceedings of the 5th Symposium on California Oak Woodlands, San Diego, CA, USA, 22–25 October 2001*; Standiford, R., McCreary, D., Eds.; United States Department of Agriculture, Forest Service, Pacific Southwest Research Station: Albany, CA, USA, 2002; pp. 733–740.
35. EPPO. Phytophthora ramorum. *EPPO Bull.* **2006**, *36*, 145–155.
36. Hansen, E.; Sutton, W.; Parke, J.; Linderman, R. *Phytophthora ramorum* and Oregon Forest Trees—One Pathogen, Three Diseases. In Proceedings of the Sudden Oak Death Science Symposium, Monterey, CA, USA, 15–18 December 2002; p. 78.
37. Davidson, J.M.; Werres, S.; Garbelotto, M.; Hansen, E.M.; Rizzo, D.M. Sudden oak death and associated diseases caused by *Phytophthora ramorum*. *Plant Health Prog.* **2003**. [CrossRef]
38. Hansen, E.M.; Kanaskie, A.; Prospero, S.; Mcwilliams, M.; Goheen, E.M.; Osterbauer, N.; Reeser, P.; Sutton, W. Epidemiology of *Phytophthora ramorum* in Oregon tanoak forests. *Can. J. For. Res.* **2008**, *38*, 1133–1143. [CrossRef]
39. DiLeo, M.V.; Bostock, R.M.; Rizzo, D.M. *Phytophthora ramorum* does not cause physiologically significant systemic injury to California bay laurel, its primary reservoir host. *Phytopathology* **2009**, *99*, 1307–1311. [CrossRef] [PubMed]
40. Davidson, J.M.; Wickland, A.C.; Patterson, H.A.; Falk, K.R.; Rizzo, D.M. Transmission of *Phytophthora ramorum* in mixed-evergreen forest in California. *Phytopathology* **2005**, *95*, 587–596. [CrossRef] [PubMed]
41. Werres, S.; Marwitz, R.; Man in't Veld, W.A.; de Cock, A.W.; Bonants, P.J.; de Weerdt, M.; Themann, K.; Ilieva, E.; Baayen, R.P. *Phytophthora ramorum* sp. nov., a new pathogen on *Rhododendron* and *Viburnum*. *Mycol. Res.* **2001**, *105*, 1155–1165. [CrossRef]
42. Davis, F.W.; Stoms, D.M.; Hollander, A.D.; Thomas, K.A.; Stine, P.A.; Odion, D.; Borchert, M.I.; Thorne, J.H.; Gray, M.V.; Walker, R.E. *The California GAP Analysis Project—Final Report*; University of California: Santa Barbara, CA, USA, 1998.
43. Davidson, J.M.; Patterson, H.A.; Rizzo, D.M. Sources of inoculum for *Phytophthora ramorum* in a redwood forest. *Phytopathology* **2008**, *98*, 860–866. [CrossRef] [PubMed]
44. Meentemeyer, R.K.; Anacker, B.L.; Mark, W.; Rizzo, D.M. Early detection of emerging forest disease using dispersal estimation and ecological niche modeling. *Ecol. Appl.* **2008**, *18*, 377–390. [CrossRef] [PubMed]
45. Cobb, R.C.; Meentemeyer, R.K.; Rizzo, D.M. Apparent competition in canopy trees determined by pathogen transmission rather than susceptibility. *Ecology* **2010**, *91*, 327–333. [CrossRef] [PubMed]
46. Eyre, C.A.; Kozanitas, M.; Garbelotto, M. Population dynamics of aerial and terrestrial populations of *Phytophthora ramorum* in a California forest under different climatic conditions. *Phytopathology* **2013**, *103*, 1141–1152. [CrossRef] [PubMed]
47. Maywald, G.F.; Sutherst, R.W. *A User's Guide to CLIMEX: A Computer Program for Comparing Climates in Ecology*; Division of Entomology, CSIRO Australia: Canberra, Australia, 1985.
48. Garbelotto, M.; Maddison, E.R.; Schmidt, D. SODmap and SODmap Mobile: Two tools to monitor the spread of sudden oak death. *For. Phytophthoras* **2014**, *4*. [CrossRef]
49. Conrad, C.C.; Hilchey, K.G. A review of citizen science and community-based environmental monitoring: Issues and opportunities. *Environ. Monit. Assess.* **2011**, *176*, 273–291. [CrossRef] [PubMed]
50. Dickinson, J.L.; Shirk, J.; Bonter, D.; Bonney, R.; Crain, R.L.; Martin, J.; Phillips, T.; Purcell, K. The current state of citizen science as a tool for ecological research and public engagement. *Front. Ecol. Environ.* **2012**, *10*, 291–297. [CrossRef]
51. Newman, G.; Wiggins, A.; Crall, A.; Graham, E.; Newman, S.; Crowston, K. The future of citizen science: Emerging technologies and shifting paradigms. *Front. Ecol. Environ.* **2012**, *10*, 298–304. [CrossRef]
52. Zandbergen, P.A. Accuracy of iPhone locations: A comparison of assisted GPS, WiFi and cellular positioning. *Trans. GIS* **2009**, *13*, 5–25. [CrossRef]

53. Hayden, K.; Ivors, K.; Wilkinson, C.; Garbelotto, M. TaqMan chemistry for *Phytophthora ramorum* detection and quantification, with a comparison of diagnostic methods. *Phytopathology* **2006**, *96*, 846–854. [CrossRef] [PubMed]

54. Mertikas, S.P. *Error Distributions and Accuracy Measures in Navigation: An Overview*; University of New Brunswick, Surveying Engineering Department: Fredericton, NB, Canada, 1985.

55. Cormack, R.M. A review of classification. *J. R. Stat. Soc. Ser. A.* **1971**, *134*, 321–367. [CrossRef]

56. Jones, H.G.; Vaughan, R.A. *Remote Sensing of Vegetation: Principles, Techniques, and Applications*; Oxford University Press: Oxford, UK, 2010.

57. Krishnamoorthy, K. *Handbook of Statistical Distributions with Applications*; CRC Press—Taylor & Francis Group: Boca Raton, FL, USA, 2016.

58. Crawley, M.J. *The R Book*, 2nd ed.; John Wiley & Sons: Chichester, UK, 2013.

59. Wagenmakers, E.J.; Farrell, S. AIC model selection using Akaike weights. *Psychon. Bull. Rev.* **2004**, *11*, 192–196. [CrossRef] [PubMed]

60. Grueber, C.E.; Nakagawa, S.; Laws, R.J.; Jamieson, I.G. Multimodel inference in ecology and evolution: Challenges and solutions. *J. Evolut. Biol.* **2011**, *24*, 699–711. [CrossRef] [PubMed]

61. Carsey, T.M.; Harden, J.J. *Monte Carlo Simulation and Resampling: Methods for Social Science*; Sage: Los Angeles, CA, USA, 2014.

62. Cohen, J.A. Coefficient of agreement for nominal scales. *Educ. Psychol. Meas.* **1960**, *20*, 37–46. [CrossRef]

63. Hosmer, D.W.; Lemeshow, S. *Applied Logistic Regression*; Wiley: New York, NY, USA, 1989.

64. McHugh, M.L. Interrater reliability: The kappa statistic. *Biochem. Med.* **2012**, *22*, 276–282. [CrossRef]

65. Mitchell, A. *The Esri Guide to GIS Analysis: Geographic Patterns & Relationships*; ESRI Press: Redlands, CA, USA, 1999; Volume 1.

66. Silverman, B.W. *Density Estimation for Statistics and Data Analysis*; Chapman and Hall: London, UK, 1986.

67. Venables, W.N.; Ripley, B.D. *Modern Applied Statistics with S*; Springer: New York, NY, USA, 2011.

68. Conover, W.J.; Iman, R.L. Rank transformations as a bridge between parametric and nonparametric statistics. *Am. Stat.* **1981**, *35*, 124–129. [CrossRef]

69. Blaker, H. Confidence curves and improved exact confidence intervals for discrete distributions. *Can. J. Stat.* **2000**, *28*, 783–798. [CrossRef]

70. Hijmans, R.J.; Cameron, S.E.; Parra, J.L.; Jones, P.G.; Jarvis, A. Very high resolution interpolated climate surfaces for global land areas. *Int. J. Climatol.* **2005**, *25*, 1965–1978. [CrossRef]

71. PRISM Climate Group, Oregon State University. Available online: http://prism.oregonstate.edu (accessed on 15 January 2017).

72. Jarvis, A.; Reuter, H.I.; Nelson, A.; Guevara, E. Hole-Filled SRTM for the Globe Version 4, Available from the CGIAR-CSI SRTM 90 m Database. 2008. Available online: http://srtm.csi.cgiar.org (accessed on 15 January 2017).

73. Lamsal, S.; Cobb, R.C.; Cushman, J.H.; Meng, Q.; Rizzo, D.M.; Meentemeyer, R.K. Spatial estimation of the density and carbon content of host populations for *Phytophthora ramorum* in California and Oregon. *For. Ecol. Manag.* **2011**, *262*, 989–998. [CrossRef]

74. Riley, S.J.; DeGloria, S.; Elliot, R.A. A Terrain ruggedness index that quantifies topographic heterogeneity. *Intermt. J. Sci.* **1999**, *5*, 23–27.

75. De Reu, J.D.; Bourgeois, J.; Bats, M.; Zwertvaegher, A.; Gelorini, V.; Smedt, P.D.; Chu, W.; Antrop, M.; Maeyer, P.D.; Finke, P.; et al. Application of the topographic position index to heterogeneous landscapes. *Geomorphology* **2013**, *186*, 39–49. [CrossRef]

76. Kennedy, H. *Introduction to 3D Data: Modeling with ArcGIS 3D Analyst and Google Earth*; John Wiley & Sons: Hoboken, NJ, USA, 2009.

77. Wilcox, R.R.; Keselman, H.J.; Kowalchuk, R.K. Can tests for treatment group equality be improved? The bootstrap and trimmed means conjecture. *Br. J. Math. Stat. Psychol.* **1998**, *51*, 123–134. [CrossRef]

78. Buckland, S.T. Monte Carlo confidence intervals. *Biometrics* **1984**, *40*, 811–817. [CrossRef]

79. Pewsey, A.; Neuhäuser, M.; Ruxton, G.D. *Circular Statistics in R*; Oxford University Press: Oxford, UK, 2013.

80. Rao, J.S. Large sample tests for the homogeneity of angular data. *Sankhya Ser. B* **1967**, *28*, 172–174.

81. Mardia, K.V. A multi-sample uniform scores test on a circle and its parametric competitor. *J. R. Stat. Soc. B Methodol.* **1972**, 102–113.

82. Martin, F.N.; Tooley, P.W.; Blomquist, C. Molecular detection of *Phytophthora ramorum*, the causal agent of Sudden Oak Death in California, and two additional species commonly recovered from diseased plant material. *Phytopathology* **2004**, *94*, 621–631. [CrossRef] [PubMed]

83. Sweets, J.A. Measuring accuracy of diagnostic systems. *Science* **1988**, *240*, 1285–1293. [CrossRef]

84. Garbelotto, M.; Davidson, J.M.; Ivors, K.; Maloney, P.E.; Hüberli, D.; Koike, S.T.; Rizzo, D.M. Non-oak native plants are main hosts for sudden oak death pathogen in California. *Calif. Agric.* **2003**, *57*, 18–23. [CrossRef]

85. DiLeo, M.V.; Bostock, R.M.; Rizzo, D.M. Microclimate impacts survival and prevalence of *Phytophthora ramorum* in *Umbellularia californica*, a key reservoir host of sudden oak death in northern California forests. *PLoS ONE* **2014**, *9*, e98195. [CrossRef] [PubMed]

86. Garbelotto, M.; Hayden, K.J. Sudden Oak Death: Interactions of the exotic oomycete *Phytophthora ramorum* with naïve North American hosts. *Eukaryot. Cell* **2012**, *11*, 1313–1323. [CrossRef] [PubMed]

87. López-Escudero, F.J.; Blanco-López, M.A. Recovery of young olive trees from *Verticillium dahliae*. *Eur. J. Plant Pathol.* **2005**, *113*, 367–375. [CrossRef]

88. Gordon, T.R.; Kirkpatrick, S.C.; Aegerter, B.J.; Fisher, A.J.; Storer, A.J.; Wood, D.L. Evidence for the occurrence of induced resistance to pitch canker, caused by *Gibberella circinata* (anamorph *Fusarium circinatum*), in populations of *Pinus radiata*. *For. Pathol.* **2010**, *41*, 227–232. [CrossRef]

89. Midi, H.; Sarkar, S.K.; Rana, S. Collinearity diagnostics of binary logistic regression model. *J. Interdiscip. Math.* **2010**, *13*, 253–267. [CrossRef]

90. Gibbon, J.D.; Pratt, J.W. P-values: Interpretation and methodology. *Am. Stat.* **1975**, *29*, 20–25. [CrossRef]

91. Kendrick, B. *The Fifth Kingdom*, 3rd ed.; Focus Publishing: Newburyport, MA, USA, 2000.

92. Goralka, R.J.; Langenheim, J.H. Analysis of foliar monoterpenoid content in the California Bay Tree, *Umbellularia californica*, among populations across the distribution of the species. *Biochem. Syst. Ecol.* **1995**, *23*, 439–448. [CrossRef]

93. Griffin, D.; Anchukaitis, K.J. How unusual is the 2012–2014 California drought? *Geophys. Res. Lett.* **2014**, *41*, 9017–9023. [CrossRef]

94. Garrett, K.A.; Dendy, S.P.; Frank, E.E.; Rouse, M.N.; Travers, S.E. Climate change effects on plant disease: Genomes to ecosystems. *Annu. Rev. Phytopathol.* **2006**, *44*, 489–509. [CrossRef] [PubMed]

95. Ansari, A.Q.; Loomis, W.E. Leaf temperatures. *Am. J. Bot.* **1959**, *46*, 713–717. [CrossRef]

96. Fuchs, M. Infrared measurement of canopy temperature and detection of plant water-stress. *Theor. Appl. Climatol.* **1990**, *42*, 253–261. [CrossRef]

97. Englander, L.; Browning, M.; Tooley, P.W. Growth and sporulation of *Phytophthora ramorum* in vitro in response to temperature and light. *Mycologia* **2006**, *98*, 365–373.

98. Tooley, P.W.; Browning, M.; Berner, D. Recovery of *Phytophthora ramorum* following exposure to temperature extremes. *Plant Dis.* **2008**, *92*, 431–437. [CrossRef]

99. Browning, M.; Englander, L.; Tooley, P.W.; Berner, D. Survival of *Phytophthora ramorum* hyphae after exposure to temperature extremes and various humidities. *Mycologia* **2008**, *100*, 236–245. [CrossRef] [PubMed]

100. Meentemeyer, R.K.; Moody, A.; Franklin, J. Landscape-scale patterns of shrub-species abundance in California chaparral—The role of topographically mediated resource gradients. *Plant Ecol.* **2001**, *156*, 19–41. [CrossRef]

101. Diffenbaugh, N.S.; Pal, J.S.; Giorgi, F.; Gao, X. Heat stress intensification in the Mediterranean climate change hotspot. *Geophys. Res. Lett.* **2007**, *34*, L11706. [CrossRef]

102. Di Castri, F. An ecological overview of the five regions of the world with a Mediterranean climate. In *Biogeography of Mediterranean Invasions*; Groves, R.H., di Castri, F., Eds.; Cambridge University Press: Cambridge, UK, 1991; pp. 3–15.

103. Mark, D.M.; Aronson, P.B. Scale-dependent fractal dimensions of topographic surfaces: An empirical investigation, with applications in geomorphology and computer mapping. *Math. Geol.* **1984**, *16*, 671–683. [CrossRef]

104. Swanson, F.J.; Kratz, T.K.; Caine, N.; Woodmansee, R.G. Landform effects on ecosystem patterns and processes. *BioScience* **1988**, *38*, 92–98. [CrossRef]

105. Catani, F.; Segoni, S.; Falorni, G. An empirical geomorphology-based approach to the spatial prediction of soil thickness at catchment scale. *Water Resour. Res.* **2010**, *46*, W05508. [CrossRef]

106. Swiecki, T.J.; Bernhardt, E.A. Increasing distance from California bay laurel reduces the risk and severity of *Phytophthora ramorum* canker in coast live oak. In *General Technical Report PSW-GTR-214, Proceedings of the Sudden Oak Death Third Science Symposium, Santa Rosa, California, 5-9 March 2007*; Frankel, S.J., Kliejunas, J.T., Palmieri, K.M., Eds.; United States Department of Agriculture, Forest Service, Pacific Southwest Research Station: Albany, CA, USA, 2008; pp. 181–194.

107. Hüberli, D.; Garbelotto, M. *Phytophthora ramorum* is a generalist plant pathogen with differences in virulence between isolates from infectious and dead-end hosts. *For. Pathol.* **2012**, *42*, 8–13. [CrossRef]

108. Giordano, L.; Gonthier, P.; Lione, G.; Capretti, P.; Garbelotto, M. The saprobic and fruiting abilities of the exotic forest pathogen *Heterobasidion irregulare* may explain its invasiveness. *Biol. Invasions* **2014**, *16*, 803–814. [CrossRef]

forests

MDPI

Article

Dispersal and Propagule Pressure of *Botryosphaeriaceae* Species in a Declining Oak Stand is Affected by Insect Vectors

Tiziana Panzavolta [1], Andrea Panichi [2], Matteo Bracalini [1], Francesco Croci [1], Beatrice Ginetti [1], Alessandro Ragazzi [1], Riziero Tiberi [1] and Salvatore Moricca [1,*]

[1]   Department of Agrifood Production and Environmental Sciences, Plant Pathology and Entomology Division, University of Florence, Piazzale delle Cascine, 28, 50144 Florence, Italy; tpanzavolta@unifi.it (T.P.); matteo.bracalini@unifi.it (M.B.); francesco.croci@unifi.it (F.C.); beatrice.ginetti@gmail.com (B.G.); alessandro.ragazzi@unifi.it (A.R.); riziero.tiberi@unifi.it (R.T.)
[2]   Fondazione Edmund Mach, Via E. Mach, 1, 38010 S. Michele all'Adige, 38010 Trento, Italy; andrea.panichi@fmach.it
*   Correspondence: salvatore.moricca@unifi.it; Tel.: +39-055-275-5864

Academic Editors: Carl Beierkuhnlein and Andreas Bolte
Received: 28 April 2017; Accepted: 23 June 2017; Published: 28 June 2017

Abstract: Many biotic and abiotic factors contribute to the onset of oak decline. Among biotic agents, a variety of fungi and insects cause extensive disease and insect outbreaks in oak forests. To date, research on fungus-insect interactions in Mediterranean forest ecosystems is still scarce and fragmentary. In this study, we investigated the assemblage of endophytic mycobiota and insect pests occurring in a declining oak stand, with the aim to explore if, and to what extent, the insect species were active vectors of fungal propagules. It emerged that some known latent pathogens of the *Botryosphaeriaceae* family, namely *Botryosphaeria dothidea*, *Diplodia corticola*, *Diplodia seriata*, *Dothiorella sarmentorum*, and *Neofusicoccum parvum* were isolated at high frequency from physiologically-impaired trees. In addition, propagules of these fungi were isolated from five insects, two of which (*Cerambyx welensii* and *Coraebus fasciatus*) are main oak pests. The life-history strategies of these fungi and those of wood-boring beetles were strikingly interconnected: both the fungi and beetles exploit drought-stressed trees and both occur at high frequency during hot, dry periods. This synchronicity increased their chance of co-occurrence and, consequently, their probability of jointly leading to oak decline. If these interactions would be confirmed by future studies, they could help to better understand the extensive decline/dieback of many Mediterranean forest ecosystems.

Keywords: *Botryosphaeriaceae*; *Cerambyx welensii*; *Coraebus fasciatus*; oak decline; climate warming; pathogen occurrence; transport vectors

1. Introduction

A widespread decline of oak forests has been observed in several parts of the world in the last four decades [1–3]. The phenomenon turned particularly severe in those geographic areas that are more exposed to global warming effects [4]; specifically, in the Mediterranean basin, a well-known climate change hotspot [5], where climate fluctuations are having a profound impact on forest ecosystems. Here, the repeated occurrence of heat extremes, accompanied by a decrease in precipitation and thus prolonged summer drought, has caused substantial heat and water stress to tree vegetation, resulting in their physiological impairment, stunted growth, dieback and, in some instances, mortality [6]. Climate-driven changes, besides having exacerbated the vulnerability of the trees, have also modified the dynamics of forest insects and pathogens, dramatically increasing the likelihood of attacks by these damaging agents [7,8].

Surveys carried out in an attempt to clarify the etiology of decline/dieback episodes at specific sites in several Mediterranean countries have not identified any single cause as being responsible for most of the events [3]. Rather, a plethora of factors, namely extended drought, exceptional weather events (e.g., rainstorms and windstorms accompanied by hail damage, branch rupture and severe flooding), inappropriate forest management, wood-boring insects, as well as oomycete and fungal attacks have been advocated from time to time as factors predisposing, inciting or contributing to the decline [9]. One drawback of many investigations, however, was that they were carried out by specialists from single, separate disciplines (e.g., climatologists, plant pathologists, entomologists, botanists, etc.) whose backgrounds led them to broach only one possible cause at a time, without a comprehensive, holistic approach to the problem. The result was in many instances a disjointed, and often incomplete, framework, which made it impossible to individuate the intertwined causes of tree declines.

As concerns the parasitic component of oak forest ecosystems, a number of pathogenic fungi have been recognized as having a prominent role in oak tree decline and mortality [10], although only after long time lags beyond the onset of the problem. For instance, it has been proven that some mycobiota, which normally occupy nonpathogenically internal host tissues, under conditions of stress (mainly water deficit) assume pathogenic behaviour, aggressively colonizing host tissues and sporulating profusely over the plant surfaces, thus spreading pervasively into oak stands [11,12]. Some of these fungi have been found together with insects on the same trees, often on the same organs [13–15] and even, sometimes, in the same tissue niches. Since the role of insects in vehiculating microbial pathogens has been amply ascertained, in this investigation we studied the correlations between occurrence of endophytic fungi and insect colonization on declining trees in an oak stand of central Italy. If the insect-mediated dispersal of important fungal pathogens in the stand were established, it would prove insects' contribution to the incidence and severity of tree diseases and decline.

2. Materials and Methods

2.1. Study Area

The study area was located at Alberese (municipality of Grosseto) in the Maremma Regional Park (42°37′03″ N, 11°06′47″ E) (Tuscany, Italy). More specifically, it is a roughly 90-year-old, mixed oak forest, with 15–25 m tall trees, composed mainly of *Quercus pubescens* Willd. (roughly 60%), *Quercus cerris* L. (roughly 20%), *Quercus suber* L. (roughly 20%), and with sporadic (less than 1%) *Quercus ilex* L. individuals. The Mediterranean undergrowth, typical of this environment, is very scarce in the study area, having the forest long been used for pasture. Many of the oaks occurring in the area showed symptoms of decline, such as exudates on the trunk, bark cankers, dead branches and twigs, and beetle exit holes. The study area has a typical Mediterranean climate with hot, dry summers (Figure 1).

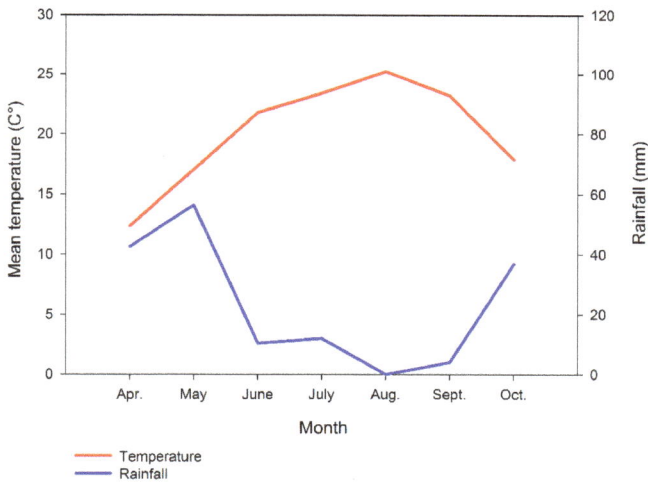

Figure 1. Meteorological data from the Alberese (Grosseto-Italy) meteorological weather station in 2015. Data supplied by the Hydrological Regional Sector (SIR) of Tuscany.

2.2. Plant Sample Collection

Tree sampling was performed according to the specific composition of the population. To this purpose, 10 trees were selected: six *Q. pubescens*, two *Q. cerris*, and two *Q. suber*. Samplings were carried out once a month from April 2015 to October 2015 on the same selected oak trees. During each sampling, four current-year twigs (2–4 cm in diameter) were collected from each of these trees at the height of 2.50 m around the crown, one from each cardinal point.

2.3. Fungal Isolation from Plants

Plant samples were taken to the laboratory within 12 h from collection. They were sterilized by immersion in 10% H_2O_2 for three minutes, then they were washed twice with sterilized distilled water, and dried on sterile tissue paper. A 2-cm-long tissue piece was removed from each twig, excluding the outer bark. From each piece 15 wood samples were removed (roughly 5 mm long and 1 mm thick) and placed, in groups of five, in three Petri dishes (90 mm in diameter) containing Potato Dextrose Agar (PDA) medium with 1 g of agar. Thus, for each sampled twig three Petri dishes were obtained. The plates were incubated at 20 °C in darkness for a week. Fungal colonies growing from the wood samples were isolated and subcultured on Oak Leaf Agar (OLA) [16,17]. Whole colonies were observed with an optical microscope and identified by macro and micro-morphological analyses, using the keys provided by Booth [18], Gams [19], Carmichael [20], Sutton [21], and Von Arx [22]. The isolation frequency of each fungus species per each month was calculated using the following formula:

$$F_i = (N_i/N_t) \times 100 \tag{1}$$

where F_i is the fungus species frequency, N_i is the number of times the species was isolated and N_t is the total number of wood samples placed in PDA.

2.4. Molecular Identification of Isolates

DNA-based identification was necessary for discriminating among related taxa whose micro-morphological characteristics alone proved inconclusive for species determination. Hyphal-tip-derived, fresh cultures were incubated under darkness for one week on MEA medium.

Genomic DNA was extracted [23] and the rDNA-ITS region was PCR-amplified using the universal primers ITS1 (5'-TCCGTAGGTGAACCTGCGG) and ITS4 (5-TCCTCCGCTTATTGATATGC) [24]. PCR cycling conditions and subsequent amplicon sequencing were as in Moricca et al. [15]. Sequences were processed in the GenBank database [25], with a BLAST search for the highest identities that was used for the identification of taxa at the species level, considering a minimum threshold of 98% [26].

2.5. Insect Collection

The study area was surveyed every two weeks, from April to October 2015, to monitor xylophagous insect presence. Different sampling methods were employed with the aim to collect living, possibly just-emerged, insects to be utilized for fungal isolation. Sweep nets were used directly on host plants or on flowers, since many adult xylophagous beetles have a flower-visiting behaviour. Some twilight or nocturnal flying beetles were collected by attracting them with light onto a vertical white sheet. Attacked twigs, branches or stems were debarked to find adult insects and, finally, oak branches and twigs were sampled and the eggs reared in the laboratory until the adults emerged. All sampled specimens were taken to the laboratory within six hours from collection in sterile plastic containers.

In addition, to have information about insect frequency in the area during the study period, suspended bait traps were also employed. Five Lindgren funnel traps and five bottle traps were located in the study area at least 100 m from each other. Lindgren funnel traps are generally used for trapping saproxylic beetles, which are attracted by their shape and their black color. In addition, in our study they were lured with ethanol, which is known to attract saproxylic beetles. Bottle traps were laced with a mixture of red wine, banana and sugar, which lures beetles. Both kinds of traps were hung on tree branches at about 3–4 m of height. Every two weeks their bait was changed and they were checked, with insects captured being collected separately for date and trap. All sampled specimens were taken to the laboratory to be identified.

2.6. Fungal Isolation from Insects

Living insects collected (excluding trapped ones) were analyzed in the laboratory. After their identification, the insects were left to walk on PDA Petri dishes, partially following the methodology used by Sabbatini Peverieri [27]. Then, each beetle was surface-washed by vortexing for 1 min in 300 μL of sterile distilled water with 1% of Tween-80 detergent. The resulting solutions were used to inoculate the PDA medium. All the Petri dishes were incubated at 20 °C in darkness for five days. Emerging fungal colonies were subcultured in a pure OLA medium. After one week of growth, the colonies were observed under an optical microscope and identified by analysing their macro- and micro-morphological characters, and by the DNA-based method described above.

2.7. Data Analysis

The significance of the data was determined with ANOVA, after the percent data were arcsin transformed. The differences in fungal isolation frequency were examined for significance using the Duncan's New Multiple Range Test.

3. Results

In the study area throughout the June–September 2015 period, rainfall was below 11 mm, and in August there was no rain. As regards temperatures, the means reached their maximum value in August, exceeding 25 °C. In addition, from June to September temperatures were always over 20 °C (Figure 1).

3.1. Recovered Fungal Taxa

A number of fungi, belonging to 21 different genera, were isolated from the oak samples collected during the study period (April 2015–October 2015). Of these, 13 were identified to the species level by coupling both conventional and molecular identification. Recovered taxa included common, ubiquitous contaminants, such as: *Alternaria* sp., *Aspergillus* sp. and *Cladosporium* sp. Five harmful species of the *Botryosphaeriaceae* family, frequently associated with woody hosts [28], heavily colonized the sampled plant material. An array of more or less common microbial inhabitants of Mediterranean oak forests, namely *Camarosporium* sp., *Candida* sp., *Cephalosporium* sp., *Fusarium solani* (Mart.) Sacc., *Gliocladium* sp., *Gonatorrhodiella* sp., *Pestalotiopsis versicolor* (Speg.) Steyaert, *Pollularia* sp., *Rhizoctonia solani* J.G. Kühn, *Rhizopus* sp., *Trichoderma viride* Pers., *Ulocladium consortiale* (Thüm.) E.G. Simmons, *Verticillium dahliae* Kleb., were also isolated from the oak tree tissues. The charcoal canker agent *Biscogniauxia mediterranea* (De Not.) Kuntze was too found infecting oaks, though its occurrence in the stand resulted almost negligible (only 1.0% and 3.5% isolation frequency in September and October, respectively).

Following this initial screening of the endophytic assemblage, we narrowed our investigation to botryosphaeriaceous fungi, owing to: (i) the overwhelming prevalence (higher isolation frequencies) of this group of fungi in the stand; (ii) their prominent role in the onset of the decline of tree species worldwide, especially on those trees already weakened by environmental stress factors [28]. The five botryosphaeriaceous species isolated in this study were: *Botryosphaeria dothidea* (Moug.) Ces. and De Not, *Diplodia corticola* A.J.L. Phillips, A. Alves and J. Luque, *Diplodia seriata* De Not., *Dothiorella sarmentorum* (Fr.) A.J.L. Phillips, A. Alves and J. Luque, and *Neofusicoccum parvum* (Pennycook and Samuels) Crous, Slippers and A.J.L. Phillips (Table 1). These fungi were found throughout the whole study period; however, their isolation frequency turned out significantly different among months (Table 2), increasing gradually during the growing season (Table 1). Higher values were recorded from June onwards for *D. sarmentorum*, from July onwards for *B. dothidea* and from August onwards for *D. corticola*, *D. seriata* and *N. parvum*. More specifically, the highest isolation frequency (%) for each fungal species, with exception of *N. parvum*, was recorded in October (Table 1).

Table 1. Percentage isolation frequencies of fungi of the *Botryosphaeriaceae* family recovered from oak samples in the studied woodland in Marina di Alberese (Grosseto-Italy). Values in columns followed by the same letter do not differ significantly ($p \leq 0.01$, Duncan's New Multiple Range Test); isolation frequencies below the 1% threshold are not reported. Standard deviation is in parentheses.

Month of Collection	Frequency of Isolation (%) (600 Monthly Samplings)				
	Fungus Species				
	Botryosphaeria dothidea	*Diplodia corticola*	*Diplodia seriata*	*Dothiorella sarmentorum*	*Neofusicoccum parvum*
April	1.75 (0.65)	3.37 a (3.24)	3.25 a (1.35)	1.37 a (0.20)	1.25 a (0)
May	3.75 b (2.25)	4.50 a (3.59)	3.50 a (1.80)	1.89 a (0.33)	1.50 a (1.10)
June	6.90 c (1.12)	4.75 a (1.98)	3.78 a (1.00)	3.12 b (0.78)	1.87 a (0.90)
July	8.78 d (0.72)	4.90 a (3.59)	4.25 a (2.10)	3.12 b (1.23)	2.00 a (0.20)
August	9.00 d (2.20)	13.89 b (0.65)	6.75 b (3.08)	3.25 b (0.50)	3.00 b (1.25)
September	10.25 d (0.12)	15.25 bc (5.82)	7.65 b (1.80)	3.50 b (1.30)	3.25 b (2.00)
October	12.50 e (0.10)	16.00 c (1.67)	11.97 c (0.89)	8.97 c (5.95)	3.50 b (1.35)

Table 2. Analysis of variance (1-way ANOVA) on isolation frequency (percent data arcsin transformed).

Variability	df	Deviation	Variance	F
Total	13	1575.21		
Between pathogens	4	915.07	228.76	57.04 **
Between months	6	648.11	108.01	26.93 **
Error	3	12.03	4.01	

** Significant at $p < 0.01$.

3.2. Recovered Insect Species

During the study period, we collected various xylophagous and non-xylophagous insect species; however, non-xylophagous taxa, or those which feed only on decaying deadwood, were excluded. Therefore, the xylophagous species belonging to the following taxa were considered: Buprestidae, Cerambycidae, and Curculionidae Scolytinae. Only xylophagous species feeding on oaks were taken into consideration: the Cerambycidae family was the most numerous, with 14 species, compared with Buprestidae (eight species) and Scolytinae (four species) (Table 3).

Table 3. Xylophagous Coleoptera feeding on oaks collected in the studied woodland in Marina di Alberese (Grosseto-Italy) and their association with botryosphaeriaceous fungi.

Coleoptera Species Collected	Botryosphaeriaceae Species Isolated	Number of Insects Tested	Isolation Frequency (%)
Buprestidae			
Acmaeoderella adspersula (Illiger, 1803)		3	0
Acmaeoderella flavofasciata (Piller and Mitterpacher, 1783)		8	0
Anthaxia millefolii polychloros Abeille de Perrin, 1894		15	0
Anthaxia scutellaris Gene, 1839		15	0
Anthaxia thalassophila Abeille de Perrin, 1900	Botryosphaeria dothidea	6	33.33
Anthaxia umbellatarum (Fabricius, 1787)		11	0
Coraebus fasciatus (Villers, 1789)	Diplodia seriata	12	33.33
Latipalpis plana (Olivier, 1790)		3	0
Cerambycidae			
Callimus abdominalis (Olivier, 1795) *		0	NA
Cerambyx welensii (Küster, 1845)	Diplodia corticola	8	37.50
Chlorophorus sartor (Müller, 1766)	Diplodia seriata	4	25.00
Chlorophorus glabromaculatus (Goeze, 1777) *		0	NA
Deilus fugax (Olivier, 1790)		1	0
Deroplia genei (Aragona, 1830) *		0	NA
Niphona picticornis Mulsant, 1839 *		0	NA
Phymatodes testaceus (Linnaeus, 1758)		13	0
Pseudosphegesthes cinerea (Laporte and Gory, 1836) *		0	NA
Purpuricenus kaehleri (Linnaeus, 1758)	Diplodia seriata	23	8.69
Stenopterus rufus (Linnaeus, 1767)		9	0
Stictoleptura cordigera (Fuessly, 1775)		2	0
Stictoleptura rufa (Brulle, 1832) *		0	NA
Trichoferus holosericeus (Rossi, 1790)		12	0
Curculionidae: Scolytinae			
Xyleborus dispar (Fabricius, 1792) *		0	NA
Scolytus intricatus (Ratzeburg, 1837) *		0	NA
Xyleborinus saxesenii Ratzeburg, 1837 *		0	NA
Xyleborus monographus (Fabricius, 1792) *		0	NA

* Insect species which could not be analyzed for fungus association. NA = not assigned.

3.3. Fungus-Insect Associations

Botryosphaeriaceous fungi were isolated from collected beetles. Fungal propagules of *B. dothidea*, *D. seriata* and *D. corticola* were found on five insect species: two Buprestidae, *A. thalassophila* and *C. fasciatus*, and three Cerambycidae, *C. welensii*, *C. sartor*, and *P. kaehleri* (Table 3). *D. seriata* was the predominant fungal species, being associated with three different insect species. With the exception of *P. kaehleri*, whose proportion of specimens bearing fungal propagules (*D. seriata*) was 8.69%, all the other insects revealed percentages ranging from 25% (*C. sartor/Diplodia seriata*), to 37.5% (*C. welensii/D. corticola*), with *A. thalassophila/B. dothidea* and *C. fasciatus/D. seriata* which presented intermediate values (33.33%). *C. fasciatus* larvae bore galleries in live branches (under 8 cm in diameter) of stressed oaks. Subsequent branch death occurs once larvae reach maturity, because at that time they bore annular galleries under the bark of the branches [29]. Generally, *C. welensii* colonizes oaks in a state of physiological decline [30]. However, its attacks to young plants in good vegetative condition are becoming more and more frequent [30,31]. *A. thalassophila*, *C. sartor*, and *P. kaehleri* feed on various broadleaved trees; nevertheless, these are not considered oak pests, as they bore galleries on the wood of dead trees or on the deadwood of living trees, particularly dead branches [32,33].

These five insect species were captured from June to August. During June–July, they were the most frequent xylophagous beetles, representing more than 43% of all species collected with suspended bait traps. In June only *A. thalassophila* and *P. kaehleri* were captured (Table 4). While the former was collected only in that month, *P. kaehleri* captures increased during the growing season, reaching its peak in August. Considering the whole study period, *P. kaehleri* was the most frequent species, representing almost 11% of all collected beetle species. Anyway, buprestid beetles deserve special mentions, because traps used in our study are not particularly attractive to some of these species; therefore, they would be more numerous if other trapping methods were used.

Table 4. Catching frequencies (%) of xylophagous beetles resulted associated with *Botryosphaeriaceae* species in the oak forest in Marina di Alberese (Grosseto-Italy). Percentages were calculated on the total number of xylophagous beetles collected with suspended bait traps.

Insect Species	Month of Collection							
	April	May	June	July	August	September	October	Total
Xylophagous Beetles Associated with Fungi								
Anthaxia thalassophila	0	0	13.04	0	0	0	0	4.83
Coraebus fasciatus	0	0	0	18.99	0	0	0	4.53
Cerambyx welensii	0	0	0	6.33	14.71	0	0	4.53
Chlorophorus sartor	0	0	0	2.53	5.88	0	0	1.81
Purpuricenus kaehleri	0	0	5.80	15.19	23.53	0	0	10.88
Total frequency	0	0	17.39	43.04	44.12	0	0	26.59
Bark-beetles	0	0	0	27.85	27.94	62.50	0	15.41
Total number of xylophagous beetles	4	26	138	79	68	16	0	331

However, not all the species collected were taken into consideration, because some of them were captured only with traps; therefore, they were not used for fungal isolation. This was particularly important for bark beetles, which were collected only with suspended bait traps. The four bark-beetle species trapped are considered key species in oak decline and they were rather frequent in the study area (Table 4), being more than 15% of all the collected specimens. This was particularly true in September, when about 62% of captures consisted of these four bark-beetle species.

4. Discussion

In our study area, xylophagous beetles associated with declining oaks were found. Specifically, the two wood-boring beetles, *C. welensii* and *C. fasciatus*, were rather frequent in summer catches. Other wood-boring beetles feeding on oaks were also collected, but, mostly, they are not considered pests, being species that feed only on deadwood. In contrast, the bark beetles caught during the study could play a key role in oak decline, since both their aggressiveness and their oviposition behaviour enhance their function as fungal vectors. In fact, adults carrying fungal propagules bore galleries under the oak bark, directly infecting the colonized trees. In addition, two of the bark beetles caught,

S. intricatus and *X. dispar*, have already been associated with fungal pathogens inhabiting the woody tissues [34]. However, since the bark beetles in our study were captured only with suspended bait traps, they were not used for fungal isolation, consequently their role as fungal vectors in the study area was not investigated.

Botryosphaeriaceae fungi were isolated from five beetle species, *C. welensii* and *C. fasciatus* included. Previous studies demonstrated that *C. welensii* emergence holes are entry ways for fungal pathogens inside trees [35]; however, information about its role as a fungal vector is not available. Here, *D. corticola* was isolated from specimens of this insect species. *D. seriata*, instead, was found on *C. fasciatus* adults. This species has already been associated with *Botryosphaeriaceae* fungi, Tiberi et al. [36] having isolated *D. mutila* from adults of this buprestid. These two beetles, differing from bark beetles, lay eggs without penetrating the oak bark; therefore, their importance in the spread of fungal infection may appear less important. Nonetheless, they do bring fungal propagules into direct contact with oak trees, since the egg-laying activity of *C. welensii* includes probing the bark with its ovipositor, as well as laying eggs into bark crevices and pruning wounds [37]. Thus, propagules are highly likely to come into contact with suitable entry sites. As regards *C. fasciatus*, females usually lay eggs near buds of young branches, although, they also oviposit around wounds [29]. In that case, they would play the same important role as *C. welensii*. In addition, these two wood-boring beetles are attracted by stressed trees, particularly drought-stressed ones, which are their preferred hosts, carrying with them propagules of *Botryosphaeriaceae* fungi, which likewise prefer physiologically-impaired trees. It appears evident that the development of *Botryosphaeriaceae*-related diseases is increasingly more likely, and potentially more severe when beetles and fungi co-occur on the same host [38].

An extremely hot and dry summer occurred during our study period. August had no rains and temperatures reached their highest values, and this could have affected the biology of both fungi and beetles. *D. corticola* and *D. seriata* were isolated more frequently from August to October, just after the driest period. It is known that members of the *Botryosphaeriaceae* family, especially some *Diplodia* species [39], are thermophilic or thermotolerant; thus, these fungi become more aggressive when temperatures rise, a fact that coincides also with a greater drought stress to trees [15]. High temperatures also favour *C. welensii* and *C. fasciatus*, as they are thermophilic species too [40].

According to previous studies focusing on other species, co-occurrence of insects and fungi shows a strong seasonality [41]. Fungal isolation frequency is known to be season-dependent [42]. This because the physiological status of the tissues influences fungal growth and sporulation, being linked to availability of carbon for the fungi. It is, however, also true that prolonged summer droughts lead to plant carbon starvation and reduced ability to counteract the attack by biotic stressors like insects and fungi [6,43]. Consequently, drought-stressed trees may become more suitable to these biotic agents, increasing their population abundance [44,45]. Accordingly, *C. welensii* and *C. fasciatus* emerged in the study area from July to August, exactly during the driest and hottest period of the year and when propagule pressure of the fungi in the stand was substantially high. The final outcome is insect and pathogen outbreaks often causing extended tree mortality [46].

The botryosphaeriaceous fungi isolated in this study are well known endophytes and latent pathogens, with a cosmopolitan host range and wide geographical distribution [28,47]. These microorganisms are able to aggressively attack the host plants when these undergo physiological stress and to induce a variety of disease symptoms [28]. Although some of these species (e.g., *D. corticola* and *N. parvum*) have in recent years come strongly to the fore in several regions of the world [48–50], their infection biology, in part because of their sometimes inconspicuous occurrence, has long been neglected [51,52]. As a matter of fact, the life history strategies of many of these taxa remain, even today, partially unexplored. Few studies have investigated, for example, to what extent these opportunistic fungi are transmitted in the woods by insect vectors [36,53]. This paper aims to partly fill this gap.

5. Conclusions

Several lines of reasoning suggest that the investigated xylophagous insects may well have a role in the dispersal of fungal species: (i) the thermotolerance or thermophily of members of both groups of parasites. These traits increase the chances of transportation of fungi by insects, being both insect population density and propagule pressure of fungi higher during hotter years; (ii) their synchronicity in their occurrence and activity, coinciding with the drought of summer months; and last but not least, (iii) the isolation of fungal propagules from the body of some of the insect species. From an epidemiological perspective, it is also worth noting that beetles, besides increasing fungal dispersal and propagule pressure, bring fungi to stressed oaks precisely during the time when these are most susceptible.

However, while it is evident that insects are effectively carrying fungal propagules and that environmental stress is the first driver of tree weakening, the causal interconnections between environmental variables and the fungus-insect-tree tripartite interactions are difficult to prove. That's because many factors may contribute to generate a more complex framework, which escapes analyses of temporal and spatial co-occurrence. Tree decline, for instance, is usually a long-term process, during which fungi may take advantage of impaired tree defenses and at the same time affect tree's response to environmental stress. Furthermore, fungi may be spreading from last year's growth inside the tree twigs, blurring the temporal aspects. To clarify this aspect, it would be interesting to repeat this research to ascertain whether the fungus-insect associations found here are stable and repeated over the years. If these harmful interactions were confirmed, they would provide a more plausible explanation for the extensive mortality of some Mediterranean forest stands whose etiology seemed uncertain. In fact, a single factor of damage (fungi or insects) alone did not explain in many cases the extent and gravity of the observed decline/dieback phenomena. In this connection, it would also deserve investigating whether other microorganisms (bacterial agents), responsible for the more recent and emerging AOD (Acute Oak Decline) syndrome [54], may also be involved.

Author Contributions: A.R., R.T., S.M. and T.P. designed the experiments and analysed the data. A.P., F.C., B.G. and M.B. contributed with field experiments and data collection. S.M. and T.P. wrote the manuscript.

Conflicts of Interest: The authors declare no conflict of interest.

References

1. Brasier, C.M.; Scott, J.K. European oak declines and global warming: A theoretical assessment with special reference to the activity of *Phytophthora cinnamomi*. *EPPO Bull.* **1994**, *24*, 221–232. [CrossRef]
2. Ragazzi, A.; Vagniluca, S.; Moricca, S. European expansion of oak decline: Involved microorganisms and methodological approaches. *Phytopathol. Mediterr.* **1995**, *34*, 207–226.
3. Thomas, F.M.; Blanck, R.; Hartmann, G. Abiotic and biotic factors and their interactions as causes of oak decline in Central Europe. *For. Pathol.* **2002**, *32*, 277–307. [CrossRef]
4. Sturrock, R.N.; Frankel, S.J.; Brown, A.V.; Hennon, P.E.; Kliejunas, J.T.; Lewise, K.J.; Worrallf, J.J.; Woods, A.J. Climate change and forest diseases. *Plant Pathol.* **2011**, *60*, 49–133. [CrossRef]
5. Diffenbaugh, N.S.; Pal, J.S.; Giorgi, F.; Gao, X. Heat stress intensification in the Mediterranean climate change hotspot. *Geophys. Res. Lett.* **2007**, *34*, 1–6. [CrossRef]
6. Allen, C.D.; Macalady, A.K.; Chenchouni, H.; Bachelet, D.; McDowell, N.; Vennetier, M.; Kitzberger, T.; Rigling, A.; Breshears, D.D.; Hogg, E.H.T.; et al. A global overview of drought and heat-induced tree mortality reveals emerging climate change risks for forests. *For. Ecol. Manag.* **2010**, *259*, 660–684. [CrossRef]
7. Moricca, S.; Linaldeddu, B.T.; Ginetti, B.; Scanu, B.; Franceschini, A.; Ragazzi, A. Endemic and emerging pathogens threatening cork oak trees: Management options for conserving a unique forest ecosystem. *Plant Dis.* **2016**, *100*, 2184–2193. [CrossRef]
8. Tiberi, R.; Branco, M.; Bracalini, M.; Croci, F.; Panzavolta, T. Cork oak pests: A review of insect damage and management. *Ann. For. Sci.* **2016**, *73*, 219–232. [CrossRef]
9. Manion, P.D. *Tree Disease Concepts*, 2nd ed.; Prentice-Hall: Englewood Cliffs, NJ, USA, 1991.

10. Halmschlager, E. Endophytic fungi and oak decline. In *Recent Advances in Studies on Oak Decline, Proceedings of An International Congress, Selva di Fasano, Brindisi, Italy, 13–18 September 1992*; Luisi, N., Lerario, P., Vannini, A., Eds.; Dipartimento di Patologia Vegetale, Università Degli Studi: Bari, Italy, 1993; pp. 77–83.

11. Kowalski, T.; Kehr, R.D. Endophytic fungi colonization of branch bases in several forest tree species. *Sydowia* **1992**, *44*, 137–168.

12. Halmschlager, E.; Butin, H.; Donaubauer, E. Endophytische pilze in blättern und zweigen von *Quercus petraea*. *Eur. J. For. Pathol.* **1993**, *23*, 51–63. [CrossRef]

13. Ragazzi, A.; Moricca, S.; Capretti, P.; Dellavalle, I.; Mancini, F.; Turco, E. Endophytic fungi in *Quercus cerris*: Isolation frequency in relation to phenological phase, tree health and the organ affected. *Phytopathol. Mediterr.* **2001**, *40*, 165–171. [CrossRef]

14. Ragazzi, A.; Moricca, S.; Capretti, P.; Dellavalle, I.; Turco, E. Differences in composition of endophytic mycobiota in twigs and leaves of healthy and declining *Quercus* species in Italy. *For. Pathol.* **2003**, *33*, 31–38. [CrossRef]

15. Moricca, S.; Ginetti, B.; Ragazzi, A. Species- and organ-specificity in endophytes colonizing healthy and declining Mediterranean oaks. *Phytopathol. Mediterr.* **2012**, *51*, 587–598. [CrossRef]

16. Cohen, S.D. Technique for large scale isolation of *Discula umbrinella* and other foliar endophytic fungi from *Quercus* species. *Mycologia* **1999**, *91*, 917–922. [CrossRef]

17. Moricca, S.; Ragazzi, A. The holomorph *Apiognomonia quercina*/*Discula quercina* as a pathogen/endophyte in oak. In *Endophytes of Forest Trees*; Pirttilä, A.M., Frank, A.C., Eds.; Springer: Dordrecht, The Netherlands, 2011; pp. 47–66.

18. Booth, C. *The Genus Fusarium*; Commonwealth Mycological Institute: Kew, UK, 1971.

19. Gams, W. *Cephalosporium-Artige Schimmelpilze (Hyphomycetes)*; Gustav Fischer Verlag: Stuttgart, Germany, 1971.

20. Charmichael, J.W.; Kendrich, W.B.; Conner, I.L.; Sigler, L. *Genera of Hyphomycetes*; The University of Alberta Press: Edmonton, AB, Canada, 1980.

21. Sutton, B.C. *The Coelomycetes Fungi Imperfecti with Pycnidia, Acervuli and Stromata*; Commonwealth Mycological Institute: Kew, UK, 1980.

22. Von Arx, J.A. *Plant Pathogenic Fungi*; Cramer, J., Ed.; Beihefte zur Nova Hedwigia: Berlin, Germany, 1987.

23. Moricca, S.; Raddi, P.; Borja, I.; Vendramin, G.G. Differentiation of *Seiridium* species associated with virulent cankers on cypress in the Mediterranean region by PCR-SSCP. *Plant Pathol.* **2000**, *49*, 774–781. [CrossRef]

24. White, T.J.; Bruns, T.; Lee, S.; Taylor, J. Amplified and direct sequencing of fungal ribosomal RNA genes for phylogenies. In *PCR Protocols: A Guide to Methods and Applications*; Innis, M.A., Gelfand, D.H., Sninsky, J.J., White, T.J., Eds.; Academic: San Diego, CA, USA, 1990; pp. 315–322.

25. Benson, D.A.; Cavanaugh, M.; Clark, K.; Karsch-Mizrachi, I.; Lipman, D.J.; Ostell, J.; Sayers, E.W. GenBank. *Nucleic Acids Res.* **2013**, *41*, D36–D42. [CrossRef] [PubMed]

26. Sánchez Márquez, S.; Bills, G.F.; Zabalgogeazcoa, I. Diversity and structure of the fungal endophytic assemblages from two sympatric coastal grasses. *Fungal Divers.* **2008**, *33*, 87–100.

27. Sabbatini Peverieri, G.; Villari, C.; Tiberi, R.; Capretti, P. Occurrence of fungal root rot diseases on pine trees in Tuscany and its relationship with *Tomicus destruens*. *J. Plant Pathol.* **2005**, *87*, 304.

28. Slippers, B.; Wingfield, M.J. *Botryosphaeriaceae* as endophytes and latent pathogens of woody plants: Diversity, ecology and impact. *Fungal Biol. Rev.* **2007**, *21*, 90–106. [CrossRef]

29. Jurc, M.; Bojović, S.; Komjanc, B.; Krč, J. Xylophagous entomofauna in branches of oaks (*Quercus* spp.) and its significance for oak health in the Karst region of Slovenia. *Biologia* **2009**, *64*, 130–138. [CrossRef]

30. Torres-Vila, L.M.; Sanchez-González, Á.; Ponce-Escudero, F.; Martín-Vertedor, D.; Ferrero-García, J.J. Assessing mass trapping efficiency and population density of *Cerambyx welensii* Küster by mark-recapture in dehesa open woodlands. *Eur. J. For. Res.* **2012**, *131*, 1103–1116. [CrossRef]

31. Sallé, A.; Nageleisen, L.; Lieutier, F. Bark and wood boring insects involved in oak declines in Europe: Current knowledge and future prospects in a context of climate change. *For. Ecol. Manag.* **2014**, *328*, 79–93. [CrossRef]

32. Curletti, G. *I Buprestidi d'Italia*; Museo Civico di Scienze Naturali: Brescia, Italy, 1994.

33. Contarini, E. Elenco faunistico commentato (check-list) dei Cerambicidi (Coleoptera Xylophytophaga) del Parco Naturale della Vena del Gesso romagnola. *Quad. Stud. Nat. Romagna* **2014**, *40*, 39–65.

34. Tiberi, R.; Panzavolta, T.; Bracalini, M.; Ragazzi, A.; Ginetti, B.; Moricca, S. Interactions between insects and fungal pathogens of forest and ornamental trees. *Ital. J. Mycol.* **2016**, *45*, 54–65. [CrossRef]

35. Martín, J.; Cabezas, J.; Buyolo, T.; Patón, D. The relationship between *Cerambyx* spp. damage and subsequent *Biscogniauxia mediterranum* infection on *Quercus suber* forests. *For. Ecol. Manag.* **2005**, *216*, 166–174. [CrossRef]
36. Tiberi, R.; Ragazzi, A.; Marianelli, L.; Peverieri Sabbatini, P.; Roversi, P.F. Insects and fungi involved in oak decline in Italy. In *IOBC/WPRS Bulletin, Working Group "Integrated Protection in Oak Forests", Proceedings of the Meeting at Oeiras, Lisbonne, Portugal, 1–4 October 2001*; Villemant, C., Sousa, E., Eds.; IOBC: Zürich, Switzerland, 2002; Volume 25, pp. 67–74.
37. Torres-Vila, L.M.; Mendiola-Diaz, F.J.; Conejo-Rodríguez, Y.; Sánchez-González, Á. Reproductive traits and number of matings in males and females of *Cerambyx welensii* (Coleoptera: Cerambycidae) an emergent pest of oaks. *Butt. Entomol. Res.* **2016**, *106*, 292–303. [CrossRef] [PubMed]
38. Belhoucine, L.; Bouhraoua, R.T.; Meijer, M.; Houbraken, J.; Harrak, M.J.; Samson, R.A.; Equihua-Martinez, A.; Pujade-Villar, J. Mycobiota associated with *Platypus cylindrus* (Coleoptera: Curculionidae, Platypodidae) in cork oak stands of North West Algeria, Africa. *Afr. J. Microbiol. Res.* **2011**, *5*, 4411–4423. [CrossRef]
39. Schumacher, J.; Heydeck, P.; Dahms, C. Increasing endangerment of forests by pathogenic thermophilic fungi-demonstrated by the example of the microfungus *Diplodia pinea* (DESM.) Kickx on *Pinus*. In *Julius-Kühn-Archiv, Proceedings of the 57th German Plant Protection Conference, Berlin, Germany, 6–9 September 2010*; Julius Kühn Institut, Bundesforschungsinstitut für Kulturpflanzen: Quedlinburg, Germany, 2010; Volume 428, p. 372.
40. Cárdenas, A.M.; Gallardo, P. The effect of temperature on the preimaginal development of the jewel beetle, *Coraebus florentinus* (Coleoptera: Buprestidae). *Eur. J. Entomol.* **2012**, *109*, 21–28. [CrossRef]
41. Brown, N.; Jeger, M.; Kirk, S.; Williams, D.; Xu, X.; Pautasso, M.; Denman, S. Acute Oak Decline and *Agrilus biguttatus*: The co-occurrence of stem bleeding and D-shaped emergence holes in Great Britain. *Forests* **2017**, *8*, 87. [CrossRef]
42. Ragazzi, A.; Moricca, S.; Capretti, P.; Dellavalle, I. Endophytic presence of *Discula quercina* on declining *Quercus cerris*. *J. Phytopathol.* **1999**, *147*, 437–440. [CrossRef]
43. Ragazzi, A.; Moricca, S.; Dellavalle, I. Water Stress and the Development of Cankers by *Diplodia mutila* on *Quercus robur*. *J. Phytopathol.* **1999**, *147*, 425–428. [CrossRef]
44. Desprez-Loustau, M.L.; Marcais, B.; Nageleisen, L.M.; Piou, D.; Vannini, A. Interactive effects of drought and pathogens in forest trees. *Ann. For. Sci.* **2006**, *63*, 597–612. [CrossRef]
45. Haavik, L.J.; Billings, S.A.; Guldin, J.M.; Stephen, F.M. Emergent insects, pathogens and drought shape changing patterns in oak decline in North America and Europe. *For. Ecol. Manag.* **2015**, *354*, 190–205. [CrossRef]
46. Jactel, H.; Petit, J.; Desprez-Loustau, M.L.; Delzon, S.; Piou, D.; Battisti, A.; Koricheva, J. Drought effects on damage by forest insects and pathogens: A meta-analysis. *Glob. Chang. Biol.* **2012**, *18*, 267–276. [CrossRef]
47. Sieber, T.N. Endophytic fungi in forest trees: Are they mutualists? *Fungal Biol. Rev.* **2007**, *21*, 75–89. [CrossRef]
48. Sakalidis, M.L.; Slippers, B.; Wingfield, B.D.; Hardy, G.E.St.J.; Burgess, T.I. The challenge of understanding the origin, pathways and extent of fungal invasions: Global populations of the *Neofusicoccum parvum-N. ribis* species complex. *Divers. Distrib.* **2013**, *19*, 873–883. [CrossRef]
49. Dreaden, T.J.; Shin, K.; Smith, J.A. First report of *Diplodia corticola* causing branch cankers on live oak (*Quercus virginiana*) in Florida. *Plant Dis.* **2011**, *95*, 1027. [CrossRef]
50. Lynch, S.C.; Eskalen, A.; Zambino, P.; Scott, T. First report of bot canker caused by *Diplodia corticola* on coast live oak (*Quercus agrifolia*) in California. *Plant Dis.* **2010**, *94*, 1510. [CrossRef]
51. Alves, A.; Correia, A.; Luque, J.; Phillips, A.J.L. *Botryosphaeria corticola* sp. nov. on *Quercus* species, with notes and description of *Botryosphaeria stevensii* and its anamorph *Diplodia mutila*. *Mycologia* **2004**, *96*, 598–613. [CrossRef] [PubMed]
52. Alves, A.; Linaldeddu, B.T.; Deidda, A.; Scanu, B.; Phillips, A.J.L. The complex of *Diplodia* species associated with *Fraxinus* and some other woody hosts in Italy and Portugal. *Fungal Divers.* **2014**, *67*, 143–156. [CrossRef]

53. Hanso, M.; Drenkhan, R. *Diplodia pinea* is a new pathogen on Austrian pine (*Pinus nigra*) in Estonia. *Plant Pathol.* **2009**, *58*, 797. [CrossRef]
54. Denman, S.; Plummer, S.; Kirk, S.; Peace, A.; McDonald, J.E. Isolation studies reveal a shift in the cultivable microbiome of oak affected with Acute Oak Decline. *Syst. Appl. Microbiol.* **2016**, *39*, 484–490. [CrossRef] [PubMed]

![forests logo] *forests*

MDPI

Review

Effects of Host Variability on the Spread of Invasive Forest Diseases

Simone Prospero [1] and Michelle Cleary [2,*]

[1] Swiss Federal Institute for Forest, Snow and Landscape Research WSL, Zuercherstrasse 111, CH-8903 Birmensdorf, Switzerland; simone.prospero@wsl.ch

[2] Swedish University of Agricultural Sciences, Southern Swedish Forest Research Centre, Sundsvägen 3, SE-23053 Alnarp, Sweden

* Correspondence: Michelle.Cleary@slu.se; Tel.: +46-40-415181

Academic Editors: Matteo Garbelotto and Paolo Gonthier
Received: 1 February 2017; Accepted: 11 March 2017; Published: 15 March 2017

Abstract: Biological invasions, resulting from deliberate and unintentional species transfers of insects, fungal and oomycete organisms, are a major consequence of globalization and pose a significant threat to biodiversity. Limiting damage by non-indigenous forest pathogens requires an understanding of their current and potential distributions, factors affecting disease spread, and development of appropriate management measures. In this review, we synthesize innate characteristics of invading organisms (notably mating system, reproduction type, and dispersal mechanisms) and key factors of the host population (namely host diversity, host connectivity, and host susceptibility) that govern spread and impact of invasive forest pathogens at various scales post-introduction and establishment. We examine spread dynamics for well-known invasive forest pathogens, *Hymenoscyphus fraxineus* (T. Kowalski) Baral, Queloz, Hosoya, comb. nov., causing ash dieback in Europe, and *Cryphonectria parasitica*, (Murr.) Barr, causing chestnut blight in both North America and Europe, illustrating the importance of host variability (diversity, connectivity, susceptibility) in their invasion success. While alien pathogen entry has proven difficult to control, and new biological introductions are indeed inevitable, elucidating the key processes underlying host variability is crucial for scientists and managers aimed at developing effective strategies to prevent future movement of organisms and preserve intact ecosystems.

Keywords: disease spread; invasive pathogens; host connectivity; host diversity; *Cryphonectria parasitica*; *Hymenoscyphus fraxineus*; biological invasions

1. Introduction

Increased connectivity and globalization have greatly accelerated the frequency and magnitude of biological invasions around the globe by facilitating the long-distance movement of species into regions outside their historical distribution range. The current rate of non-indigenous species movement, resulting from human mediated intervention, is unprecedented [1], and has been the main accelerator driving the increase in novel encounters between host plants and pathogens, and the probability of invasive species emergence [2]. Biosecurity policies adopted by most countries for preventing new introductions (e.g., inspections at ports of entry, phytosanitary certification) can have positive effects, but, unless efforts are further strengthened and better coordinated internationally, biological invasions will inevitably continue [3].

Recent reviews have highlighted the escalating rate of exotic pathogen species introductions affecting forest trees in North America and Europe [4–6], mainly as a result of changes in trade practices surrounding the global movement of live plants and plant materials [7,8]. In some cases, alien forest pathogens (i.e., those that are nonnative, introduced from distant countries) have caused large-scale transformations of native ecosystems and changed the ecological dynamics through local and regional extinction of native host species. For example, the host-specialist pathogen *Cryphonectria parasitica*, the causal agent of chestnut blight, has dramatically reduced populations of *Castanea dentata* (Marsh.) Borkh. in North America where it had a defined role as a keystone species in ecosystem structure and function [9]. Generalist pathogens are thought to be better invaders than specialists due to their non-selective ability of seeking out new hosts in a new environment. For example, the oomycete plant pathogen *Phytophthora ramorum* (Werres, De Cock & Man in't Veld), has a broad host range affecting both natural (forests) and semi-natural (urban green area) environments and hardy nursery stock. In western USA, *P. ramorum* has largely influenced the California oak woodlands landscape through diffuse impacts on a large number of host species [10,11], causing lethal stem infections on 'canker hosts' (mainly *Notholithocarpus densiflorus* (Hook. & Arn.) Manos, Cannon & S.H. Oh and *Quercus* spp.), and non-lethal foliar infections on 'foliar hosts' (e.g., *Umbellularia californica* (Hook. & Arn.) Nutt.).

In the strictest sense, 'spread' (also synonymous with the terms 'dissemination' and 'dispersal') refers to the movement of individuals either through random diffusion or directed dispersal such that they expand beyond the frontier of their geographic range [12]. The expansion phase is generally described by the change in range distance as a function of time [13]. Pathogens become 'invasive' when they acquire a competitive advantage in a new area following the disappearance of natural obstacles to their proliferation including native antagonists, and successfully adapt to new conditions [14]. Most experimental studies (including modeling) on the spread of invasive pathogens focus on agricultural systems, humans, and animals. Given the importance of invasive pathogens currently causing global threats to biodiversity, a deeper understanding of spread mechanisms and dynamics of spread can improve the ability to predict and manage impacts of biological invasions affecting forests and urban landscapes. Though several recent reviews of invasive forest pathogens have highlighted human activity and international trade as major determinants of invasiveness [5,6], surprisingly few have focused specifically on factors driving disease spread of invasive pathogens across various scales from a local stand level to a broader landscape level.

The goal of this review is to discuss mechanisms underlying the invasion process for alien pathogens and disease spread, post-introduction and establishment. We present a framework for local, regional, and continental-scale factors governing the spread and impact of invasive forest pathogens. Specifically, we recapitulate characteristics and modes of pathogen reproduction, pathogen dispersal, and parasitic specialization, in relation to spread and epidemiology. We provide an overview of the main factors affecting variability in disease spread: host diversity, host connectivity, and host susceptibility and use one historical and one recent example of invasive pathogens that have significantly impacted forests to illustrate their invasion success and spread dynamics in that context. Finally, we discuss the way-forward in which a deeper understanding of the factors promoting disease spread across local and continental scales can help address the global ecological and societal challenges of managing novel disease epidemics.

1.1. Novel Environments, Novel Hosts

Biological invasions are a special type of range expansion [15] that can be defined by a series of discrete, consecutive phases which include: 'Arrival'—single or multiple arrivals of a nonindigenous species at one or more points of entry into a new locale/environment, and 'Establishment'—whereby arriving populations start to reproduce *in situ*, surpassing barriers of initial extinction (geographic, environment, reproductive, antagonists) [16–18].

After arrival and initial invasion of a new host ('spillover'), production of transmission stages within the new host, and subsequent establishment, invasive forest pathogens can then 'Spread' expanding their range of occupied territory into new hospitable areas. This ability to spread is highly dependent on environmental suitability, resource availability, and the ability to adapt and naturalize [12]. Strong filtering exists between the different phases; successful invasion is a rare event such that only a small fraction of alien species survive to pass through and establish themselves in a novel environment. Both host availability (as a suitable ecological niche) and environmental conditions (extremes of temperature, moisture, and UV radiation) in the new location place strong limitations on a pathogen's survival, its ability to reproduce and disperse, and subsequently spread [19,20]. These factors act as a strong selection filter leading to rapid adaptation to new environmental conditions and rapid evolution and exploitation of novel hosts [20]. Understanding novel forest pathogen introductions and the factors driving invasion success requires a deeper understanding of the invasion sequence that is conditioned by local or long-distance transport mechanisms from their native habitat to a novel environment, as well as environmental conditions and ecological factors determining an organism's survival and reproducibility, and any population and community effects affecting their dynamics across a range of spatial and temporal scales (see Figure 1).

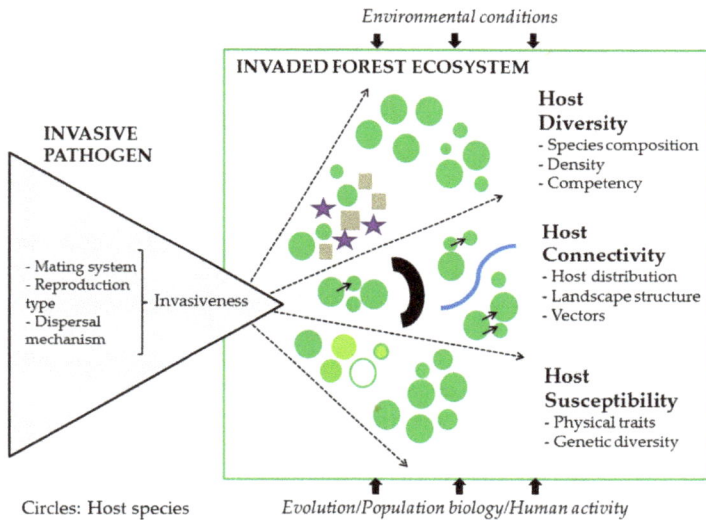

Figure 1. Local, regional, and continental-scale factors governing the spread and impact of invasive forest pathogens *post-introduction and establishment* are affected by three main factors: host diversity, host connectivity, and host susceptibility. Following arrival and establishment, invasiveness is inherently affected by organismal traits (mating system, reproduction type, and dispersal mechanisms). *Host diversity* is mainly affected by plant species richness (density/composition) creating a dilution effect of pathogen impacts on the ecosystem. Non-hosts, competent hosts, and less competent hosts will have variable effects on their ability to intercept inoculum and subsequently reduce pathogen spread. *Host connectivity* is largely influenced by the distribution of available host species; more or less aggregated. Spatial heterogeneity of hosts becomes important for vector-induced pathogens. Variations in landscape structure (topography, natural geographic/environmental barriers, forest fragmentation) will largely influence spread dynamics on the landscape level. *Host susceptibility* to invasive pathogens is influenced by physical traits (size, age, morphology), the random presence of other (potentially antagonistic) organisms, environmental and site factors, and host genetic background. Intraspecific genetic diversity (mixtures of host genotypes) offers the best insurance against invasive pathogens through a dilution effect on inoculum production/deposition and the likelihood that some hosts will possess effective mechanisms to resist or minimize damage caused by invasives. Evolutionary and environmental factors, as well as continuous pressures caused by human activity will influence spread dynamics over time. Understanding how host variability is affected by *host diversity, connectivity*, and *susceptibility* will improve our ability to predict disease spread on the landscape and potential consequences to ecosystem services.

The most accurate records documenting the spread of forest pathogens can be derived from field mapping of invasion fronts over successive years from a defined geographic area, or from a time series of aerial photos. The kind and quality of data that is used as a measure of increasing abundance (presence/absence), the scale of mapping, and range size is important for extrapolating information about spread rates [21]. For example, ground surveys and geographic information system (GIS) mapping of Dutch elm-diseased and elm bark beetles-attacked trees on the geographically isolated island of Gotland allowed for tracking patterns of disease spread in relation to management strategies to reduce disease prevalence [22]. For many invading forest pathogens, the key to understanding dispersal is by measuring the human transport process through imported plant material via random checks in quarantine facilities from nurseries [23,24], or on vehicles [25]. For example, the spread of *P. ramorum* has been demonstrated using various approaches such as environmental niche models [26], risk assessment maps based on host distribution [27], landscape structure [28,29], multi-scale patterns of human activity [30], and trade networks [31]. When patterns of past spread are missing, future spread may be forecasted with simulations using either a mechanistic model or extrapolations (e.g., [32,33]).

1.2. Pathogen Invasiveness Affected by Species Traits

Certain species traits can favor invasiveness of pathogens and their spread and subsequent impact on forest ecosystems [34]. To some extent the plasticity of pathogen traits allows for some level of pre-adaption prior to an introduction. Janzen [35] described 'ecological fitting' as the interaction of an organism with its biotic and abiotic environment in a way that indicates a shared evolutionary history, when the organismal traits relevant to the interaction actually evolved elsewhere under different environmental conditions. Ecological fitting of novel forest pathogens depends in part on the ability of certain organismal traits to be co-opted for novel functions [36], and is largely influenced by phenotypic plasticity, correlated trait evolution, and phylogenetic conservatism [37].

Among the most important traits affecting spread ability for invasive forest pathogens is the mating system [38], namely for its ability to generate more virulent strains [34] and also to adapt to newly encountered host species in a new environment [20], and the type of reproduction (e.g., polycyclic pathogens complete their lifecycle multiple times throughout the growing season). Bazin et al. [39] showed that invasion dynamics of an introduced population are largely affected by the rate of asexuality. It is generally assumed that purely asexual organisms may exhibit lower invasion success compared to other organisms with, for example, mixed mating systems due to their inability to generate new sets of meiotic progeny which can rapidly adapt to the new host and environment [40]. In addition, asexual spores are dispersed mainly over short distances or at the plant level. Several examples, however, show that clonality does not necessarily reduce invasiveness. For instance, new populations of *C. parasitica* in Europe are frequently founded by one or a few genotypes (e.g., [41]). Similarly, Laurel wilt disease, which in the USA is threatening communities of native plant species in the family Lauraceae, is caused by the clonal ascomycete fungus *Raffaelea lauricola* T.C. Harr., Fraedrich & Aghayeva 2008 [42].

Other important species traits include: spore shape, which affects the release, transport, and deposition of inoculum (especially for aerial pathogens), and long-distance dispersal mechanisms (e.g., mito- or meiospores mediated by wind, running water, or vectors versus rain splash only) [34]. Other traits affecting spread, but perhaps of lesser importance than sexuality, spore shape, and dispersal mechanisms, include pluricellular spores which can facilitate survival in stressful environments, and abiotic niche characteristics such as the climate in the area of origin and the pathogen's optimal temperature (as a proxy of climate-matching) [34].

Propagative spores are produced in many different ways, in for example, sporangia or simple or complex conidiophores in or on ascocarps and basidiocarps, by budding or fusion, within pycnidia, perithecia, or other various types [43]. The number of spores produced per unit area of infected leaf tissue can be enormous, (e.g., a relatively small apothecium can produce several million spores). The duration and periodicity of sporulation is as important as the number of spores produced [44]. Some pathogens produce an almost continuous crop of propagules, while others may have sporophores that bear several successive propagules (e.g., *Phytophthora* spp.). This capacity to produce a steady stream of infectious propagules over a prolonged period of time is advantageous to the pathogen [43]. Following initial primary infection on a suitable susceptible host plant, pathogens may undergo secondary spread to the same or new host plant species within the same location or to new hosts in another location. The transmission of a pathogen is highly dependent on its ability to produce numerous spores or infectious propagules under favorable environmental conditions that are dispersed and then deposited in a viable condition on a susceptible host plant under conditions conductive for infection. In many cases, pathogens also possess mechanisms to survive between periods of unfavorable environmental conditions (through formation of resting structures).

In nature, organisms can move or be transferred over short and large distances. Stratified dispersal, a two-scale dispersal process of combined short-distance, continuous dispersal, and discontinuous, long-range dispersal, is a major driver of spread dynamics [45]. Where long-distance dispersal events are normally rare, but facilitated by inadvertent human transport, they can cause much greater rates of spread than that which would normally occur with short-range dispersal since populations jump well ahead of the advancing invasion front [13]. Subsequently, isolated populations become established far from the moving population front, and will grow and eventually coalesce with the source population to significantly advance the population front.

Plant pathogens typically disseminate through direct transmission and indirect (passive) transmission (Table 1). With direct transmission, pathogen dispersal occurs intrinsically alongside seeds (germinative) and/or other plant parts (vegetative). Indirect transmission may be autonomous, by wind, water, insect, or mammalian vectors, and human-mediated. Many pathogens are dispersed by more than one mechanism. The importance of understanding these organismal traits and how they affect spread dynamics is essential for preventative and predictive actions. For example, during Pest Risk Assessment (PRA) the likelihood of pathogen spread within an importing country or region considers such factors like the dispersal potential as it relates to the pathogen's reproductive potential (rated by the presence of multiple generations per year or growing season, and the relative number of offspring or propagules per generation), the pathogen's inherent mobility (e.g., rapid movement), and external dispersal facilitation modes (e.g., the presence of natural barriers or enemies, and dissemination enhanced by wind, water, vectors, or human assistance).

Table 1. Dispersal mechanisms of plant pathogens.

Mechanism	Description
Autonomous transmission	This dispersal mechanism is characterized by continuous and persistent growth of hyphal strands that can migrate independently through the soil from plant to plant, quite characteristic of soil fungi (e.g., *Armillaria* spp.). Dispersal can range from a few cm to several meters per year.
Wind	Most fungal pathogens that produce spores or conidia externally on host surfaces are easily carried by wind currents (e.g., downy and powdery mildews, rust fungi). Fungal spores behave as inert particles, with terminal velocities ranging from about 0.05 to 2.5 cm per second, with the larger spores falling more rapidly than smaller ones. Turbulence redistributes spores and affects their progressive dilution with increasing distance from its source [43]. With normal wind and turbulence conditions, spores can travel large distances (from several hundred meters to kilometers)
Water	Except in the case of streams or rivers that may carry inoculum, water is usually less effective than wind for long-distance dissemination. Rain splash or splatter during heavy rains can locally distribute inoculum on or around the same or neighboring plants. Similarly, rain or irrigation water that moves either through the soil or on the soil surface can disseminate pathogen propagules.
Vectors	Some plant pathogens cannot be directly transferred from one plant to another and require a completely unrelated species to act as a vector. Many insects have piercing and sucking mouthparts that penetrate the plant surface and facilitate the transmission and inoculation of host plants (e.g., Dutch elm disease caused by *Ophiostoma ulmi* or *O. novo-ulmi* growing within the egg galleries of *Scolytus* bark beetles, contaminating emerging adults; Laurel wilt caused by *Raffaelea lauricola* transmitted by species of ambrosia beetles). Thus, vector-transmitted pathogens are usually transferred to the host with great efficiency and play a major role in the infection lifecycle. Some insects cause wounding of plant tissue through which plant pathogens can enter secondarily. Other vectors of pathogens may include nematodes or mammals that may transmit diseases both externally and internally. Most vectors of forest pathogens are usually, but not always, insects, and are sometimes referred to as alternate hosts or as having 'hitchhiking' dispersal. Hitchhiking dispersal is favored by typical fungal features such as inconspicuousness, and the production of numerous small propagules [46]. For vector-dependent fungi, if no alternate host exists, the infection cycle is broken.
Human	Plant diseases are often dispersed through human-mediated, extra-range dispersal typically through transportation of infected propagative material (e.g., seed, nursery stock, timber, plant products, or soil), or through mishandling or contamination of healthy plants or plant parts during cultivation practices. Organismal spread may be complicated by multiple introductions (genotypes) from multiple sources to multiple locations [7,47].

2. Host Factors

2.1. Diversity

Most studies recognize that plant diversity can affect disease prevalence and spread through the direct effect of host density (plant species richness) on the transmission of plant pathogens and the role that plant diversity has in influencing host density through, for example, competition (e.g., [48,49]). Although most plant species are susceptible to infection and damage by one or more pathogens, some species or individuals may exhibit different degrees of susceptibility [50]. For example, within the genus *Fraxinus*, there is large variation between species in susceptibility to the ash dieback pathogen *Hymenoscyphus fraxineus* [51,52] (see Section 3.1 *Case studies*), and large genotypic effects in susceptibility among individuals of European ash (*Fraxinus excelsior* L.) [53]. Similarly, *R. lauricola* affects several members of the laurel family (Lauraceae) in the USA though lethal damage is most prominent on only a few native hosts including red bay (*Persea borbonia* (L.) Spreng. *sensu stricto* [42,54]).

In ecological communities, a high diversity of plant species usually contributes to maintaining the functional integrity of the ecosystem ('insurance hypothesis'; e.g., [55]). As different species show varying responses to a specific pathogen, a high species diversity will act as a sort of buffer, diluting the effects of the pathogen on the ecosystem ('diversity-disease hypothesis' or 'dilution hypothesis'; [56,57]). When pathogen transmission is density dependent or where the host range is narrow, biodiversity can alter infection prevalence through a change in the absolute abundance of important hosts and any associated vectors [58]. For example, a non-host species may reduce the probability of encountering hosts and therefore lessening opportunities for healthy susceptible individuals to become infected [56]. Similarly, in the case of an environmentally transmitted disease, added species abundance can reduce the probability that contact between individuals lead to transmission, thereby leading to the idea of encounter reduction [56,59]. In contrast, if added species function as alternative sources of infection, or serve as a source for increased vector activity, disease prevalence may subsequently increase [56]. Thus, host biodiversity can largely influence disease spread through interspecific variability in suitable host species. At the same time, a mixture of species including non-hosts, competent hosts (i.e., with a high effectiveness of passing on infection), and less competent hosts, may intercept inoculum and also reduce the spread of a pathogen [48].

Monocultures of host species, despite a few exceptions, are highly susceptible to epidemics of invasive pathogens because of the lack of intraspecific host diversity [49,50]. Moreover, hosts in high quantity and density, as observed in monocultures, mean reduced distance to which inoculum must traverse to spread between plants, consequently increasing pathogen transmission [59]. This is particularly relevant for competent hosts on which the pathogen may readily produce inoculum. In humans, when transmission of a specialist pathogen is density-dependent, theory predicts a minimal density of the host population below which the pathogen becomes extinct ('crowd disease', [60]). In mixed communities, density-dependent disease dynamics confer an advantage to uncommon species which benefit from a lower enemy pressure ('rare-species advantage'; [61]) and may, therefore, increase in incidence. Examples where host density has affected the spread of an invasive forest pathogen include the beech bark disease involving the exotic beech scale insect *Cryptococcus fagisuga* Lind. and the exotic fungus *Neonectria coccinea* var. *faginata* (Pers.) Fr. in North America. Morin et al. [62] showed that about 100 years after its first detection the disease had invaded most regions where American beech (*Fagus grandifolia* Ehrh.) is a dominant component of stands, but not the regions where the host occurs at low densities. Noteworthy, a study conducted in northern Maine showed that the disease epidemic is also influenced by climatic conditions (e.g., winter temperatures) which affect the survival of the insect vector [63]. A host-density dependent disease dynamic was also revealed using a model by Hatala et al. [64] for the invasive white pine blister rust *Cronartium ribicola* J.C. Fisch in the forests of the Greater Yellowstone Ecosystem. In general, these examples among others suggest that the risk of disease is lower if the competent host for the invasive pathogen comprises a small fraction of the overall diverse host community.

The above discussed points may, however, not strictly apply to invasive pathogens infecting a wide range of host species (i.e., polyphagous or generalist pathogens). In such case, the "insurance hypothesis" can fail and hosts driving the epidemics seem to be decisive. Weste et al. [65] showed that in the major types of forest and woodland of the Grampians (Western Australia) the activity of the generalist root pathogen *Phytophthora cinnamomi* Rands. in certain cases resulted in important changes in species composition and community structure. The high functional diversity of the local ecosystems could not prevent the spread of this invasive oomycete. In mixed forests of coastal California, the spread of *P. ramorum*, the causal agent of Sudden Oak Death, is driven by foliar hosts, on which this polyphagous pathogen sporulates [66]. These so-called 'reservoir hosts' showing only cryptic or asymptomatic infections but acting as a source of inoculum for other hosts may be difficult to predict and can also play a major role in the epidemic of invasive pathogens. This, for example, is also the case for the invasive *Rhododendron ponticum* and *P. ramorum* and *P. kernoviae* in the UK [67], and for non-beech hardwood hosts (e.g., *Acer rubrum* L.) and bark beech disease in North America [68].

2.2. Connectivity

The spatial structure of the host population influences the spread of invasive pathogens both at the local (forest stand) and regional (landscape) scale [69,70]. In a forest stand, the epidemic spread rates are driven by the contacts between infected and healthy hosts or between vectors and infected/healthy hosts [71]. Both types of contact strongly depend on host density, which, in mixed forests, may vary significantly. Frequently, hosts show a clumped (aggregated) distribution, with a 'patch phase' (i.e., higher than average host density) and a 'gap phase' (i.e., lower than average host density) [59]. For pathogens spreading via active vectors (e.g., insects), the spatial heterogeneity of hosts can hinder the advance of the vector and/or physically separate the pathogen from the vector and consequently limit the spread of the disease [72,73]. Similarly, if the host heterogeneity scales at distances over which a pathogen is transmitted, a spatial variation in host density may constrain the spread of pathogens that rely on passive vectors (e.g., wind, water).

At the landscape scale, invasive pathogens generally first colonize areas with continuous forests and then, eventually, isolated (scattered) forest stands or trees (e.g., chestnut blight, see Section 3.2 *Case studies*). A scattered distribution of hosts does not always allow them to escape infection. For instance, although butternut (*Juglans cinerea* L.) in North America usually occurs as scattered individuals or in small groups in deciduous and mixed forests, it could not escape infection by the invasive canker pathogen *Ophiognomonia clavigignenti-juglandacearum* (N.B. Nair, Kostichka & J.E. Kuntze) Broders & Boland, which was dispersed over longer distances by beetle vectors and infected seeds [74]. Similarly, a fragmented distribution of some white pines (e.g., *Pinus strobiformis* Engelm.) in western North America could only retard, but not prevent their infection by the white pine blister rust fungus *C. ribicola* [75]. When hosts are scattered, landscape connectivity (i.e., how the landscape structure facilitates or impedes the disease spread among patches [76]) plays an important role for disease spread. Such connectivity is strongly related to landscape structure, which, as shown by Real and Biek [77] for rabies, can present two possibilities; namely individuals aggregated over a uniform landscape or individuals assorted by the environment into different spatial locations. This second possibility most likely applies to trees whose distribution is mainly shaped by environmental features [78]. Although most studies on the influence of landscape heterogeneity on disease dynamics have focused on agricultural systems, frequently using simulation models (e.g., [76]), in recent years the interest in how landscape features affect the spread of forest pathogens has increased considerably (landscape pathology; [70]). Several investigations have dealt with the spread of *P. ramorum* in coastal California. For example, Condeso and Meentemeyer [28] showed that the effect of forest fragmentation on disease severity is scale-dependent. In another study, Ellis et al. [79] demonstrated that environmental variables were relatively more important than landscape connectivity in shaping the spatial pattern of Sudden Oak Death. Filipe et al. [29] found that host-free barriers would contain the spread of *P. ramorum* for a significantly long time only if combined with additional buffers (e.g., topographic conditions).

In a landscape still not completely colonized by an invasive pathogen, the connectivity may be increased by additional introductions, which also have the potential to introduce new genotypes [80,81], or by insect vectors that may actively or passively spread the pathogen over long distances. *Fusarium circinatum* Nirenberg and O'Donnell, the causal agent of pitch canker, is a potentially dangerous invasive pathogen in pine (*Pinus* sp.) forests in Europe. Möykkynen et al. [82] modeled the rate of spread of *F. circinatum* as a function of several factors, among which included host distribution and flight distance of insect vectors. Their model showed that because of the short distance at which spores are dispersed and the fragmentation of pine forests, unless there will be new introductions, the pathogen will most likely not spread to northern Europe. In a broader sense, one of the main reasons for the increasing number of invasive pathogens which are spread around the planet is the high global connectivity through transportation and trade networks [5–7].

2.3. Susceptibility

Intraspecific variability in host susceptibility is a main source of heterogeneity controlling the trajectory of a disease epidemic [83]. The susceptibility of a host to a specific pathogen's transmission, maintenance, and proliferation is influenced by physical host traits (e.g., size, morphology), external factors (e.g., availability of nutrients, local microclimate, topography), and, of course, the genetic host background [49,84]. Tree size, which often correlates with tree age, may show a contradictory effect on host susceptibility. Frequently, the impact of invasive pathogens is greater on larger trees, probably because of the larger contact area available for the pathogen and its vectors and/or traits correlated with age (e.g., vigor; [57,85]). A study by Jules et al. [25] indicated that in Oregon and California large Port Orford cedars (*Chamaecyparis lawsoniana* (A. Murray) Parl.) located in close proximity of streams and roads were more likely to be killed by the invasive oomycete *Phytophthora lateralis* Tucker & Milbrath. According to that study, high susceptibility was due to the larger root systems and the position of the trees which allowed the roots to reach the water and, thus, be a good target for waterborne zoospores. On the other hand, in Europe the impact of oak (*Quercus* sp.) powdery mildew caused mainly by the invasive fungus *Erysiphe alphitoides* Griffon & Maublanc seems to be more severe on seedlings, particularly in natural regeneration, than on adult trees [86]. Age-related susceptibility has also been documented for *Ophiostoma novo-ulmi* (Brasier) causing Dutch elm disease on *Ulmus* spp., whereby younger trees, possessing smaller diameter vessels, are generally more tolerant [87].

Host population genetic diversity appears to play an important role in buffering populations against disease epidemics [88]. In fact, empirical observations and modeling studies indicate that in a population of a host species intraspecific genetic diversity represents the best insurance against pathogens [89]. According to Garrett and Mundt [90], a mixture of host genotypes reduces a disease in three ways: first, the presence of less susceptible/resistant genotypes dilutes the inoculum; second, less susceptible/resistant genotypes represent a physical barrier for inoculum deposition; and third, the potential for induced host resistance is increased. Monocultures of genetically similar or identical (clones) trees are usually highly susceptible to invasive pathogens (the so-called 'monoculture effect' earlier mentioned; [91,92]). Numerous examples worldwide support this general assumption, including plantations of *Pinus radiata* D. Don in the Southern hemisphere affected by needle blight (*Dothistroma septosporum* (Dorog.) Morelet) [93] or *Eucalyptus* spp. plantations in south-east Asia which are susceptible to leaf, bud, and shoot blight caused by *Teratosphaeria destructans* (M.J. Wingf. & Crous) [94]. Increased host susceptibility to infectious parasites may also be due to reduced individual-level and population-level genetic heterozygosity, which may increase the occurrence of inbreeding [95,96].

Usually, in the initial phase of an invasive pathogen epidemic, either resistance in the host population is completely lacking or resistant host genotypes are at too low of a frequency to reduce the effects of the pathogen [97,98]. Successively, the continued exposure to the pathogen will select for less susceptible host genotypes, which will increase in frequency. However, if the impact of an invasive pathogen is extreme with large-scale mortality within a relatively short-time frame, host resistance may never evolve due to rapid elimination of the host species. In forest trees, large differences in generation times may also be a disadvantage for developing resistance to invasive pathogens. As stated by Aegerter and Gordon [97], for such a directional selection to be successful, young individuals in a population have to be challenged by a pathogen after the stage where the physiological mechanisms for resistance can be operative. The same authors showed that in *Pinus radiata* different mechanisms of resistance against *Fusarium circinatum* are active in seedlings and adult trees. Still, benefits deriving from a genetically diverse host population may also depend on the genetic diversity of the pathogen population [99].

3. Case Studies

Following initial introduction and establishment, spread rates for alien forest pathogens are typically slower than alien insect pests due in part to dispersal mechanisms. The following two examples of invasive forest pathogens serve to illustrate their spread post-introduction and establishment and their effects on local ecosystems.

3.1. Ash Dieback

The invasion of the ash dieback pathogen *Hymenoscyphus fraxineus* (Table 2) into Europe and its subsequent spatial spread to most European countries throughout the natural distribution range of native *Fraxinus* species can been characterized by relatively slow spread dispersal by airborne spores, coupled with some few pulsed events involving long-distance establishment via anthropogenic means within or ahead of the invasion boundary namely through—then, unknowing or naïve movement of nursery stock from commercial nurseries within or outside already-infested areas [100,101]. Ash dieback has had variable recognizable consequences, partially driven by both the known and still unknown or unrecognized importance of the species from a forestry or nature conservations perspective. Mitchell et al. [102] suggest that the disease could have wide-ranging ecological implications particularly for obligate-associated organisms to ash and the indirect effect of the disease on nutrient cycling in woodland ecosystems. Baseline information on tree species contributions to ecosystem functions is necessary in order to determine actual short-term impacts on light penetration, nitrogen cycling, and primary production, long-term impacts involving interactions with other abiotic or biotic stress factors, and any compensatory effects of other tree species in the post-epidemic/decline phase.

Spread rates for ash dieback have been difficult to accurately track due to the lag time in reporting disease presence in various countries. However, where good records have been kept, mean spread rates have ranged between 30 and up to 75 km per year [103]. Regular monitoring of disease presence at the onset of its introduction to Norway initially gave an annual mean spread rate of 30 km, but then it increased rapidly in subsequent years averaging more than 50 km per year [104]. In Sweden, the disease was observed already in 2001 in only a few places and within just a few years, the occurrence of ash dieback was reported on trees of all age classes throughout the natural distribution range of *F. excelsior* in that country [105]. By 2010, ash was added to the Swedish Red-List with vulnerable status, and recently elevated to critically endangered status considered to be at high risk of extinction in the wild.

Some factors contributing to variable rates of disease spread among countries may be due to large differences in natural geographic barriers that may limit natural dispersal, the length of growing (and hence sporulation) seasons, fragmented distribution and density of the host species throughout Europe, between and within-season fluctuations in optimal climate conditions, and stratified dispersal also involving the import or transfer of diseased plants which resulted in large jumps ahead of the advancing infection front. Within the nearly 20 years since damage was first reported in Lithuania and Poland, the pathogen has spread throughout most of western and eastern Europe where native *Fraxinus* spp. are growing. Within this region, the large majority of native ash species (especially *F. excelsior*) are highly susceptible, and some non-native species planted within the zone of infestation also exhibit moderate to high susceptibility. Fortunately, a small proportion (<5%) of the natural population has shown better tolerance to the disease, which offers the potential to revitalize and restore forest and urban landscapes through breeding for resistance [53].

3.2. Chestnut Blight

Chestnut blight caused by the fungus *Cryphonectria parasitica* (Table 2) is most likely the best-known example of an invasive forest disease in Europe and North America. After its first discovery, the pathogen rapidly colonized the whole geographic range of the susceptible host species in the two continents. In eastern North America, the fungus spread at a rate of more than 30 km per year throughout the approx. 800,000 km^2 of native American chestnut forests [106]. In some stands, *C. dentata* accounted for more than 50% of the basal area of standing trees and was the canopy forming species [107]. Within 50 years, this species was confined to the understory, with significant ecological and economic consequences [108]. This dramatic course of the epidemic was favored by the high susceptibility of American chestnut to the introduced pathogen [109]. Nevertheless, it is still not clear whether the low genetic diversity of *C. dentata* compared to that of the congeneric species is the result or also a cause of the species decline due to chestnut blight [110]. Recent population genetic analyses [111] showed that the initial introduction of *C. parasitica* into North America occurred from the main Japanese island Honshu. Later on, the fungus was also introduced from other Japanese regions, China, and Korea. Introductions into Europe occurred both from North America (into Italy) and Asia (into south-western France) [111]. Although European chestnut is slightly less susceptible to *C. parasitica* than the American chestnut, the spread of the pathogen on the European continent was also rapid. Thirty years after its first detection, the disease was reported in the main chestnut growing areas of Europe and to date chestnut blight is widespread throughout most of the distribution range of *C. sativa* [112,113]. The presence of important geographic barriers and the sometimes scattered distribution of chestnut stands could only slow down, but not stop the spread of the pathogen. For instance, in Switzerland chestnut blight was first observed south of the Alps in 1948 [41]. About forty years later, despite the potential barrier represented by the Alps and the adoption of quarantine regulations, the disease also appeared in the scattered stands north of the Alps. The host connectivity in Europe has likely increased through several introduction events targeting different regions, which unintentionally propagated *C. parasitica* via infected plant material.

Due to the unexpected appearance and spread of natural hypovirulence within the *C. parasitica* population, however, consequences of chestnut blight in Europe were less dramatic than in North America. European chestnut stands, in spite of a high disease incidence, have survived the epidemics and are still successfully fulfilling their important ecological and cultural functions. Nevertheless, differences in the success of biocontrol by hypovirulence exist among chestnut-growing regions, which may be explained by several factors, including the diversity of the local *C. parasitica* population in terms of vegetative compatibility (vc) types, the subtype of the occurring hypovirus, the presence of adequate vectors, management practices, and variable environmental conditions [113,114].

In eastern North America, biological control has failed nearly completely due to the high vc type diversity of the local *C. parasitica* populations and the high susceptibility of *C. dentata* to the pathogen [114]. To increase the chance of establishment of the hypovirus, transgenic *C. parasitica* strains which transmit the hypovirus also to the sexual spores (ascospores), as well as super hypovirus donor strains, have been recently used [115,116]. The success of these fungal strains in the field, however, still needs to be demonstrated. From the host side, blight-resistant chestnut trees have been obtained by backcrossing the resistance to *C. parasitica* of the Chinese chestnut into the genome of the American chestnut [117]. Recently, a chestnut tree with an increased resistance to chestnut blight was also created using a transgenic approach [118].

Table 2. Main characteristics of the two invasive forest pathogens causing ash dieback (*Hymenoscyphus fraxineus*) and chestnut blight (*Cryphonectria parasitica*).

Common Name	Ash Dieback [1]	Chestnut Blight [2]
Causal agent	*Hymenoscyphus fraxineus* (T. Kowalski) Baral, Queloz, Hosoya, comb. nov. (Family *Helotiaceae*, Order Helotiales, Class Leotiomycetes)	*Cryphonectria parasitica* (Murr.) Barr (Family *Cryphonetriaceae*, Order Diaporthales, Class Sordariomycetes)
Host species	Major: genus *Fraxinus*, in particular European ash (*F. excelsior*), Narrow-leafed ash (*F. angustifolia*); Black ash (*F. nigra*); Green ash (*F. pennsylvanica*); White ash (*F. americana*); Tianshan ash (*F. sogdiana*), and Blue ash (*F. quadrangulata*) Minor [3]: Manna ash (*F. ornus*); Chinese ash (*F. chinensis*), Manchurian ash (*F. mandschurica*), Texas ash (*F. albicans*), Oregon ash (*F. latifolia*), Spaeth's ash (*F. platypoda*), Pumpkin ash (*F. profunda*), and Velvet ash (*F. veluntina*).	Major: genus *Castanea*, in particular American chestnut (*C. dentata*), European chestnut (*C. sativa*), Japanese chestnut (*C. crenata*), and Chinese chestnut (*C. mollissima*); Minor (incidental): *Quercus* spp., *Acer* spp., *Carpinus betulus*, *Castanea pumila*
Symptoms	Necrotic lesions on leaflets expanding preferentially along leaf veins, and on leaf rachises; leaf wilting; premature leaf abscission; necrosis of buds, perennial necrotic lesions (cankers) on the bark of twigs, branches, and stem; brown discoloration of xylem; dieback of crown; prolific formation of epicormics shoots; basal stem necrosis	Perennial necrotic lesions (cankers) on the bark of above-ground woody parts (stem, branches) of host plants. The plant part distal to the infection point may wilt.
Spread mechanism	Sexual spores (ascospores) for short- (local), medium-, and potentially long-distance spread; Over long distances via movement of latently infected plants or plant material	Over short distances mainly via splash dispersed asexual spores (conidia); Over long distances via sexual ascospores or latently infected plants or plant material.
Mating system	Random mating, heterothallic, outcrossing	Mixed with outcrossing and self-fertilization occurring at variable frequencies.
Native range	Eastern Asia (China, Korea, Japan, Far East Russia)	Eastern Asia (China, Japan, Korea).
Invaded range	Europe	North America, Europe.
First detection	Early 1990s (Lithuania, Poland)	1904 (North America), 1938 (Europe).
Introduction pathway	Primary pathway for introduction through nursery stock of latently infected plants (e.g., Manchurian ash *F. mandshurica*) for planting	Most likely infected plants for planting (probably *Castanea crenata*).
Primary dispersal pathway	Wind-dispersed spores (seasonal—between June and September); movement of infected plants or plant material.	Spores (spontaneous dispersal), infected plant material (e.g., plants for planting, wood with bark).
Mean dispersal rates	30 km per year (Norway) 50–60 km per year (Italy) 75 km per year (from east Poland to Switzerland)	In North America, 37 km per year. In Europe (Italy), 29 km per year.
Control	Breeding for resistance through traditional screening and selection of disease tolerant genotypes from wild populations. Chemical and biological treatments have been tried on an experimental basis with varied efficacy.	North America: breeding for resistance by backcrossing blight-resistant Chinese chestnut into the American chestnut genome; Transgenic fungal strains and chestnut trees. Europe: biocontrol by a mycovirus in the family Hypoviridae (*Cryphonectria* hypovirus 1, CHV-1) which reduces both virulence and sporulation of the infected fungal strain (phenomenon called hypovirulence); Hybrids between *C. sativa* and *C. crenata*.

[1] Based on [51,52,103,119–129]; [2] Based on [41,106,108,111–113,115–118]; [3] Based on known reports and confirmation of pathogen presence on symptomatic trees and/or few to no disease symptoms but support development of ascomata on leaf rachises.

4. "The Way Forward"

The study of invasive species epidemiology will continue to be important in the future as new introductions associated with anthropogenic activity and novel plant-organismal interactions will inevitably continue to cause disease epidemics around the globe. Despite recent advances in our understanding of pathways for biological introductions and factors contributing to emerging invasive diseases, much remains to be learned. Here, we mention some future challenges and research priorities which are strongly related to the spread of invasive forest pathogens.

4.1. Host Diversity Threshold

As mentioned above, diversity of plant species can strongly influence the spread of invasive forest pathogens by modifying the relative abundance of host species. In particular, a high species diversity negatively affects the spread of pathogens with a limited host range or whose transmission is density dependent. Thus, we could imagine that in a forest a threshold of tree species diversity may exist, above which an invasive pathogen may not be able to become established and spread. Determining the diversity threshold of a given pathosystem toward specific invasive pathogens would be of great value for estimating the potential vulnerability of ecosystems. An eventual high vulnerability could be reduced by introducing new, possibly non-host or less susceptible, tree species. Although this approach may be difficult to apply to already existing natural forests, it could be adopted when establishing new artificial forests. As shown by studies conducted with agricultural crops (e.g., [73,130]), not only plant diversity but also the spatial organization of the host species may influence the spread of pathogens. Therefore, when introducing additional plant diversity, host spatial heterogeneity should also be created.

A similar 'diversity threshold' approach could also target the intraspecific genetic diversity of the host. Although it is known that a high genetic variation within a specific host represents the best insurance against pathogens, only a few studies have examined the rate of spatial spread of invasive forest pathogens as a function of the genetic diversity of the host population. In particular, almost no information is available about how much host genetic diversity is needed for the effects to be realized, specifically for the spread of a pathogen to be reduced and less susceptible host genotypes to emerge.

4.2. Spread Dynamics

Predicting the spread dynamics (e.g., when the invader will arrive at a specific location, from where it will arrive) of an invasive pathogen after its establishment in a new area is of importance for developing and applying effective management strategies. In recent years, network theory has started to be used to model the epidemiology of plant pests and diseases (e.g., [131]). For example, Ferrari et al. [132] developed a dynamic network model for analyzing the spread of the invasive phloem-boring pest *Adelges tsugae* Annand (hemlock woolly adelgid) in the eastern United States. They showed that this kind of model, which allows connections to change through time, can provide valuable information about the spatio-temporal dynamics of invasion processes. Harwood et al. [31] developed a simulation model of the *P. ramorum* epidemic development in the UK which included information on the spatial distribution of hosts and a realistic network of plant trade. For several reasons (e.g., heterogeneity of the landscape, patchiness of host distribution, and environmental conditions; [131]), however, network theory is still poorly applied to invasive pathogens in forest ecosystems [133]. In the future, approaches combining network analysis with landscape genetic analyses [134], monitoring data (e.g., host distribution in the target region), and realistic network data (e.g., plant trade) should become a standard tool for predicting (and reconstructing) the spread of invasive plant diseases.

4.3. How to Slow the Spread

The eradication of invasive forest pathogens post-establishment is difficult and has been rarely achieved [135]. For example, in the case of ash dieback, photographic evidence suggests the disease likely had reached an epidemic level by the time conspicuous symptoms were formally documented and had been noticed by foresters in the mid-1990s. Therefore, slowing the spread may be a more realistic objective. A premise for such a goal to be reached is the early detection of a pathogen by optimizing the monitoring strategies (for a modeling approach, see e.g., [136]). Once an invasive pathogen is detected, the ability to actually slow down its spread and the type of management strategy to be adopted will vary between pathogen species due to differences in their infection biology. However, any attempts to minimize the spread and impact of invasive pathogens must ideally consider all factors that influence their dispersal, population growth, or a combination of both. For example, long distance dispersal events (frequently mediated by human activity) bringing a pathogen ahead of the actual advancing front of the epidemic, even if difficult to predict, are an important determinant of spread rates. Hence, even smaller efforts aimed at minimizing long-distance dispersal can greatly reduce pathogen spread [137]. Among possible approaches to be considered, potential strategies might include optimizing pathogen detection/diagnostics and biosurveillance monitoring [138], reducing landscape connectivity and susceptibility by adopting adequate land-use strategies (e.g., promoting increased diversity in mixed plantations instead of monocultures), or, in particular for soil borne pathogens, careful management of routine operations, recreation activities, and machinery traffic in and around known infested areas.

4.4. Evolution of Invasive Forest Pathogens

Invading populations are faced with new environmental conditions to which they have to respond rapidly [139,140]. In the case of invasive forest pathogens, changes in the selective regime (e.g., increased selection for adapted genotypes) compared to the native range, combined with intrinsic characteristics (e.g., genetic diversity) of the invading population may influence the dynamic of an epidemic. It is easy to predict that adaptations in response to invasions will also affect the dispersal ability of a pathogen, thereby introducing uncertainty in the prediction of the rate of its range expansion. Additional uncertainty may be provided by the ability of some pathogens to overcome the so-called 'host species barrier' and infect a new host ('host jump'; [141]), which could accelerate the spread and amplify the consequences of an invasion. Similarly, hybridization events of an established invasive pathogen with native or other similar invasive species [46] could originate new, rapidly-spreading invasive species. Unfortunately, both host jumps and hybridization events of invasive species are particularly difficult to predict. Of importance is also the fact that invading pathogen populations can also induce evolutionary changes in their host populations. For example, virulent host genotypes may select for less susceptible host genotypes. Last, but not least, the coevolution of an invasive pathogen with its new host will also be influenced by the environmental conditions.

5. Conclusions

Understanding the host variability as it is affected by host diversity, host connectivity, and host genetics (Figure 1) will improve our ability to predict invasion success and potential consequences to local ecosystem health and services—such as carbon sequestration, nutrient cycling—on which humans rely. Quite often data related to host species characteristics under various local driver values (e.g., gradients of temperature, moisture, elevation, human activities) is either lacking altogether, is fragmented, or has emerged slowly because sampling is too local and often not adequate enough to apply across a landscape-level [2].

Further complicating our ability to predict and manage disease epidemics is 'episodic temporal variation', where the success of a new invader only coincides with, for example, a disturbance, or major rainfall event [21]. In addition, changing land-use patterns and changes in climate can and will continue to influence the range expansion of native species affecting the spread and impact of introduced non-native species [2]. Continued efforts on the avoidance of new introductions and recognizing the human factor affecting the emergence of pests and diseases, interdisciplinary research on emerging invasive diseases, and creative strategies for the mitigation of plant disease impacts in natural and semi-natural ecosystems, is warranted [6,142].

Acknowledgments: Funding was provided by The Swedish Research Council FORMAS, Carl Tryggers Stiftelse för Vetenskaplig Forskning and Kungl. Skogs- och Lantbruksakademien.

Author Contributions: M.C. and S.P. contributed equally to this work.

Conflicts of Interest: The authors declare no conflict of interest. The funding sponsors had no role in the design or writing of this review manuscript.

References

1. Ricciardi, A. Are Modern Biological Invasions an Unprecedented Form of Global Change? *Conserv. Biol.* **2007**, *21*, 329–336. [CrossRef] [PubMed]
2. Crowl, T.A.; Cris, T.O.; Parmenter, R.R.; Belovsky, G.; Lugo, A.E. The spread of invasive species and infectious disease as drivers of ecosystem change. *Front Ecol. Environ.* **2008**, *6*, 238–246. [CrossRef]
3. Lovett, G.M.; Weiss, M.; Liebhold, A.M.; Holmes, T.P.; Leung, B.; Lambert, K.F.; Orwig, D.A.; Campbell, F.T.; Rosenthal, J.; McCullough, D.G.; et al. Nonnative forest insects and pathogens in the United States: Impacts and policy options. *Ecol. Appl.* **2016**, *26*, 1437–1455. [CrossRef] [PubMed]
4. Aukema, J.E.; McCullough, D.G.; Von Holle, B.; Liebhold, A.J.; Britton, K.O.; Frankel, S.J. Historical accumulation of non-indigenous forest pests in the continental United States. *BioScience* **2010**, *60*, 886–897. [CrossRef]
5. Fisher, M.C.; Henk, D.A.; Briggs, C.; Brownstein, J.S.; Madoff, L.; McCraw, S.L.; Gurr, S. Emerging fungal threats to animal, plant and ecosystem health. *Nature* **2012**, *484*, 186–194. [CrossRef] [PubMed]
6. Santini, A.; Ghelardini, L.; de Pace, C.; Desprez-Loustau, M.L.; Capretti, P.; Chandelier, A.; Cech, T.; Chira, D.; Diamandis, S.; Gaitniekis, T.; et al. Biogeographical, patterns and determinants of invasion by forest pathogens in Europe. *New Phytol.* **2013**, *197*, 238–250. [CrossRef] [PubMed]
7. Liebhold, A.M.; Brockerhoff, E.G.; Garrett, L.J.; Parke, J.L.; Britton, K.O. Live plant imports: The major pathway for forest insect and pathogen invasions of the US. *Front. Ecol. Environ.* **2012**, *10*, 135–143. [CrossRef]
8. Ghelardini, L.; Pepori, A.L.; Luchi, N.; Capretti, P.; Santini, A. Drivers of emerging fungal diseases of forest trees. *For. Ecol. Manag.* **2016**, *381*, 235–246. [CrossRef]
9. Anagnostakis, S.L. Chestnut blight: The classical problem of an introduced pathogen. *Mycologia* **1987**, *79*, 23–37. [CrossRef]
10. Hansen, E.M.; Kanaskie, A.; Prospero, S.; McWilliams, M.; Goheen, E.M.; Osterbauer, N.; Reeser, P.; Sutton, W. Epidemiology of *Phytophthora ramorum* in Oregon tanoak forests. *Can. J. For. Res.* **2008**, *38*, 1133–1143. [CrossRef]
11. Rizzo, D.M.; Garbelotto, M. Sudden oak death: Endangering California and Oregon forest ecosystems. *Front. Ecol. Environ.* **2003**, *1*, 197–204. [CrossRef]
12. Hui, C.; Krug, R.M.; Richardson, D.M. Modelling spread in invasion ecology: A synthesis. In *Fifty Years of Invasion Ecology: The Legacy of Charles Elton*, 1st ed.; Richardson, D.M., Ed.; Wiley-Blackwell: Oxford, UK, 2011; pp. 329–343.
13. Shigesada, N.; Kawasaki, K.; Takeda, Y. Modelling stratified diffusion in biological invasions. *Am. Nat.* **1995**, *146*, 229–251. [CrossRef]
14. Valéry, L.; Fritz, H.; Lefeuvre, J.C.; Simberloff, D. In search of a real definition of the biological invasion phenomenon itself. *Biol. Invasions* **2008**, *10*, 1345–1351. [CrossRef]
15. Wilson, J.R.U.; Dormontt, E.E.; Prentis, P.J.; Lowe, A.J.; Richardson, D.M. Biogeographic concepts define invasion biology. *TRENDS Ecol. Evol.* **2009**, *24*, 586. [CrossRef]
16. Lockwood, J.L.; Hoopes, M.F.; Marchetti, M.P. *Invasion Ecology*, 2nd ed.; Wiley Blackwell: West Sussex, UK, 2013.
17. Smith, C.S.; Lonsdale, W.M.; Fortune, J. When to ignore advice: Invasion predictions and decision theory. *Biol. Invasions* **1999**, *1*, 89–96. [CrossRef]
18. Mack, R.N.; Simberloff, D.; Lonsdale, W.M.; Evans, H.; Clout, M.; Bazzaz, F. Biotic invasions: Causes, epidemiology, global consequences, and control. *Ecol. Appl.* **2000**, *10*, 689–710. [CrossRef]
19. Aylor, D.E. Spread of plant disease on a continental scale: Role of aerial dispersal of pathogens. *Ecology* **2003**, *84*, 1989–1997. [CrossRef]
20. Parker, I.M.; Gilbert, G.S. The Evolutionary Ecology of Novel Plant-Pathogen Interactions. *Annu. Rev. Ecol. Evol. Syst.* **2004**, *35*, 675–700. [CrossRef]
21. Hastings, A.; Cuddington, K.; Davies, K.F.; Dugaw, C.J.; Elmendorf, S.; Freestone, A.; Harrison, S.; Holland, M.; Lambrinos, J.; Malvadkar, U.; et al. The spatial spread of invasions: New developments in theory and evidence. *Ecol. Lett.* **2005**, *8*, 91–101. [CrossRef]
22. Menkis, A.; Östbrant, I.-L.; Wågström, K.; Vasaitis, R. Dutch elm disease on the island of Gotland: Monitoring disease vector and combat measures. *Scand. J. For. Res.* **2016**, *31*, 237–241. [CrossRef]

23. Goss, E.M.; Larsen, M.; Vercauteren, A.; Werres, S.; Heungens, K.; Grünwald, N.J. *Phytophthora ramorum* in Canada: Evidence for migration within North America and from Europe. *Phytopathology* **2011**, *101*, 166–171. [CrossRef] [PubMed]

24. Goss, E.M.; Larsen, M.M.; Chastagner, G.A.; Givens, D.R.; Grünwald, N.J. Population genetic analysis infers migration pathways of *Phytophthora ramorum* in US nurseries. *PLoS Pathol.* **2009**, *5*, e1000583. [CrossRef] [PubMed]

25. Jules, E.W.S.; Kauffman, M.J.; Ritts, W.D.; Carrol, A.L. Spread of an invasive pathogen over a variable landscape: A nonnative root rot on Port Orford cedar. *Ecology* **2002**, *83*, 3167–3181. [CrossRef]

26. Kelly, M.; Gueo, Q.; Liu, D.; Shaari, D. Modelling the risk for a new invasive forest disease in the United States: An evaluation of five environmental niche models. *Comput. Environ. Urban* **2007**, *31*, 689–710. [CrossRef]

27. Meentemeyer, R.; Rizzo, D.; Mark, W.; Lotz, E. Mapping the risk of establishment and spread of sudden oak death in California. *For. Ecol. Manag.* **2004**, *200*, 195–214. [CrossRef]

28. Condeso, T.E.; Meentemeyer, R.K. Effects of landscape heterogeneity on the emerging forest disease sudden oak death. *J. Ecol.* **2007**, *95*, 364–375. [CrossRef]

29. Filipe, J.A.N.; Cobb, R.C.; Meentemeyer, R.K.; Lee, C.A.; Valachovic, Y.S.; Cook, A.R.; Rizzo, D.M.; Gilligan, C.A. Landscape Epidemiology and Control of Pathogens with Cryptic and Long-Distance Dispersal: Sudden Oak Death in Northern Californian Forests. *PLoS Comput. Biol.* **2012**, *8*, e1002328. [CrossRef] [PubMed]

30. Cushman, J.H.; Meentemeyer, R.K. Multi-scale patterns of human activity and the incidence of an exotic forest pathogen. *J. Ecol.* **2008**, *96*, 766–776. [CrossRef]

31. Harwood, T.D.; Xu, X.; Pautasso, M.; Jeger, M.J.; Shaw, M.W. Epidemiological risk assessment using linked network and grid based modelling: *Phytophthora ramorum* and *Phytophthora kernoviae* in the UK. *J. Ecol. Model.* **2009**, *220*, 3353–3361. [CrossRef]

32. Gilbert, M.; Liebhold, A. Comparing methods for measuring the rate of spread of invading populations. *Ecography* **2010**, *33*, 809–817. [CrossRef]

33. Tisseuil, C.; Gryspeirt, A.; Lancelot, R.; Pioz, M.; Liebhold, A.; Gilbert, M. Evaluating methods to quantify spatial variation in the velocity of biological invasions. *Ecography* **2016**, *39*, 409–418. [CrossRef]

34. Philibert, A.; Desprez-Loustau, M.-L.; Fabre, B.; Frey, P.; Halkett, F.; Husson, C.; Lung-Escarmant, B.; Marçais, B.; Robin, C.; Vacher, C.; et al. Predicting invasion success of forest pathogenic fungi from species traits. *J. Appl. Ecol.* **2011**, *48*, 1381–1390. [CrossRef]

35. Janzen, D.H. On Ecological Fitting. *Oikos* **1985**, *45*, 308–310. [CrossRef]

36. Brooks, D.R.; McLennan, D.A. *The Nature of Diversity: An Evolutionary Voyage of Discovery*; University of Chicago Press: Chicago, IL, USA, 2002.

37. Agosta, S.J.; Klemens, J.A. Ecological fitting by phenotypically flexible genotypes: Implications for species associations, community assembly and evolution. *Ecol. Lett.* **2008**, *11*, 1123–1134. [CrossRef] [PubMed]

38. Brown, A.H.D.; Burdon, J.J. Mating systems and colonizing success in plants. In *Colonization, Succession and Stability*; Cary, A.J., Crawley, M.J., Edwards, P.J., Eds.; Blackwell Scientific Publications: Oxford, UK, 1987; pp. 115–131.

39. Bazin, E.; Mathe-Hubert, H.; Facon, B.; Carlier, J.; Ravigné, V. The effect of mating system on invasiveness: Some genetic load may be advantageous when invading new environments. *Biol. Invasions* **2014**, *16*, 875–886. [CrossRef]

40. McDonald, B.A.; Linde, C. Pathogen population genetics, evolutionary potential, and durable resistance. *Annu. Rev. Phytopathol.* **2002**, *40*, 349–379. [CrossRef] [PubMed]

41. Prospero, S.; Rigling, D. Invasion genetics of the chestnut blight fungus *Cryphonectria parasitica* in Switzerland. *Phytopathology* **2012**, *102*, 73–82. [CrossRef] [PubMed]

42. Pisani, C.; Ploetz, R.C.; Stover, E.; Ritenour, M.; Scully, B. Laurel Wilt in Avocado: Review of an Emerging Disease. *Int. J. Plant Biol. Res.* **2015**, *3*, 1043.

43. Mehrotra, R.S.; Aggarwal, A. Dispersal of plant pathogens. In *Plant Pathology*, 2nd ed.; Tata McGraw-Hill Publishing Co. Ltd.: New Delhi, India, 2005; pp. 199–211.

44. Mehrotra, R.S.; Aggarwal, A. Dispersal of plant pathogens. In *Fundamentals of Plant Pathology*; Tata McGraw-Hill Publishing Co. Ltd.: New Delhi, India, 2013; Chapter 9.

45. Hengeveld, R. *Dynamics of Biological Invasions*; Chapman & Hall: London, UK, 1989.

46. Gladieux, P.; Feurtey, A.; Hood, M.E.; Snirc, A.; Clavel, J.; Dutech, C.; Roy, M.; Giraud, T. The population biology of fungal invasions. *Mol. Ecol.* **2015**, *24*, 1969–1986. [CrossRef] [PubMed]
47. Pyšek, P.; Jarošik, V.; Hulme, P.E.; Kühn, I.; Wild, J.; Arianoutsou, M.; Bacher, S.; Chiron, F.; Didziulis, V.; Essl, F.; et al. Disentangling the role of environmental and human pressures on biological invasions. *Proc. Natl. Acad. Sci. USA* **2010**, *107*, 12157–12162.
48. Keesing, F.; Belden, L.K.; Daszak, P.; Dobson, A.; Harvell, C.D.; Holt, R.D.; Hudson, P.; Jolles, A.; Jones, K.E.; Mitchell, C.E.; et al. Impacts of biodiversity on the emergence and transmission of infectious diseases. *Nature* **2010**, *468*, 647–652. [CrossRef] [PubMed]
49. Ostfeld, R.S.; Keesing, F. Effects of Host Diversity on Infectious Disease. *Annu. Rev. Ecol. Evol. Syst.* **2012**, *43*, 157–182. [CrossRef]
50. Pautasso, M.; Holdenrieder, O.; Stenlid, J. Susceptibility to Fungal Pathogens of Forests Differing in Tree Diversity. In *Forest Diversity and Function: Temperate and Boreal Systems, Ecological Studies*; Scherer-Lorenzen, M., Körner, C., Schulze, E.-D., Eds.; Springer: New York, NY, USA, 2005; Volume 176, pp. 263–289.
51. Cleary, M.; Nguyen, D.; Marčiulynienė, D.; Berlin, A.; Vasaitis, R.; Stenlid, J. Friend or foe? Biological and ecological traits of the European ash dieback pathogen *Hymenoscyphus fraxineus* in its native environment. *Sci. Rep.* **2016**, *6*, 21895. [CrossRef] [PubMed]
52. Nielsen, L.R.; McKinney, L.V.; Hietala, A.M.; Kjær, E.D. The susceptibility of Asian, European and North American *Fraxinus* species to the ash dieback pathogen *Hymenoscyphus fraxineus* reflects their phylogenetic history. *Eur. J. Forest. Res.* **2016**. [CrossRef]
53. McKinney, L.V.; Neilsen, L.R.; Collinge, D.B.; Thomsen, I.M.; Hansen, J.K.; Kjær, E.J. The ash dieback crisis: Genetic variation in resistance can prove a long-term solution. *Plant Pathol.* **2014**, *63*, 485–499. [CrossRef]
54. Hughes, M.A.; Smith, J.A.; Ploetz, R.C.; Kendra, P.E.; Mayfield, A.E., III; Hanula, J.L.; Hulcr, J.; Stelinski, L.L.; Cameron, S.; Riggins, J.J.; et al. Recovery plan for laurel wilt on redbay and other forest species caused by *Raffaelea lauricola* and disseminated by *Xyleborus glabratus*. *Plant Health Prog.* **2015**. [CrossRef]
55. Mitchell, C.E.; Tilman, D.; Groth, J.V. Effects of grassland plant species diversity, abundance, and composition on foliar fungal disease. *Ecology* **2002**, *83*, 1713–1726. [CrossRef]
56. Keesing, F.; Holt, R.D.; Ostfel, R.S. Effects of species diversity on disease risk. *Ecol. Lett.* **2006**, *9*, 485–498. [CrossRef] [PubMed]
57. Haas, S.E.; Cushman, J.H.; Dillon, W.W.; Rank, N.E.; Rizzo, D.M.; Meentemeyer, R.K. Effects of individual, community, and landscape drivers on the dynamics of a wildland forest epidemic. *Ecology* **2016**, *97*, 649–660. [CrossRef] [PubMed]
58. Lacroix, C.; Jolles, A.; Seabloom, E.W.; Power, A.G.; Mitchell, C.E.; Borer, E.T. Non-random biodiversity loss underlies predictable increases in viral disease, prevalence. *J. R. Soc. Interface* **2014**, *11*, 20130947. [CrossRef] [PubMed]
59. Burdon, J.J.; Chilvers, G.A. Host density as a factor in plant disease ecology. *Annu. Rev. Phytopathol.* **1982**, *20*, 143–166. [CrossRef]
60. Wolfe, N.D.; Dunavan, C.P.; Diamond, J. Origins of major human infectious diseases. *Nature* **2007**, *447*, 279–283. [CrossRef] [PubMed]
61. Parker, I.M.; Saunders, S.; Bontrager, M.; Weitz, A.P.; Hendricks, R.; Magarey, R.; Suiter, K.; Gilbert, G.S. Phylogenetic structure and host abundance drive disease pressure in communities. *Nature* **2015**, *520*, 542–544. [CrossRef] [PubMed]
62. Morin, R.S.; Liebhold, A.M.; Tobin, P.C.; Gottschalk, K.W.; Luzader, E. Spread of beech bark disease in the eastern United States and its relationship to regional forest composition. *Can. J. For. Res.* **2007**, *37*, 726–736. [CrossRef]
63. Kasson, M.T.; Livingston, W.H. Relationships among beech bark disease, climate, radial growth response and mortality of American beech in northern Maine, USA. *For. Pathol.* **2012**, *42*, 199–212. [CrossRef]
64. Hatala, J.A.; Dietze, M.C.; Crabtree, R.L.; Kendall, K.; Six, D.; Moorcroft, P.R. An ecosystem-scale model for the spread of a host-specific forest pathogen in the Greater Yellowstone Ecosystem. *Ecol. Appl.* **2011**, *21*, 1138–1153. [CrossRef] [PubMed]
65. Weste, G.; Brown, K.; Kennedy, J.; Walshe, T. *Phytophthora cinnamomi* infestation—A 24 year study of vegetation change in forests and woodlands of the Grampians, Western Victoria. *Aust. J. Bot.* **2002**, *50*, 247–274. [CrossRef]

66. Davidson, J.M.; Patterson, H.A.; Wickland, A.C.; Fichtner, E.J.; Rizzo, D.M. Forest type influences transmission of *Phytophthora ramorum* in California oak woodlands. *Phytopathology* **2001**, *101*, 492–501. [CrossRef] [PubMed]

67. Purse, B.V.; Graeser, P.; Searle, K.; Edwards, C.; Harris, C. Challenges in predicting invasive reservoir hosts of emerging pathogens: Mapping *Rhododendron ponticum* as a foliar host for *Phytophthora ramorum* and *Phytophthora kernoviae* in the UK. *Biol. Invasions* **2013**, *15*, 529–545. [CrossRef]

68. Kasson, M.T.; Livingston, W.H. Spatial distribution of *Neonectria* species associated with beech bark disease in northern Maine. *Mycologia* **2009**, *101*, 190–195. [CrossRef] [PubMed]

69. Park, A.W.; Gubbins, S.; Gilligan, C.A. Extinction times for closed epidemics: The effects of host spatial structure. *Ecol. Lett.* **2002**, *5*, 747–755. [CrossRef]

70. Holdenrieder, O.; Pautasso, M.; Weisberg, P.J.; Lonsdale, D. Tree diseases and landscape processes: The challenge of landscape pathology. *TRENDS Ecol. Evol.* **2004**, *19*, 446–452. [CrossRef] [PubMed]

71. Dwyer, G. On the spatial spread of insect viruses: Theory and experiment. *Ecology* **1992**, *73*, 479–494. [CrossRef]

72. Burdon, J.J.; Jarosz, A.M.; Kirby, G.C. Pattern and patchiness in plant-pathogen interactions: Causes and consequences. *Ann. Rev. Ecol. Syst.* **1989**, *20*, 119–136. [CrossRef]

73. Caraco, T.; Duryea, M.C.; Glavanakov, S.; Maniatty, W.; Szymanski, B.K. Host Spatial Heterogeneity and the Spread of Vector-Borne Infection. *Theor. Popul. Biol.* **2001**, *59*, 185–206. [CrossRef] [PubMed]

74. Broders, K.; Boraks, A.; Barbison, L.; Brown, J.; Boland, G.J. Recent insights into the pandemic disease butternut canker caused by the invasive pathogen *Ophiognomonia clavigignenti-juglandacearum*. *For. Pathol.* **2015**, *45*, 1–8.

75. Brar, S.; Tsui, C.K.M.; Dhillon, B.; Bergeron, M.-J.; Joly, D.L.; Zambino, P.J.; El-Kassaby, Y.A.; Hamelin, R.C. Colonization History, Host Distribution, Anthropogenic Influence and Landscape Features Shape Populations of White Pine Blister Rust, an Invasive Alien Tree Pathogen. *PLoS ONE* **2015**, *10*, e0127916. [CrossRef] [PubMed]

76. Papaïx, J.; Touzeau, S.; Monod, H.; Lannou, C. Can epidemic control be achieved by altering landscape connectivity in agricultural systems? *Ecol. Model.* **2014**, *284*, 35–47. [CrossRef]

77. Real, L.A.; Biek, R. Spatial dynamics and genetics of infectious diseases on heterogeneous landscapes. *J. R. Soc. Interface* **2007**, *4*, 935–948. [CrossRef] [PubMed]

78. Castello, J.D.; Leopold, D.J.; Smallidge, P.J. Pathogens patterns, and processes in forest ecosystems. *Bioscience* **1995**, *45*, 16–24. [CrossRef]

79. Ellis, A.M.; Váklavík, T.; Meentemeyer, R.K. When is connectivity important? A case study of the spatial pattern of sudden oak death. *Oikos* **2010**, *119*, 485–493. [CrossRef]

80. Garnas, J.R.; Auger-Rozenberg, M.-A.; Roques, A.; Bertelsmeier, C.; Wingfield, M.J.; Saccaggi, D.L.; Roy, H.E.; Slippers, B. Complex patterns of global spread in invasive insects: Eco-evolutionary and management consequences. *Biol. Invasions* **2016**, *18*, 935–952. [CrossRef]

81. Dutech, C.; Fabreguettes, O.; Capdevielle, X.; Robin, C. Multiple introductions of divergent genetic lineages in an invasive fungal pathogen, *Cryphonectria parasitica*, in France. *Heredity* **2010**, *105*, 220–228. [CrossRef] [PubMed]

82. Möykkynen, T.; Capretti, P.; Pukkala, T. Modelling the potential spread of *Fusarium circinatum*, the causal agent of pitch canker in Europe. *Ann. For. Sci.* **2015**, *72*, 169–181. [CrossRef]

83. Jules, E.S.; Carroll, A.L.; Garcia, A.M.; Steenbock, C.M.; Kauffman, M.J. Host heterogeneity influences the impact of a non-native disease invasion on populations of a foundation tree species. *Ecosphere* **2014**, *5*, 1–17. [CrossRef]

84. Jarosz, A.M.; Burdon, J.J. The effect of small scale environmental changes on disease incidence and severity in a natural plant pathogen interaction. *Oecologia* **1988**, *75*, 78–81. [CrossRef]

85. Smith, J.P.; Hoffman, J.T. Site and stand characteristics related to white pine blister rust in high-elevation forests of Southern Idaho and Western Wyoming. *West. N. Am. Nat.* **2001**, *61*, 409–416.

86. Marçais, B.; Desprez-Loustau, M.-L. European oak powdery mildew: Impact on trees, effects of environmental factors, and potential effects of climate change. *Ann. For. Sci.* **2014**, *71*, 633–642. [CrossRef]

87. Solla, A.; Martín, J.A.; Ouellette, G.B.; Gil, L. Influence of plant age on symptom development in *Ulmus minor* following inoculation by *Ophiostoma novo-ulmi*. *Plant Dis.* **2005**, *89*, 1035–1040. [CrossRef]

88. King, K.C.; Lively, C.M. Does genetic diversity limit disease spread in natural host populations? *Heredity* **2012**, *109*, 199–203. [CrossRef] [PubMed]
89. Burdon, R.D. Genetic diversity and disease resistance: Some considerations for research, breeding, and deployment. *Can. J. For. Res.* **2001**, *31*, 596–606. [CrossRef]
90. Garrett, K.A.; Mundt, C.C. Epidemiology in mixed populations. *Phytopathology* **1999**, *89*, 984–990. [CrossRef] [PubMed]
91. Pilet, F.; Chacon, G.; Forbes, G.A.; Adrivon, D. Protection of susceptible potato cultivars against late blight in mixtures increases with decreasing disease pressure. *Phytopathology* **2006**, *96*, 777–783. [CrossRef] [PubMed]
92. Zhu, Y.; Chen, H.; Fan, J.; Wang, Y.; Li, Y.; Chen, J.; Fan, J.X.; Yang, S.; Hu, L.; Leung, H.; et al. Genetic diversity and disease control in rice. *Nature* **2000**, *406*, 718–722. [CrossRef] [PubMed]
93. Barnes, I.; Wingfield, M.J.; Carbone, I.; Kirisits, T.; Wingfield, B.D. Population structure and diversity of an invasive pine needle pathogen reflects anthropogenic activity. *Ecol. Evol.* **2014**, *4*, 3642–3661. [CrossRef] [PubMed]
94. Andjic, V.; Dell, B.; Barber, P.; Hardy, G.; Wingfield, M.; Burgess, T. Plants for planting; indirect evidence for the movement of a serious forest pathogen, *Teratosphaeria destructans*, in Asia. *Eur. J. Plant Pathol.* **2011**, *131*, 49–58. [CrossRef]
95. Dwyer, G.; Elkinton, J.S.; Buonaccorsi, J.P. Host Heterogeneity in susceptibility and disease dynamics: Tests of a mathematical model. *Am. Nat.* **1997**, *150*, 685–707. [CrossRef] [PubMed]
96. Spielman, D.; Brook, B.W.; Briscoe, D.A.; Frankham, R. Does inbreeding and loss of genetic diversity decrease disease resistance. *Conserv. Genet.* **2004**, *5*, 439–448. [CrossRef]
97. Aegerter, B.J.; Gordon, T.R. Rates of pitch canker induced seedling mortality among *Pinus radiate* families varying in levels of genetic resistance to *Gibberella circinata* (anamorph *Fusarium circinatum*). *For. Ecol. Manag.* **2006**, *235*, 14–17. [CrossRef]
98. Sniezko, R.A. Resistance breeding against nonnative pathogens in forest trees—Current successes in North America. *Can. J. Plant Pathol.* **2006**, *28*, S270–S279. [CrossRef]
99. Ganz, H.H.; Ebert, D. Benefits of host genetic diversity for resistance to infection depend on parasite diversity. *Ecology* **2010**, *91*, 1263–1268. [CrossRef] [PubMed]
100. Freer-Smith, P.H.; Webber, J.F. Tree pests and diseases: The threat to biodiversity and the delivery of ecosystem services. *Biodivers. Conserv.* **2015**, 1–15. [CrossRef]
101. Heuch, J. What lessons need to be learnt from the outbreak of Ash Dieback Disease, *Chalara fraxinea* in the United Kingdom? *Arboric. J.* **2014**, *36*, 32–44. [CrossRef]
102. Mitchell, R.J.; Beaton, J.K.; Bellamy, P.E.; Broome, A.; Chetcuti, J.; Eaton, S.; Ellis, C.J.; Gimona, A.; Harmer, R.; Hester, A.J.; et al. Ash dieback in the UK: A review of the ecological and conservation implications and potential management options. *Biol. Conserv.* **2014**, *175*, 95–109. [CrossRef]
103. Gross, A.; Holdenrieder, O.; Pautasso, M.; Queloz, V.; Sieber, T.N. *Hymenoscyphus pseudoalbidus*, the causal agent of European ash dieback. *Mol. Plant Pathol.* **2014**, *15*, 5–21. [CrossRef] [PubMed]
104. Børja, I.; Timmermann, V.; Hietala, A.M.; Tollefsrud, M.M.; Nagy, N.E.; Vivian-Smith, A.; Cross, H.; Sønstebø, H.J.; Myking, T.; Solheim, H. Ash dieback in Norway—Current situation. In *Dieback of European Ash (Fraxinus spp.)—Consequences and Guidelines for Sustainable Management*; Vasaitis, R., Enderle, R., Eds.; SLU Service/Repro: Uppsala, Sweden, 2017; pp. 166–175.
105. Cleary, M.; Nguyen, D.; Stener, L.-G.; Stenlid, J.; Skovsgaard, J.-P. Ash and ash dieback in Sweden: A review of disease history, current status, pathogen and host dynamics, host tolerance and management options in forests and landscapes. In *Dieback of European Ash (Fraxinus spp.)—Consequences and Guidelines for Sustainable Management*; Vasaitis, R., Enderle, R., Eds.; SLU Service/Repro: Uppsala, Sweden, 2017; pp. 195–208.
106. Roane, M.K.; Griffin, G.J.; Elkins, J.R. *Chestnut Blight, Other Endothial Diseases, and the Genus Endothia*; American Phytopathological Society: St. Paul, MN, USA, 1986.
107. Braun, E.L. *Deciduous Forests of Eastern North America*; Hafner: New York, NY, USA, 1950.
108. Jacobs, D.F. Toward development of silvical strategies for forest restoration of American chestnut (*Castanea dentata*) using blight-resistant hybrids. *Biol. Conserv.* **2007**, *137*, 497–506. [CrossRef]
109. Graves, A.H. Relative blight resistance in species and hybrids of *Castanea*. *Phytopathology* **1950**, *40*, 1125–1131.
110. Dane, F.; Lang, P.; Huang, H.; Fu, Y. Intercontinental genetic divergence of *Castanea* species in eastern Asia and eastern North America. *Heredity* **2003**, *91*, 314–321. [CrossRef] [PubMed]

111. Dutech, C.; Barres, B.; Bridier, J.; Robin, C.; Milgroom, M.G.; Ravigne, V. The chestnut blight fungus world tour: Successive introduction events from diverse origins in an invasive plant fungal pathogen. *Mol. Ecol.* **2012**, *21*, 3931–3946. [CrossRef] [PubMed]

112. Heiniger, U.; Rigling, D. Biological control of chestnut blight in Europe. *Annu. Rev. Phytopathol.* **1994**, *32*, 581–599. [CrossRef]

113. Rigling, D.; Prospero, S. *Cryphonectria parasitica*, the causal agent of chestnut blight: Invasion history, population biology and disease control. *Mol. Plant Pathol.* **2017**. [CrossRef] [PubMed]

114. Milgroom, M.G.; Cortesi, P. Biological control of chestnut blight with hypovirulence: A critical analysis. *Annu. Rev. Phytopathol.* **2004**, *42*, 311–338. [CrossRef] [PubMed]

115. Chen, B.S.; Choi, G.H.; Nuss, D.L. Mitotic stability and nuclear inheritance of integrated viral cDNA in engineered hypovirulent strains of the chestnut blight fungus. *EMBO J.* **1993**, *12*, 2991–2998. [PubMed]

116. Zhang, D.X.; Nuss, D.L. Engineering super mycovirus donor strains of chestnut blight fungus by systematic disruption of multilocus *vic* genes. *Proc. Natl. Acad. Sci. USA* **2016**, *113*, 2062–2067. [CrossRef] [PubMed]

117. Diskin, M.; Steiner, K.C.; Hebard, F.V. Recovery of American chestnut characteristics following hybridization and backcross breeding to restore blight-ravaged *Castanea dentata*. *For. Ecol. Manag.* **2006**, *223*, 439–447. [CrossRef]

118. Newhouse, A.E.; Polin-McGuigan, L.D.; Baier, K.A.; Valletta, K.E.; Rottmann, W.H.; Tschaplinski, T.J.; Maynard, C.A.; Powell, W.A. Transgenic American chestnuts show enhanced blight resistance and transmit the trait to T1 progeny. *Plant Sci.* **2014**, *228*, 88–97. [CrossRef] [PubMed]

119. Hosoya, T.; Otani, Y.; Furuya, K. Materials for the fungus flora of Japan. *Trans. Mycol. Soc. Jpn.* **1993**, *34*, 429–432.

120. Drenkhan, R.; Hanson, M. New host species for *Chalara fraxinea*. *New Dis. Rep.* **2010**, *22*, 16. [CrossRef]

121. Cleary, M.R.; Arhipova, N.; Gaitnieks, T.; Stenlid, J.; Vasaitis, R. Natural infection of *Fraxinus excelsior* seeds by *Chalara fraxinea*. *For. Pathol.* **2013**, *43*, 83–85.

122. Luchi, N.; Montecchio, L.; Santini, A. Situation with ash in Italy: Stand characteristics, health condition, ongoing work and research needs. In *Interim Report from Chalara fraxinea, FRAXBACK Meeting in Vilnius, 13–14 November 2012*; Mainprize, N., Hendry, S., Weir, J., Eds.; Forestry Commission: Bristol, UK, 2012; pp. 25–26.

123. Zhao, Y.-J.; Hosoya, T.; Baral, H.-O.; Hosaka, K.; Kakishima, M. *Hymenoscyphus pseudoalbidus*, the correct name for *Lambertella albida* reported from Japan. *MycoTaxon* **2013**, *122*, 25–41. [CrossRef]

124. Chandelier, A.; Gerarts, F.; San Martin, G.; Herman, M.; Delahaye, L. Temporal evolution of collar lesions associated with ash dieback and the occurrence of *Armillaria* in Belgian forests. *For. Pathol.* **2016**. [CrossRef]

125. Chandelier, A.; Helson, M.; Dvorak, M.; Gischer, F. Detection and quantification of airborne inoculum of *Hymenoscyphus pseudoalbidus* using real-time PCR assays. *Plant Pathol.* **2014**, *63*, 1296–1305. [CrossRef]

126. Zheng, H.-D.; Zhuang, W.-Y. *Hymenoscyphus albidoides* sp. nov. and *H. pseudoalbidus* from China. *Mycol. Prog.* **2014**, *13*, 625–638. [CrossRef]

127. Han, J.-G.; Shrestha, B.; Hosoya, T.; Lee, K.-H.; Sung, G.-H.; Shin, H.-D. First report of the ash dieback pathogen *Hymensocyphus fraxineus* in Korea. *Mycobiology* **2014**, *42*, 391–396. [CrossRef] [PubMed]

128. Kirisits, T.; Schwanda, K. First definite report of natural infection of *Fraxinus ornus* by *Hymenoscyphus fraxineus*. *For. Pathol.* **2015**, *45*, 430–432.

129. Marçais, B.; Husson, C.; Godart, L.; Caël, O. Influence of site and stand factors on *Hymenoscyphus fraxineus*-induced basal lesion. *Plant Pathol.* **2016**. [CrossRef]

130. Park, A.W.; Gubbins, S.; Gilligan, C.A. Invasion persistence of plant parasites in a spatially structured host population. *Oikos* **2001**, *94*, 162–174. [CrossRef]

131. Jeger, M.J.; Pautasso, M.; Holdenrieder, O.; Shaw, M.K. Modelling disease spread and control in networks: Implications for plant sciences. *New Phytol.* **2007**, *114*, 279–297. [CrossRef] [PubMed]

132. Ferrari, J.R.; Preisser, E.L.; Fitzpatrick, M.C. Modeling the spread of invasive species using dynamic network models. *Biol. Invasions* **2014**, *16*, 949–960. [CrossRef]

133. Moslonka-Lefebvre, M.; Finley, A.; Dorigatti, I.; Dehnen-Schmutz, K.; Harwood, T.; Jeger, M.J.; Xu, X.; Holdenrieder, O.; Pautasso, M. Networks in plant epidemiology: From genes to landscapes, countries, and continents. *Phytopathology* **2011**, *101*, 392–403. [CrossRef] [PubMed]

134. Biek, R.; Real, L.A. The landscape genetics of infectious disease emergence and spread. *Mol. Ecol.* **2010**, *19*, 3515–3531. [CrossRef] [PubMed]

135. Desprez-Loustau, M.-L. The alien fungi of Europe. In *Handbook of Alien Species in Europe, DAISIE. Invading Nature, Series in Invasion Ecology*; Drake, J.A., Ed.; Springer: Berlin, Germany, 2009; Volume 3, pp. 15–28.
136. Horie, T.; Haight, R.G.; Homans, F.R.; Venette, R.C. Optimal strategies for the surveillance and control of forest pathogens: A case study with oak wilt. *Ecol. Econ.* **2013**, *86*, 78–85. [CrossRef]
137. Liebhold, A.M.; Tobin, P.C. Exploiting the Achilles Heels of Pest Invasions: Allee Effects, Stratified Dispersal and Management of Forest Insect Establishment and Spread. *N. Zeal. J. For. Sci.* **2010**, *40*, S25–S33.
138. Homans, F.; Horie, T. Optimal detection strategies for an established invasive pest. *Ecol. Econ.* **2011**, *70*, 1129–1138. [CrossRef]
139. Hänfling, B.; Kollmann, J. An evolutionary perspective of biological invasions. *TRENDS Ecol. Evol.* **2002**, *17*, 545–546. [CrossRef]
140. Lee, C.E. Evolutionary genetics of invasive species. *TRENDS Ecol. Evol.* **2002**, *17*, 386–391. [CrossRef]
141. Woolhouse, M.E.J.; Haydon, D.T.; Antia, R. Emerging pathogens: The epidemiology and evolution of species jumps *TRENDS Ecol. Evol.* **2005**, *20*, 238–244.
142. Desprez-Loustau, M.-L.; Courtecuisse, R.; Robin, C.; Husson, C.; Moreau, P.-A.; Blancard, D.; Selosse, M.-A.; Lung-Escarmant, B.; Piou, D.; Sache, I. Species diversity and drivers of spread of alien fungi (*sensu lato*) in Europe with a particular focus on France. *Biol. Invasions* **2010**, *12*, 157–172. [CrossRef]

forests

MDPI

Article

Pre-Infection Stages of *Austropuccinia psidii* in the Epidermis of *Eucalyptus* Hybrid Leaves with Different Resistance Levels

Renata Ruiz Silva [1], André Costa da Silva [2], Roberto Antônio Rodella [3], José Eduardo Serrão [4], José Cola Zanuncio [5] and Edson Luiz Furtado [1,*]

[1] Departamento de Produção Vegetal, Faculdade de Ciências Agronômicas–UNESP, Botucatu 18610-307, São Paulo, Brasil; renatatibs@gmail.com
[2] Departamento de Fitopatologia, Universidade Federal de Viçosa, Viçosa 36570-900, Minas Gerais, Brasil; andrec_agro@yahoo.com.br
[3] Departamento de Botânica, Instituto de Biociências de Botucatu–UNESP, Botucatu 18618-000, São Paulo, Brasil; rodella@ibb.unesp.br
[4] Departamento de Biologia Geral, Universidade Federal de Viçosa, Viçosa 36570-900, Minas Gerais, Brasil; jeserrao@ufv.br
[5] Departamento de Entomologia/BIOAGRO, Universidade Federal de Viçosa, Viçosa 36570-900, Minas Gerais, Brasil; zanuncio@ufv.br
* Correspondence: elfurtado@fca.unesp.br; Tel.: +55-31-3899-2924

Academic Editors: Matteo Garbelotto and Paolo Gonthier
Received: 25 July 2017; Accepted: 20 September 2017; Published: 17 October 2017

Abstract: Rust is a major *Eucalyptus* spp. disease, which is especially damaging for early-stage plants. The aim of this study was to verify the pre-infection process of *Austropuccinia psidii* (*A. psidii*) in the leaves of three phenological stages of *Eucalyptus* clones with different resistance levels. Plants from the hybrids of *Eucalyptus urophylla* × *Eucalyptus grandis* (*E. grandis*) with variable levels of resistance to this disease were used. The pathogen was inoculated in vitro on abaxial leaf discs of first, third, and fifth leaf stages and maintained under conditions suitable for disease development. Subsequently, samples from these discs were collected 24 and 120 h after inoculation and processed using scanning electron microscopy analysis. No symptoms were seen in any leaf stage of the resistant clone. Additionally, a low incidence of *A. psidii* germination (1.3–2%) and appressoria (0–0.5%) in three leaf stages was observed. However, the first leaf stage of the susceptible clone presented germination of large numbers of urediniospores (65%) with appressoria (55%) and degradation of the cuticle and wax. From the third stage, the percentage of germinated urediniospores (<15%) and appressoria (<2%) formation of this clone decreased. Protrusions on the leaf surface, associated with the pathogen, were observed on the first and third leaf stages of the resistant clone and on the fifth stage of the susceptible clone, suggesting a possible defensive plant reaction.

Keywords: eucalypt rust; pathogenesis; phenological stage leaves; pre-infection stage; resistance; leaf discs; scanning electron microscopy; *Puccinia psidii*

---

1. Introduction

The expansion of eucalypt plantations to new areas around the world is confronting abiotic and biotic diseases that can limit certain genetic materials of this plant [1]. Rust, caused by *Austropuccinia psidii* (*A. psidii*) (G. Winter) Beenken comb. nov., is one of the most notable diseases of *Eucalyptus* spp., and is mainly associated with its immature tissues [1–3]. This fungus infects both native and exotic Myrtaceae [4–6].

Rust is found mainly on early stage plants [1,7]. Most damage is found in nurseries, but it can also attack plants in the field. *Eucalyptus* spp. are more susceptible to *A. psidii* infection up to two years of age, due to the higher number of shoots and tender tissues, which is ideal for the establishment of this pathogen [7–9].

The initial infection events of fungi that cause rust include adherence to the cuticle and direct germ tube growth on host plant surface [10]. The *A. psidii* urediniospores often directly penetrate through the host cuticle and epidermis, between the anticlinal walls of the epidermal cells, by forming appressoria [11]. The process of intercellular colonization by the pathogen starts after its penetration through intracellular haustoria that draw nutrients from the host cells [1,7,12].

Resistant genetic plant materials are the most widely used control method to manage rust, but many cultivated eucalypt clones are susceptible to this disease. As with other rust species [13], the pathogenesis of this fungus depends on its thigmotropic response for host recognition [14]. Young leaves and shoots of susceptible eucalypt plants are infected by *A. psidii*, while mature or older leaves are resistant [1,7]. The resistance to *A. psidii* infection of older *Eucalyptus* leaves occurs at the pre-penetration stage and is due, amongst other inherent factors, to their higher wax quantity [15]. However, the resistance of young leaves in resistant plants needs to be investigated. The study of *A. psidii* initial infection processes on leaf surfaces of the clones with different resistance levels is important to understand the behavior of these eucalypt clones.

Thus, the objective of this study was to evaluate the pre-infection process of *A. psidii* in leaves at different phenological stages of eucalypt clones with different resistance levels.

2. Materials and Methods

Seedlings of rust-susceptible and resistant *Eucalyptus urophylla* S. T. Blake × *Eucalyptus grandis* (*E. grandis*) W. Hill ex Maiden "urograndis" were grown in 2 L capacity pots, containing Carolina Soil® (Carolina Soil do Brasil, Santa Cruz do Sul, RS, Brazil) substrate enriched with simple superphosphate (6 kg/m^3) and Osmocote® (Tecnutri do Brasil, Tietê, SP, Brazil) (NPK 19:06:10) and kept in a greenhouse at 20–30 °C for three months. These clones were selected based on their response to *A. psidii*, and their seedlings were supplied by Votorantim Celulose e Papel S.A, located in Jacareí, São Paulo State, Brazil. A total of 50 discs per leaf stage with a 1.5 cm diameter were cut with a punch from the first, third, and fifth leaf pairs from branches of six plants of each clone. These discs were placed on water-saturated foam in plastic trays (Figure 1A). An isolate of *A. psidii* (FCA-PP2303), obtained from the mycological collection of the Laboratory of Forest Pathology at the Universidade Estadual Paulista (UNESP-Botucatu), was inoculated on *Syzygium jambos* L. (Alston) and after sporulation of the fungus, the urediniospores were collected to prepare the inoculum [16]. An inoculum suspension in water containing 1% Tween 20 was prepared at a concentration of 9×10^4 spores/mL [17]. The concentration of spore suspension was determined with a haemocytometer. Five microliters of the urediniospores suspension were inoculated on the central part of the abaxial surface of leaf discs for easy identification of the inoculum deposition site for scanning electron microscopy (SEM) analysis. The trays with the inoculated materials were covered with a glass lid to reduce moisture loss and placed in an incubation chamber at 20 ± 1 °C in continuous darkness for the first 24 h, followed by a 12 h photoperiod at approximately 10 µM photons/s/m^2, which is ideal for rust development, according to previous studies [15,16]. Samples were taken from the first leaf stage of the susceptible clone and put in a humid chamber for 12 days in order to confirm that the inoculation method was able to cause the disease (Figure 1B). The experiment was set up using a completely randomized design, with four replications. The assays were repeated once.

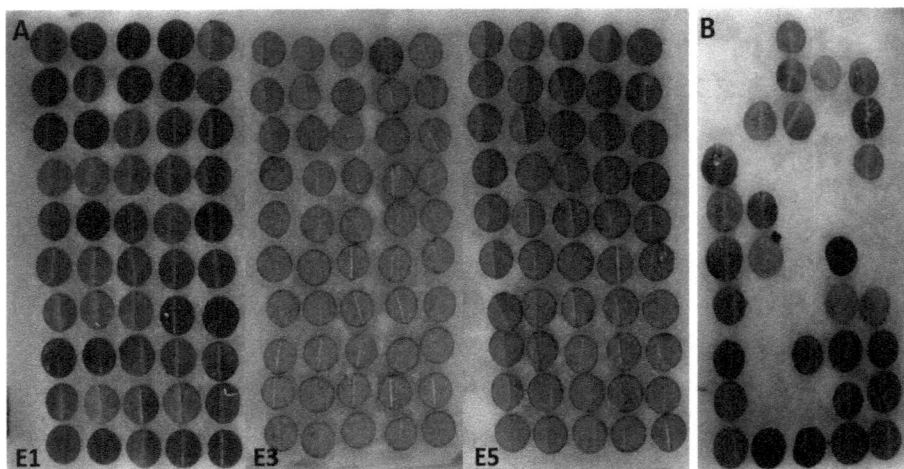

Figure 1. Leaf discs of *Eucalyptus urophylla* × *Eucalyptus grandis* (*E. grandis*) "urograndis" in the first (E1), third (E3), and fifth (E5) leaf development stages after inoculation with *Austropuccinia psidii* (**A**); leaf discs of the first leaf stage of the susceptible clone with rust symptoms, proving the efficiency of this inoculation method (**B**).

Leaf samples (5 mm^2) were collected at 24 and 120 h after pathogen inoculation for SEM. The preparation and analysis of the samples followed the previously described methodology [18] with some adaptations. Leaf samples were fixed in 2.5% glutaraldehyde in 0.1 M phosphate buffer (pH 7.3) for 24 h. Fragments used per treatment were transferred to 0.05 M cacodylate buffer and washed three times for 10 min. These fragments were placed into a 1% solution of osmium tetroxide in 0.1 M phosphate buffer (pH 7.3) for 1 h, washed with distilled water three times, and dehydrated in an acetone series (25%, 50%, 75%, 90%, and 100% (3×) for 10 min each stage). After dehydration, samples were taken at the critical point apparatus (Balzers CPD 030) to replace the acetone by CO_2 and complete drying. Specimens were mounted on aluminum stubs and covered with gold (SCD 050, Balzers) for observation under a scanning electron microscope (Philips SEM 515). The images generated at 20 kV and 9 mm working distance were digitally recorded. Percentage of urediniospores germination and appressoria formation were quantified through SEM images in leaf samples collected 24 and 120 h after inoculation. A total of 50 urediniospores were observed. The experiment was distributed as a double factorial in split plot in time. The factorial was constituted by the combination of two clones differing in resistance level × three leaf stages. The data were subjected to analysis of variance (ANOVA). When the results were significant, the average values were separated using a Tukey test at 5% probability. An AIC (Akaike Information Criteria) analysis using generalized linear models via likelihood was performed [19] because discrete counting data was used. The generalized model of normal distribution was the best fit for germination and appressoria formation data in comparison to a model of Poisson distribution. The statistical analyses were carried out using the R program.

3. Results

ANOVA showed that only the clone × leaf stage interaction was a significant source of variation (Table 1). There was no difference in the percentage of urediniospores germination and appressoria formation at 24 and 120 h after inoculation of the pathogen. A higher urediniospore germination and appressoria formation of *A. psidii* was observed in the first leaf stage of the susceptible clone (Table 2 and Figure 2). Dehydration of urediniospores and formation of germ tubes and appressoria were occasionally observed in the resistant clone at 24 h after its inoculation (Figure 2A,B). On the other hand, urediniospores germinated and formed intact germ tubes of various sizes, producing appressoria, beginning the infectious process in the susceptible clone (Figure 2C,D). In the third leaf stage, the incidence of germinated urediniospores was lower on the leaf surface of the resistant clone and appressorium formation was rarely observed (Table 2, Figure 3A,B). A low incidence of urediniospore germination with appressoria on the leaf of the susceptible clone in relation to the first leaf stage was observed (Table 2, Figure 3C,D). In the fifth leaf stage, the number of germinated urediniospores was lower in the resistant clone and no appressorium formation was observed (Table 2, Figure 4A,B). However, the susceptible clone had urediniospores with long germ tubes without appressoria (Table 2 and Figure 4C). Skin cells formed protrusions on its leaf surface (Figure 4D). The third and fifth leaf stages had higher wax quantity on the leaf surface compared to that of the first stage.

Table 1. Summary of analysis of variance.

Sources of Variation	Degrees of Freedom	Germination F_{calc}	Appressorium F_{calc}
Clone	1	554.2391 ***	336.3952 ***
Leaf stage	2	190.0900 ***	315.4363 ***
Clone × leaf stage	2	181.7854 ***	304.0159 ***
Error A	18		
Hours after inoculation	1	4.3235 [ns]	2.5946 [ns]
Clone × hours after inoculation	1	1.4118 [ns]	0.6486 [ns]
Leaf stage × hours after inoculation	2	0.0221 [ns]	0.7703 [ns]
Clone × leaf stage × hours after inoculation	2	0.2868 [ns]	0.2838 [ns]
Error B	18		

[ns] not significant, *** significant at 0.0001% probability by Tukey test.

Table 2. Percentage of *Austropuccinia psidii* urediniospore germination and germinated + appressorium formation on the first, third, and fifth leaf stages of susceptible and resistant hybrid *Eucalyptus urophylla* × *E. grandis* clones.

Leaf Stage	Germination (%)	
	Resistant Clone	Susceptible Clone
1st	2 [bA]	65.6 [aA]
3rd	1.6 [bA]	15.4 [aB]
5th	1.3 [bA]	12.9 [aB]
Appressorium (%)		
Leaf stage		
1st	0.5 [bA]	55.1 [aA]
3rd	0.1 [aA]	1.9 [aB]
5th	0 [aA]	0 [aB]

Means followed by the same lowercase in horizontal and capital letters in vertical for the variables germination and appressorium do not differ by the Tukey test, $p < 0.0001$.

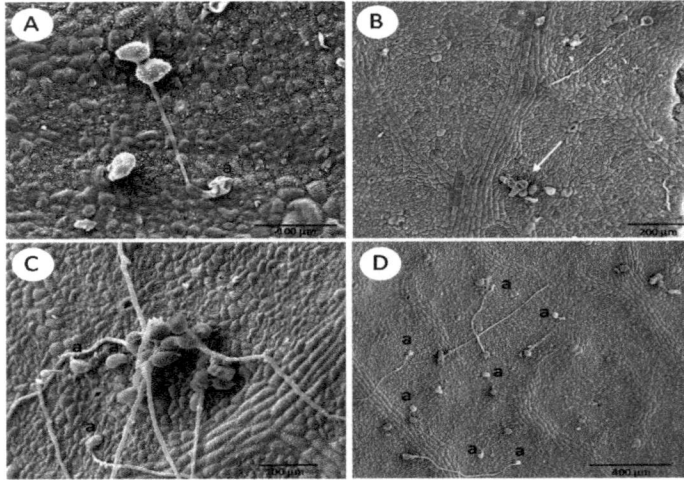

Figure 2. Abaxial surface of first leaf development stage of hybrid *Eucalyptus urophylla* × *E. grandis* "urograndis" clones 24 h after *Austropuccinia psidii* inoculation. Resistant clone with germ tubes with appressoria (a) in a dehydration state (**A**). Resistant clone presenting viable urediniospores (arrow) (**B**). Susceptible clone with various germinated urediniospores and intact appressoria (a) (**C,D**).

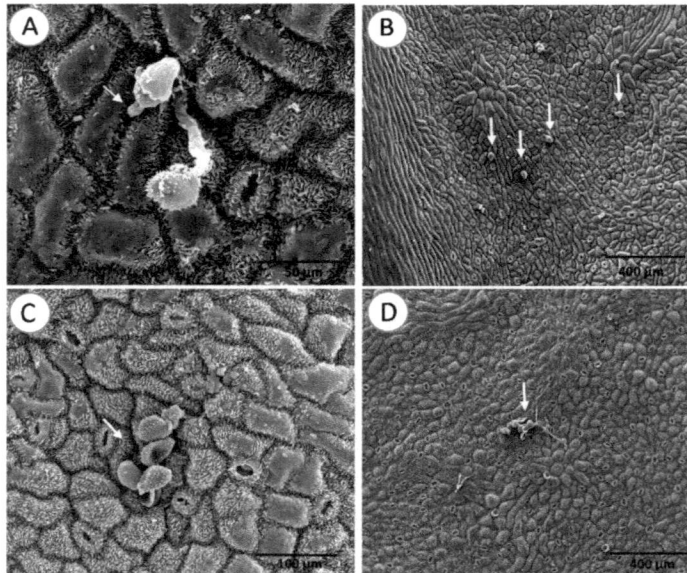

Figure 3. Abaxial leaf surface of third development stage of hybrid *Eucalyptus urophylla* × *E. grandis* "urograndis" clones 24 h after *Austropuccinia psidii* inoculation. Resistant clone with germinated urediniospores showing infrequent formation of appressoria (arrow) (**A**). Resistant clone without germination of some urediniospores (arrow) (**B**). Susceptible clone with urediniospores germinated and formation of germ tubes and appressoria (arrow) (**C,D**).

124

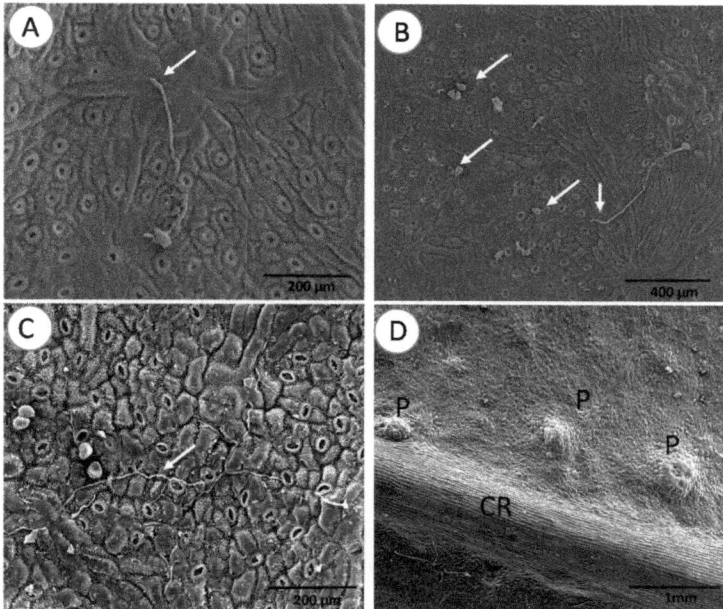

Figure 4. Abaxial surface of fifth leaf development stage of hybrid *Eucalyptus urophylla* × *E. grandis* "urograndis" clones 24 h after *Austropuccinia psidii* inoculation. Resistant clone without urediniospores germinated or germinated devoid of appressoria (arrows) (**A**,**B**). Susceptible clone showing extensive germ tube formation on the leaf surface but without penetration (**C**). Susceptible clone with protuberance formation (P) on the inoculum deposition site near the central rib (CR) (**D**).

SEM images showed the presence of protrusions on the resistant clone leaves involving urediniospores as a possible plant reaction with dehydration of appressoria and urediniospores (Figure 5A,B). These reactions were observed in all leaf stages of the resistant clone, 120 h after inoculation, and in the fifth stage of the susceptible clone, 24 h after inoculation. Cuticle and wax degradation, indicating probable onset of *A. psidii* colonization in the host, were observed in the leaves of the first leaf stage of the susceptible clone (Figure 5C,D). The resistant clone in this stage did not exhibit symptoms on the leaf surface and presented a lower incidence of appressoria (Figure 5A,B). Leaves of the third leaf stage of the resistant clone had a low quantity of germinated urediniospore and protrusions involving urediniospores (Figure 6A,B). This was characterized by protrusions on the leaf surface, similar to those observed during the first leaf stage, with low urediniospores germination. However, the susceptible clone showed extensive germ tube formation and damage by the pathogen (Figure 6C,D). The fifth leaf stage of the resistant clone had a lower quantity of urediniospore germinated, without appressoria and urediniospore wilting (Figure 7A,B). In the susceptible clone, the pathogen presented long germ tubes without appressoria and penetration in the leaves (Figure 7C,D).

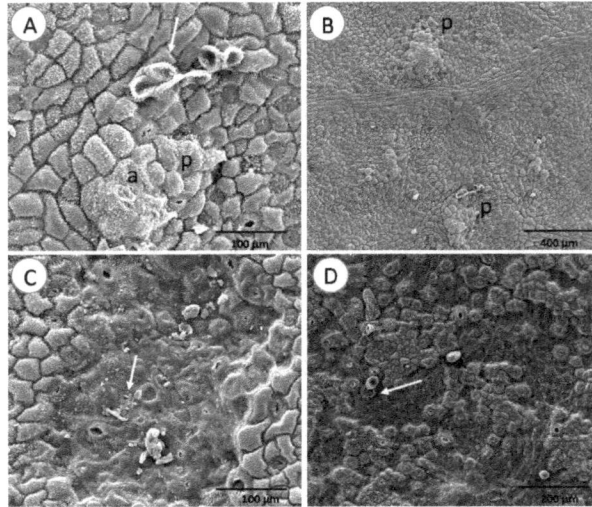

Figure 5. Abaxial surface of first leaf development stage of hybrid *Eucalyptus urophylla* × *E. grandis* "urograndis" clones 120 h after *Austropuccinia psidii* inoculation. Resistant clone with protuberances (P) on the surface of the leaf, dehydration of appressoria (a) and of spores (arrow) (**A,B**). Susceptible clone with cuticle and wax degradation (arrows) (**C,D**).

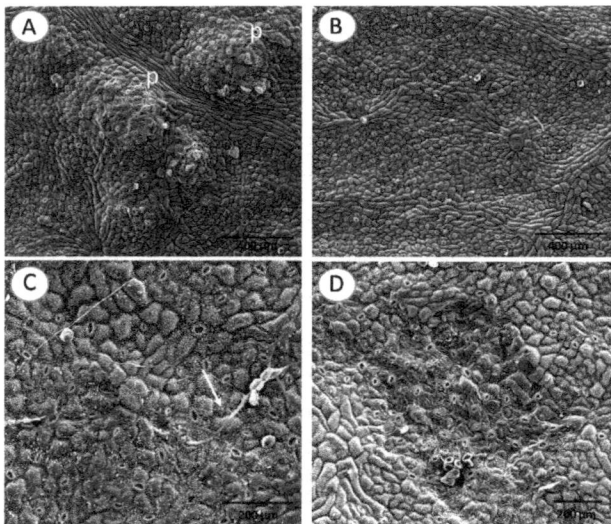

Figure 6. Abaxial surface of third leaf development stage of hybrid *Eucalyptus urophylla* × *E. grandis* "urograndis" clones 120 h after *Austropuccinia psidii* inoculation. Resistant clone with protuberances (P) on the leaf surface and low number of urediniospores germinated (**A,B**). Susceptible clone with extensive germ tube formation (arrow) and damage by the pathogen (arrow) (**C,D**).

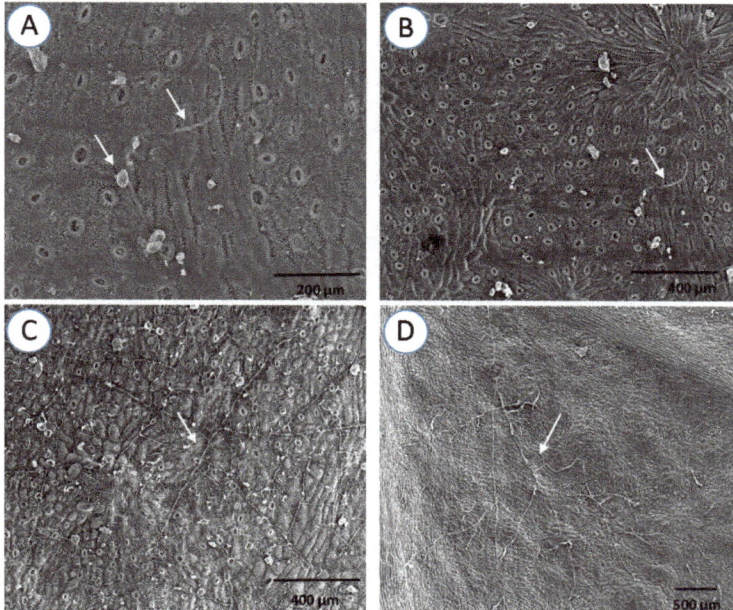

Figure 7. Abaxial surface of fifth leaf development stage of hybrid *Eucalyptus urophylla* × *E. grandis* "urograndis" clones 120 h after *Austropuccinia psidii* inoculation. Resistant clone with shriveled spores (arrow) without appressorium (arrow) (**A,B**). Susceptible clone with extensive germ tube formation without the pathogen penetration (arrows) (**C,D**).

4. Discussion

In the first leaf stage, the susceptible clone showed a higher degree of urediniospore germination with appressoria and degradation of the cuticle and wax. From the third leaf stage, the inhibition of the pre-infection fungus process in leaves of the susceptible *Eucalyptus urophylla* × *E. grandis* clone showed that there is some plant defense response related to leaf age. In the resistant clone, these mechanisms occurred from the first leaf stage. In the fifth leaf stage, the susceptible clone showed superficial growth, but without penetration of the pathogen, while the resistant clone had many inactive urediniospores. This resembles the pattern observed on *Eucalyptus grandis* susceptible to *A. psidii* [15]. The germination, appressoria formation, and penetration of this fungus gradually decreased from the first to the fourth leaf stage, but without penetrating in the fifth leaf stage [15].

In the first leaf stage of the resistant clone, dehydration of urediniospores and formation of germ tubes and appressoria were occasionally observed, as reported for the *Phakopsora pachyrhizi* (*P. pachyrhizi*) Syd. & P. Syd. urediniospores in soybean [20]. The susceptible clone already had many urediniospores germinated with appressorium followed by cuticle and wax degradation. Different to other *Puccinia* species, in that penetration occurs via stomata [21–23], the penetration of *A. psidii* occurs between the anticlinal walls of the leaf epidermis directly into the mesophyll of the leaf after appressoria formation [24]. Once formed, appressoria adhere tightly to the leaf surface and secrete extracellular enzymes, or generate physical force, or use a combination of both factors to bring about cuticle penetration [25]. The leaf surface is the first line of defense against plant invaders, where the adhesion of the pathogen occurs, followed by penetration and infection [26]. In the interaction *Puccinia recondita* f. sp. *tritici*–wheat, the resistance is not related to the pre-infectious processes at the leaf surface level [23], different to what was observed in our research.

In the fifth leaf stage of the susceptible clone, the urediniospores with long germ tubes without appressoria, and some with short germ tubes with appressoria, are similar to *P. pachyrhizi* urediniospores on soybean [20]. Germ tube extension and differentiation can occur in response to signals including surface hardness, hydrophobicity, plant signals, and surface topography [27]. The short germ tubes decrease the amount of endogenous energy required for growth, which thus can be used to penetrate the cells [24]. These results demonstrate that the fungus germinates on the surface of leaves of plants in the fifth leaf stage, mainly on those of the susceptible clone, but without penetrating or colonizing its tissues.

The presence of epicuticular waxes on leaves of the two clones, especially in the third and fifth leaf stages, agrees with that reported for some eucalypt species as an important factor for resistance to *A. psidii* infection [15]. Appressorium formation in *P. pachyrhizi* and *Phakopsora apoda* (Har. & Pat.) Mains occurred in a place with lower wax deposition on the leaf surface [20,28]. Variability in the quantity of wax on the surface may also modify fungal behavior and interfere with the infection process [29]. Recognition of the cuticle surface by the fungus, necessary for appressorium differentiation, depends on the wax distribution pattern on leaves and also on the pathogen capacity to degrade it [30]. Thus, the increase in the quantity of wax on the leaf surface may have interfered with appressorium development, as seen in the pathosystem *Hordeum chilense* Roem. et Schult.- *Puccinia hordei* G.H. Otth [31]. Surface contact is essential for appressorium induction [32] and germination and germ tube growth requires fewer stimuli than appressorium formation [33].

In the fifth leaf stage, the presence of long germ tubes and absence of appressoria and penetration by the pathogen in the susceptible clone is similar to that reported for *Alternaria solani* Sorauer on the surface of *Solanum lycopersicum* L. leaves resistant and susceptible to this pathogen [34]. The higher chitinase and peroxidase activity on older eucalypt leaves [26] can be one of the reasons for the fungal wilt structures because chitinases increase plant resistance to pathogens by catalyzing chitin polymer hydrolysis, the main components of fungal cell walls [35]. Peroxidases are also involved in numerous cellular processes including the final suberin and lignin biosynthesis steps [36] and the metabolism of phenylpropanoids [37]. In addition, peroxidases are involved in the oxidation of phenols and compounds toxic to pathogenic organisms [38] and the formation of papillae that can block fungal entry [39].

Protrusions associated with the pathogen and observed on the leaf surface of first and third leaf stages of the resistant clone and in the fifth leaf stage of the susceptible clone suggest a possible defensive plant reaction. This was also shown by the protrusions involving urediniospores and dehydration of appressoria and urediniospores. This may be due to chemical compounds produced by leaves, such as essential oils, and extracted from eucalypt leaves with high antimicrobial activity [40–43].

The inoculation method using detached eucalypt leaf discs was effective for SEM studies and for evaluating the resistance of the plant to rust, with the onset of symptoms in the first leaf stage of the susceptible eucalypt clone. The efficiency of this method had also been reported for bean [44–46] and soybean [47].

The results obtained in this research will be of great importance for the international forestry industry. From the results observed in this study, we recommend the use of the genetic materials with a lower leaf maturation period and with higher wax content in the younger leaves to be tested in the breeding programs for the control of this disease. Future studies could be conducted to analyze the anatomy and chemical composition of leaves of different phenological stages of the resistant and susceptible plants, before and after the infectious process of *A. psidii* in order to identify other possible resistance mechanisms of eucalypt plants to this disease.

5. Conclusions

Resistance and susceptibility of eucalypt clones occur in the pre-infection process. The failure of the pathogen to recognize an infection site on the leaves of the resistant clone and on old leaves of the susceptible clone and initiate appressorium formation appears to be the key factor explaining such resistance. The urediniospores germination with appressoria and degradation of the cuticle waxes had higher values in the first leaf stages of the susceptible *Eucalyptus urophylla* × *E. grandis* clone. From the third leaf stage in the susceptible clone, the germination and appressorium formation of the fungus was prevented by the defense mechanisms, while this took place in the first leaf stage of the resistant clone. Protrusions on the leaf surface were associated with the pathogen in the first and third leaf stages of the resistant clone, at 120 h after inoculation, and in the fifth leaf stage of the susceptible clone, 24 h after inoculation, suggesting a possible plant defense reaction. The results presented in this work help to explain the resistance of old and young eucalypt leaves to *A. psidii*.

Acknowledgments: The authors wish to acknowledge researcher Donizete Dias Costa of Votorantim Celulose e Papel, "Conselho Nacional de Desenvolvimento Científico e Tecnológico (CNPq)", "Coordenação de Aperfeiçoamento de Pessoal de Nível Superior (CAPES)" and "Fundação de Amparo à Pesquisa do Estado de Minas Gerais (FAPEMIG)" and "Programa Cooperativo sobre Proteção Florestal-Instituto de Pesquisas e Estudos Florestais" (PROTEF-IPEF) for financial support. Global Edico Services and Phillip John Villani (The University of Melbourne, Australia) revised and corrected the English language used in this manuscript.

Author Contributions: R.R.S. conceived and designed the study, performed the sampling, processed and analyzed the data, and wrote the paper; A.C.S. discussed results and wrote the paper; R.A.R. conceived and designed the study, and reviewed drafts of the paper; J.E.S. reviewed drafts and wrote of the paper; J.C.Z. reviewed drafts and wrote of the paper; E.L.F. conceived and designed the study, reviewed drafts of the paper, and contributed reagents/materials/analysis tools.

Conflicts of Interest: The authors declare no conflict of interest. The funding sponsors had no role in the design of the study; in the collection, analyses, or interpretation of data; in the writing of the manuscript, and in the decision to publish the results.

References

1. Alfenas, A.C.; Zauza, E.A.V.; Mafia, R.G.; Assis, T.F. *Clonagem e Doenças do Eucalipto*, 2nd ed.; Editora UFV: Viçosa, Brazil, 2009.
2. Miranda, A.C.; Moraes, M.L.T.; Tambarussi, E.V.; Furtado, E.L.; Mori, E.S.; Silva, P.H.M.; Sebbenn, A.M. Heritability for resistance to *Puccinia psidii* Winter rust in *Eucalyptus grandis* Hill ex Maiden in Southwestern Brazil. *Tree Genet. Genomes* **2013**, *9*, 321–329. [CrossRef]
3. Beenken, L. *Austropuccinia*: A new genus name for the myrtle rust *Puccinia psidii* placed within the redefined family Sphaerophragmiaceae (Pucciniales). *Phytotaxa* **2017**, *297*, 53–61. [CrossRef]
4. Figueiredo, M.B. Doenças fúngicas emergentes em grandes culturas. *Biológico* **2001**, *63*, 29–32.
5. Morin, L.; Aveyard, R.; Lidbetter, J.R.; Wilson, P.G. Investigating the host-range of the rust fungus *Puccinia psidii* sensu lato across tribes of the family Myrtaceae present in Australia. *PLoS ONE* **2012**, *7*, e35434. [CrossRef] [PubMed]
6. Yamaoka, Y. Recent outbreaks of rust diseases and the importance of basic biological research for controlling rusts. *J. Gen. Plant Pathol.* **2014**, *80*, 375–388. [CrossRef]
7. Ferreira, F.A. *Patologia Florestal–Principais Doenças Florestais do Brasil*; Sociedade de Investigações Florestais: Viçosa, Brazil, 1989.
8. Demuner, N.L.; Alfenas, A.C. Fungicidas sistêmicos para o controle da ferrugem causada por *Puccinia psidii* em *Eucalyptus cloeziana*. *Fitopatol. Bras.* **1991**, *16*, 174–177.
9. Coutinho, T.A.; Wingfield, M.J.; Alfenas, A.C.; Crous, P.W. *Eucalyptus* rust: A disease with the potential for serious international implications. *Plant Dis.* **1998**, *82*, 819–825. [CrossRef]
10. Mendgen, K.; Hahn, M.; Deising, H. Morphogenesis and mechanisms of penetration by plant pathogenic fungi. *Annu. Rev. Phytopathol.* **1996**, *34*, 367–386. [CrossRef] [PubMed]
11. Glen, M.; Alfenas, A.C.; Zauza, E.A.V.; Wingfield, M.J.; Mohammed, C. *Puccinia psidii*: A threat to the Australian environment and economy—A review. *Australas. Plant Pathol.* **2007**, *36*, 1–16. [CrossRef]
12. Bedendo, I.P. Ferrugens. In *Manual de Fitopatologia: Princípios e Conceitos*; Bergamin Filho, A., Kimati, H., Amorim, L., Eds.; Agronômica Ceres: São Paulo, Brazil, 1995; pp. 872–880.
13. Terhune, B.T.; Hoch, H. Substrate hydrophobicity and adhesion of *Uromyces* urediospores and germlings. *Exp. Mycol.* **1993**, *17*, 241–252. [CrossRef]

14. Bailey, J.A.; O'Connell, R.J.; Pring, R.J.; Nash, C. Infection strategies of *Colletotrichum* species. In *Colletotrichum: Biology, Pathology and Control*; Bailey, J.A., Jeger, M.J., Eds.; CAB International: Wallingford, UK, 1992; pp. 88–120.
15. Xavier, A.A.; da Silva, A.C.; Guimarães, L.M.S.; Matsuoka, K.; Hodges, C.S.; Alfenas, A.C. Infection process of *Puccinia psidii* in *Eucalyptus grandis* leaves of different ages. *Trop. Plant Pathol.* **2015**, *40*, 318–325. [CrossRef]
16. Ruiz, R.A.R.; Alfenas, A.C.; Ferreira, F.A. Influência da temperatura, do tempo de molhamento foliar, fotoperíodo e intensidade de luz sobre a infecção de *Puccinia psidii* em eucalipto. *Fitopatol. Bras.* **1989**, *14*, 55–61.
17. Pinto, C.S.; Costa, R.M.L.; Moraes, C.B.; Pieri, C.; Tambarussi, E.V.; Furtado, E.L.; Mori, E.S. Genetic variability in progenies of *Eucalyptus dunnii* Maiden for resistance to *Puccinia psidii*. *Crop Breed. Appl. Biotechnol.* **2014**, *14*, 187–193. [CrossRef]
18. Medice, R.; Alves, E.; Assis, R.T.; Magno, R.G., Jr.; Lopes, E.A.G.L. Óleos essenciais no controle da ferrugem asiática da soja *Phakopsora pachyrhizi* Syd. & P. Syd. *Cienc. Agrotec.* **2007**, *31*, 83–90.
19. Akaike, H. A new look at the statistical model identification. *IEEE Trans. Automat. Control* **1974**, *19*, 716–723. [CrossRef]
20. Magnani, E.B.Z.; Alves, E.; Araújo, D.V. Eventos dos processos de pré-penetração, penetração e colonização de *Phakopsora pachyrhizi* em folíolos de soja. *Fitopatol. Bras.* **2007**, *32*, 156–160. [CrossRef]
21. Hughes, F.L. Scanning electron microscopy of early infection in the uredial stage *of Puccinia sorghi* in *Zea mays*. *Plant Pathol.* **1985**, *34*, 61–68. [CrossRef]
22. Lennox, C.L.; Rijkenberg, F.H.J. Scanning electron microscopy study of infection structure formation of *Puccinia graminis* f. sp. *tritici* in host and non-host cereal species. *Plant Pathol.* **1989**, *38*, 547–556.
23. Hu, G.; Rijkenberg, F.H.J. Scanning electron microscopy of early infection structure formation by *Puccinia recondita* f. sp. *tritici* on and in susceptible and resistant wheat lines. *Mycol. Res.* **1998**, *102*, 391–399.
24. Hunt, P. Cuticular penetration by germinating uredospores. *Trans. Br. Mycol. Soc.* **1983**, *51*, 103–112. [CrossRef]
25. Pryce-Jones, E.; Carver, T.; Gurr, S.J. The roles of cellulase enzymes and mechanical force in host penetration by *Erysiphe graminis* f. sp. *hordei*. *Physiol. Mol. Plant Pathol.* **1999**, *55*, 175–182. [CrossRef]
26. Boava, L.P.; Kuhn, O.J.; Pascholati, S.F.; Di Piero, R.M.; Furtado, E.L. Atividade de quitinases e peroxidases em folhas de eucalipto em diferentes estágios de desenvolvimento após tratamento com acibenzolar-S-metil (ASM) e inoculação com *Puccinia psidii*. *Trop. Plant Pathol.* **2010**, *35*, 124–128. [CrossRef]
27. Tucker, S.L.; Talbot, N.J. Surface attachment and pre-penetration stage development by plant pathogenic fungi. *Annu. Rev. Phytopathol.* **2001**, *39*, 385–417. [CrossRef] [PubMed]
28. Adendorff, R.; Rijkenberg, F.H.J. Direct penetration from *urediospores* of *Phakopsora apoda*. *Mycol. Res.* **2000**, *104*, 317–324. [CrossRef]
29. Martin, J.T. Role of cuticle in the defense against plant disease. *Annu. Rev. Phytopathol.* **1964**, *2*, 81–100. [CrossRef]
30. Wynn, W.R.; Staples, R.C. Tropism of fungi in host recognition. In *Plant Disease Control: Resistance and Susceptibility*; Staples, R.C., Toenniessen, G.H., Eds.; Wiley: New York, NY, USA, 1981; pp. 45–69.
31. Vaz Patto, M.C.; Niks, R.E. Leaf wax layer may prevent appressorium differentiation but does not influence orientation of the leaf rust fungus *Puccinia hordei* on *Hordeum chilense* leaves. *Eur. J. Plant Pathol.* **2001**, *107*, 795–803. [CrossRef]
32. Jelitto, T.C.; Page, H.A.; Read, N.D. Role of external signals in regulating the pre-penetration phase of infection by the rice blast fungus, *Magnaporthe grisea*. *Planta* **1994**, *194*, 471–477. [CrossRef]
33. Goodman, R.N.; Kiraly, Z.; Wood, K.R. The infection process. In *The Biochemistry and Physiology of Plant Disease*; Goodman, R.N., Kiraly, Z., Wood, K.R., Eds.; University of Missouri Press: Colombia, MO, USA, 1986; pp. 1–45.
34. Araújo, J.C.A.; Matsuoka, K. Histopatologia da interação *Alternaria solani* em tomateiros resistente e suscetível. *Fitopatol. Bras.* **2004**, *29*, 268–275. [CrossRef]
35. Van Loon, L.C.; Rep, M.; Pieterse, C.M.J. Significance of inducible defense-related proteins in infected plants. *Annu. Rev. Phytopathol.* **2006**, *44*, 135–162. [CrossRef] [PubMed]
36. Whetten, R.W.; MacKay, J.J.; Sederoff, R.R. Recent advances in understanding lignin biosynthesis. *Annu. Rev. Plant Physiol. Plant Mol. Biol.* **1998**, *49*, 585–609. [CrossRef] [PubMed]

37. Vidhyasekaran, P.; Ponmalar, T.R.; Samiyappan, R.; Velazhahan, R.; Vimala, R.; Ramanathan, A. Host-specific toxin production by *Rhizoctonia solani*, the rice sheath blight pathogen. *Phytopathology* **1997**, *87*, 1258–1263. [CrossRef] [PubMed]
38. Montealegre, J.R.; Lopez, C.; Stadnik, M.J.; Henrıquez, J.L.; Herrera, R.; Polanco, R.; Di Piero, R.M.; Perez, L.M. Control of grey rot of apple fruits by biologically active natural products. *Trop. Plant Pathol.* **2010**, *35*, 271–276.
39. Godard, S.; Slacanin, I.; Viret, O.; Gindro, K. Induction of defense mechanisms in grapevine leaves by emodin- and anthraquinone-rich plant extracts and their conferred resistance to downy mildew. *Plant Physiol. Biochem.* **2009**, *47*, 827–837. [CrossRef] [PubMed]
40. Delaquis, P.J.; Stanich, K.; Girard, B.; Mazza, G. Antimicrobial activity of individual and mixed fractions of dill, cilantro, coriander and eucalyptus essential oils. *Int. J. Food Microbiol.* **2002**, *74*, 101–109. [CrossRef]
41. Vilela, G.R.; Almeida, G.S.; D'Arce, M.A.B.R.; Moraes, M.H.D.; Brito, J.O.; Silva, M.F.G.F.; Silva, S.C.; Piedade, S.M.S.; Calori-Domingues, M.A.; Gloria, E.M. Activity of essential oil and its major compound, 1,8-cineole, from *Eucalyptus globulus* Labill., against the storage fungi *Aspergillus flavus* Link and *Aspergillus parasiticus* Speare. *J. Stored Prod. Res.* **2009**, *45*, 108–111. [CrossRef]
42. Amorim, E.P.D.R.; Andrade, F.W.R.D.; Moraes, E.M.D.S.; Silva, J.C.D.; Lima, R.D.S.; Lemos, E.E.P.D. Antibacterial activity of essential oils and extracts on the development of *Ralstonia solanacearum* in banana seedlings. *Rev. Bras. Frutic.* **2011**, *33*, 392–398. [CrossRef]
43. Pereira, R.B.; Lucas, G.C.; Perina, F.J.; Resende, M.L.V.; Alves, E. Potential of essential oils for the control of brown eye spot in coffee plants. *Ciênc. Agrotec.* **2011**, *35*, 115–123. [CrossRef]
44. Bigirimana, J.; Höfte, M. Bean anthracnose: inoculation methods and influence of plant stage on resistance of *Phaseolus vulgaris* cultivars. *J. Phytopathol.* **2001**, *149*, 403–408. [CrossRef]
45. Rios, G.P.; Andrade, E.M.; Costa, J.L.S. Avaliação da resistência de cultivares e linhagens do feijoeiro comum a diferentes populações de *Uromyces appendiculatus*. *Fitopatol. Bras.* **2001**, *26*, 128–133. [CrossRef]
46. Jerba, V.F.; Rodella, R.A.; Furtado, E.L. Relação entre a estrutura foliar de feijoeiro e a pré-infecção por *Glomerella cingulata* f.sp. *phaseoli*. *Pesqui. Agropecu. Bras.* **2005**, *40*, 217–223. [CrossRef]
47. Gonçalves, E.C.P.; Centurion, M.A.P.C.; Di Mauro, A.O. Avaliação da reação de genótipos de soja ao oídio em diferentes condições. *Summa Phytopathol.* **2009**, *35*, 151–153. [CrossRef]

Article

Simulated Summer Rainfall Variability Effects on Loblolly Pine (*Pinus taeda*) Seedling Physiology and Susceptibility to Root-Infecting Ophiostomatoid Fungi

Jeff Chieppa *, Lori Eckhardt and Arthur Chappelka

School of Forestry and Wildlife Sciences, Auburn University, Auburn, AL 36849, USA;
eckhalg@auburn.edu (L.E.); chappah@auburn.edu (A.C.)
* Correspondence: jjchieppa@gmail.com; Tel.: +61-0422-851-582

Academic Editors: Matteo Garbelotto and Paolo Gonthier
Received: 26 January 2017; Accepted: 27 March 2017; Published: 30 March 2017

Abstract: Seedlings from four families of loblolly pine (*Pinus taeda* L.) were grown in capped open-top chambers and exposed to three different weekly moisture regimes for 13 weeks. Moisture regimes varied in intensity and frequency of simulated rainfall (irrigation) events; however, the total amounts were comparable. These simulated treatments were chosen to simulate expected changes in rainfall variability associated with climate change. Seedlings were inoculated with two root-infecting ophiostomatoid fungi associated with Southern Pine Decline. We found susceptibility of loblolly pine was not affected by water stress; however, one family that was most sensitive to inoculation was also most sensitive to changes in moisture availability. Many studies have examined the effects of drought (well-watered vs. dry conditions) on pine physiology and host-pathogen interactions but little is known about variability in moisture supply. This study aimed to elucidate the effects of variability in water availability, pathogen inoculation and their interaction on physiology of loblolly pine seedlings.

Keywords: rainfall patterns; *Pinus taeda*; Southern Pine Decline; *Leptographium terebrantis*; *Grosmannia huntii*

---

1. Introduction

Southern Pine Decline (SPD) is the term attributed to the premature death of *Pinus* spp. in the Southern United States due to a series of biotic and abiotic factors [1–3]. These factors include associated root pathogenic fungi (e.g., *Leptographium terebrantis* Barras and Perry and *Grosmannia huntii* (Rob-Jeffry.) Zipfel, de Beer and Wingfield, and their root-feeding beetle vectors (*Hylastes salebrosus* Eichoff, *H. tenuis* Eichoff, *Hylobius pales* Herbst., and *Pachylobius picivorus* Germar). Predisposing abiotic factors include resource stress (nutrient deficiencies, edaphic factors, and moisture stress), management strategies such as overstocking, mechanical injury and prescribed burning [4]. Studies have shown that when loblolly pine (*P. taeda*) is inoculated with *L. terebrantis*, the fungus can result in the development of lesions in the phloem and resin-soaking in the xylem [5–7]. *Grosmannia huntii*, a non-indigenous species, is a related fungal pathogen and has been reported to be more virulent in young pine seedlings when compared to *L. terebrantis* [7].

In the 1950's, Brown and McDowell [8] observed the decline of mature *P. taeda* stands in Talladega National Forest in Alabama and since then numerous studies have been performed to find causality of the decline as well as detect the phenomena at the landscape level [9–15]. Detection of SPD might be difficult as aboveground symptoms in mature trees (short chlorotic needles, sparse crowns, reduced radial growth, tree morality) occur following root damage and mortality associated with both associated insects and fungi [3]. Regardless, numerous studies have examined the virulence of root-infecting ophiostomatoid fungi on mature and juvenile families of loblolly pine among other southern *Pinus* species [6,16–18]. Since infection is dependent on the bark beetle vectors, it is important to investigate how predisposing factors (e.g., drought) that lead to root feeding, and thus fungal infection, interact with fungi associated with SPD.

Future climate change scenarios may play a significant role in the predisposing factors associated with SPD. An uncertainty with these potential developments is how much precipitation will occur in the Southern U.S. in the next 50–100 years [19,20]. One of the most important and least studied factors regarding climate change is extremes in climatic variability. For example, in 2007, the worst drought in 100 years occurred in the Southern U.S. and was followed by flooding in 2009 [21]. While changes in the intensity and frequency of summer precipitation may continue in the Southeastern U.S., there is still debate as to the underlying cause [21–23]. Another trend in precipitation patterns has been the daily variation in precipitation events where storms are occurring less frequently but are characterized by more intense rainfall for longer durations in North America [20,24,25].

A concern when considering future precipitation patterns is how forests will respond to altered drying and wetting periods [19,26]. Trees may thrive during wetter periods and experience moisture stress if evaporative losses increase during warmer, drier periods [27]. Droughts can reduce tree vigor and alter insect and pathogen physiology [28]. The effects of precipitation changes are anticipated to be unique based on both the host and pathogen physiology [29–31]. For example, mature forests would likely be tolerant of seasonal variability in rainfall frequency and magnitude [32]. The linkages between tree size and mortality due to changes in precipitation patterns are likely to be size dependent with seedlings and tall trees being most sensitive [33,34]. While tree size is likely important due to physiological constraints when under reduced available moisture [35], climate of origin may be equally important. For example, vegetation communities from more xeric sites may be more sensitive to changes in rainfall magnitude [36], while it is mesic sites that may be more sensitive to changes in frequency of precipitation [37].

The linkages between biotic and abiotic tolerance (cross-tolerance) is a useful tool to help understand how to select appropriate families/genotypes for out planting [31,38–40].In the case of root-infecting ophiostomatoid fungi and loblolly pine seedlings, the role of water regulation is likely important as inoculation can cause resin-soaking in the xylem, which has negative impacts on water movement. Lesions in the phloem can affect carbon transport, which can affect allocation and production of biomass [16,41,42]. The direct effects that water availability can have on physiological traits and productivity is also important [43,44]. Therefore, a suite of response traits (e.g., chlorophyll content, water potential, lesion length) seems appropriate for investigating interactive effects of multiple stressors.

Based on several studies [30,45,46] there are three common relationships to look for when analyzing climate-host-pathogen relationships: (1) Climate can affect the pathogen's virulence, abundance, distribution and general biology/ecology; (2) Climate can alter the host's defense, abundance, distribution and general biology/ecology; (3) Climate can change the way the host and pathogen interact, through direct and/or indirect effects. In an assessment of the effect of potential future climate change scenarios for the Southern U.S., Jones et al. [47] stated that changes in variation in water availability are important and require further investigation. Variability in water availability can cause alterations in loblolly pine vigor, resulting in biotic organisms, such as *L. terebrantis* or *G. huntii*, potentially exacerbating declines and reducing productivity. The overall goal of this study, therefore, was to elucidate the interactions of two root-infecting ophiostomatoid fungi (*L. terebrantis* and *G. huntii*) in the presence of climatic conditions similar to those predicted in the next 50 to 100 years in the Southern U.S. More specifically, our main focus was to understand how variability in water availability may affect the outcome of loblolly pine infection with the root-infecting ophiostomatoid fungi. The hypotheses tested include: (1) Loblolly pine will become more susceptible to *L. terebrantis* and *G. huntii* as variability in water availability increases and (2) Loblolly pine families selected for their tolerance to root-infecting ophiostomatoid fungi would be more tolerant to changes in water availability.

2. Materials and Methods

2.1. Study Site and Capped Open-Top Chamber

The research site (approximately 0.02 km^2 in area) is located approximately 5 km north of Auburn University Campus, Auburn, AL, USA. The site contained 24 open-top chambers (OTCs), monitoring sheds and a small laboratory. The OTCs were 4.8 m height × 4.5 m diameter aluminum framed structures with fans (1.5 horse-power or 1.1 kw motors) and chamber plastics [48]. Plastic caps were attached to each OTC to exclude ambient rainfall and permit adequate airflow [49].

Prior to the commencement of the study (March 2014), the vegetation growing in each OTC was killed with a 3% solution of glyphosate. Once dead, the vegetation was removed prior to the ground being covered with landscape fabric to prevent further unwanted vegetation growth within each OTC.

2.2. Seedlings

Bareroot 1–0 seedlings (sown in March 2013) from four commercially grown loblolly pine families were used for this study (lifted/extracted from the nursery in January 2014). We utilize the term "family" as we did not test genetic distinction between groupings. We also do not use the term ecotype because seedling parents are not from sites with distinct/contrasting ecological characteristics. Based on previous findings, two of these loblolly pine families were considered "tolerant" (T1 and T2) and two "susceptible" (S1 and S2) to root-infecting ophiostomatoid fungi [18,50]. In January 2014, 2700 seedlings (750 per family) were planted in 2.4 liter pots (1 trade gallon) with ProMix BX® peat-based potting mix (Premier Tech, Quebec, Canada). Seedlings were kept in a shade-house and watered daily for 17 weeks until being deployed into the OTCs in May 2014. The seedlings at the commencement of the study were approximately 10 months removed from sowing in the nursery.

2.3. Simulated Rainfall Treatments

To determine the longest duration the saturated potting mix could last without additional water, eight seedlings (two from each family) were placed in a greenhouse at approximately 32 °C (~90 °F). After three days of water being withheld, the potting mix became dried out and therefore, the longest period between simulated rainfall (irrigation) events was set at two days.

Three simulated precipitation treatments were used (3 replicates/treatment) over 13 weeks. The treatments were as follows: (1) 3 days week^{-1} (3D) during the experimental period; (2) 4 days week^{-1} (4D) during the experimental period; and (3) 7 days week^{-1} (7D) during the experimental period. Irrigation nozzles within each OTC were adjusted to ensure an even water distribution and flow rates within and between the chambers. The amount of water distributed was adjusted to ensure 58 minutes of watering resulted in 25.4 mm (1 inch) of precipitation. While the days of watering varied between treatments, each chamber received approximately the same amount of precipitation at the end of each week. Weekly watering values were estimated based on the 30-year (1971–2000) average precipitation for Auburn, AL. Therefore, our target for the water amount from May to August was 97, 103, 149 and 92 mm per month, respectively. Irrigation events occurred three times a day at 09:00, 12:00 and 15:00 hour. In June 2014, a 20% increase was applied to all treatment amounts/time to compensate for higher temperatures and increased airflow in the chamber. Average temperature inside the chambers was about 3–5 °C higher than ambient site temperature [51]. Monitoring throughout June indicated this adjustment approximately offset the increased evaporation of moisture in the chambers.

2.4. Inoculations

Stem inoculations were conducted as described by Nevill et al. [50] in May 2014 using the wound + inoculum method. Five inoculation treatments were used in this study: no wound (NW), wound only (W), wound + media (WM), *L. terebrantis* (LT) and *G. huntii* (GH). The *L. terebrantis* isolate (LOB-R-00-805/MYA-3316) was obtained from a *P. taeda* root exhibiting symptoms characteristic of root disease in Talladega National Forest, Oakmulgee Ranger District, AL. The *G. huntii* isolate (LLP-R-02-100/MYA-3311) was obtained from *Pinus palustris* Mill. root exhibiting symptoms characteristic of root disease in Fort Benning Military Reservation, GA. These isolates were obtained by excavating primary lateral roots, cutting the root into ~1 cm^3 sections, surface sterilizing and plating into 2% malt extract agar (MEA) and MEA with 800 mg/L cycloheximide and 200 mg/L of streptomycin sulfate [3,52]. Isolates were identified using Jacobs and Wingfield [53] by growing them in the dark and using a compound microscope. These isolates have been used in previous studies [3,17,18,54]. Long-term storage of fungi occurred on silica gel [55] at 4° C. *Leptographium terebrantis* is characterized by aerial mycelium but is very general in characteristics of related species. It can be distinguished from other *Leptographium* species by the branching of mycelium. Unlike *L. terebrantis*, *G. huntii* is readily distinguishable by the presence of serpentine-like hyphae and the presence of sexual structures (peithecia) (descriptions from Jacobs and Wingfield [53]). Control seedlings (NW, W and WM) also were plated to determine if contamination had occurred through the presence of *L. terebrantis* and/or *G. huntii*.

To inoculate seedlings, a sterile razor blade was used to cut a 15 mm vertical lesion into the bark (<1 cm depth) 5 cm above the soil line. Plugs on 2% MEA (3 mm) were placed into the wound (a single plug per seedling). Media were either sterile or colonized by cultures of *L. terebrantis* (LT) or *G. huntii* (GH). Seedling stem wounds were wrapped in cotton dampened with deionized water and then wrapped in Parafilm® to prevent desiccation of the MEA and avoid contact with other biological contaminants [18,54].

2.5. Measurements and Harvest

Root collar diameter (RCD) and height measurements were recorded for all seedlings at both the study initiation (February 2014, Table 1) and completion (August 2014) using a digital caliper and meterstick. Seedling volume increment was calculated (Volume$_{Final}$ − Volume$_{Initial}$ = Volume$_{Change}$) to determine overall growth of individual seedlings. The equation *Volume* = RCD^2 × *height* was used to estimate seedling volume/biomass in young pine seedlings (2–3 years of age) [56].

During planting in February, 40 seedlings (extra seedlings not included in the design) from each family (160 total) were destructively harvested and separated into needles (NE), shoot (SH), coarse roots (CR, >2 mm diameter) and fine roots (FR, <2 mm diameter) (Table 1). These components were placed in drying-ovens at 70 °C for 72 h. At the conclusion of the study (August 2014), two seedlings from each treatment combination per chamber were selected for final dry matter seedling biomass. Initial family dry matter averages (the 160 extra seedlings; 40 per family) for each component (needles, shoots etc.) were subtracted from the final dry matter values to estimate dry matter yield.

Table 1. Initial family dry matter, root collar diameter (RCD) and height averages and standard deviations (*n* = 160; 40/family).

Family	Needle (g)	Shoot (g)	Coarse Root (g)	Fine Root (g)	RCD (mm)	Height (mm)
S1	8.35 ± 0.87	7.54 ± 0.53	7.15 ± 0.40	6.46 ± 0.26	4.10 ± 0.72	263.06 ± 41.65
S2	8.50 ± 0.89	7.76 ± 0.62	6.73 ± 0.31	6.28 ± 0.10	4.18 ± 0.89	267.08 ± 28.12
T1	11.36 ± 1.94	8.90 ± 1.22	8.03 ± 0.99	6.58 ± 0.38	5.81 ± 1.47	283.04 ± 42.53
T2	9.22 ± 1.16	7.80 ± 0.56	6.78 ± 0.34	6.31 ± 0.13	6.29 ± 1.52	279.69 ± 35.13

S1, S2 denote loblolly pine families selected for their susceptibility to root-infecting ophiostomatoid fungi while T1, T2 denotes families selected for their tolerance.

Eleven seedlings from each treatment combination, from each of the nine OTCs, were examined for relative leaf chlorophyll (or needle greenness) using a SPAD-502 chlorophyll meter (Spectrum Tech. Inc., Plainfield, IL, USA) during the final harvest (August 2014). Needles from the first 2013 flush (previous year) were selected as they had reached physiological maturity [57]. A group of 5 to 7 needles were used for measurements. Two measurements were taken on separate parts of a plant and averaged. The same seedlings were measured for lesion characteristics. Seedlings were cut at the soil line, and stems were placed in plastic bins filled with FastGreen stain (FastGreen FCF; Sigma Chemical Co., St. Louis, MO, USA) as described by Singh et al. [18]. After 72 h, stems were removed, and the lesion length and occlusion length were measured. Lesions are the portion of the phloem colonized by *L. terebrantis* and *G. huntii*. The occlusion is the portion of the xylem that does not conduct water due to resin-soaking (thus the use of FastGreen stain). Lesion/occlusion length is presented on a per height basis (lesion length ratio) to standardize for plant growth (e.g., a 50 mm long lesion would be functionally different between plants with heights of 100 cm and 300 cm). Two 5 mm cross-sections of stem tissue from each lesion were removed from the stem and plated on malt extract agar with cyclohexamide and streptomycin sulfate for fungal re-isolation [18].

The remaining two seedlings from each combination treatment per chamber were sampled for water potential using a Scholander pressure bomb (PMS Instrument Company, Albany, OR, USA) during the final week of the experimental period. Five cm of a randomly selected lateral branch for each seedling was excised and sampled as described by Kaufmann [58]. Predawn water potential sampling occurred between 03:00 and 05:00 hour. Seedlings were sampled on watered and non-watered days. For example, those irrigated 4 day/week were measured following a day of watering and the following day after a day with no watering (herein referred to wet and dry). Seedlings irrigated daily (7D) were measured twice and randomly assigned the treatment 'watered' and 'non-watered'. This allows for the experimental design to be balanced (important for statistical analysis used, see Section 2.6) in addition to determine if seedling water potential values were significantly affected by the rainfall treatments themselves.

2.6. Data Analysis

The experimental design was a split-split-split plot with replicates at all levels: three simulated rainfall treatments replicated 3 times (9 total chambers), 4 loblolly pine families and 5 inoculation treatments produced 60 treatment combinations. Each treatment combination was replicated 15 times per chamber at the initiation of the study. Those seedlings where fungi were not recovered from re-isolations were excluded from the analysis as changes in measured characteristics cannot be attributed to the presence of the inoculated fungi. All statistical analyses were conducted using SAS (Version 9.3, SAS Institute, Inc., Cary, NC, USA) and STATISTICA (Statsoft, Inc., Tulsa, OK, USA). ANOVA *F*-test procedures followed by post hoc Tukey (Honest Significant Difference) procedures were used to determine individual treatment effects. ANOVA assumptions were verified (checked for normality) using both Kolmogorov-Smirnov and Lilliefors tests. Homogeneity of variance was inspected visually in addition to using both Levene's and Bartlett's tests [59,60]. Alpha was set at 0.05.

3. Results

3.1. Overall Results

An overview of significant interactions is presented in Table 2, including transformations utilized and number of plants used for the analysis. An overview of significant interacts for water potential data is presented in Table 3 ($p < 0.05$).

Table 2. Results of ANOVA *F*-tests.

Measurement	Transformation	*n*	Rain	Family	Inoculation	Rain × Fam	Rain × Inoc	Fam × Inoc	Rain × Fam × Inoc
SPAD	Square root	1561	***	NS	NS	NS	NS	NS	NS
Seedling volume increment	Log$_{10}$	2043	*	***	NS	***	NS	NS	NS
Total Dry Matter Yield	Log$_{10}$	298	**	***	NS	**	NS	NS	NS
Needle DMY	Square root	298	NS	***	**	***	NS	NS	NS
Shoot DMY	Square root	298	**	***	NS	*	NS	NS	NS
Coarse Root DMY	Square root	298	***	***	NS	*	NS	NS	NS
Fine Root DMY	Square root	298	***	***	**	NS	NS	NS	NS
Lesion Length/Seedling Height	Log$_{10}$	1194	NS	***	***	**	NS	**	NS
Occlusion Length/Seedling	Log$_{10}$	1161	NS	***	***	*	NS	**	NS

*: $p < 0.05$, **: $p < 0.01$, ***: $p < 0.001$; NS: not significant; Fam: Family, Inoc: Inoculation.

Table 3. Results of ANOVA *F*-tests for water potential data.

Treatments/Combinations	Predawn Water Potential
	n = 787
	Transformation: Square Root
Rain	****
Family	NS
Inoculation	NS
Dry-wet (DW)	*
Rain × Fam	***
Rain × Inoc	*
Rain × DW	NS
Fam × Inoc	NS
Fam × DW	NS
Inoc × DW	NS
Rain × Fam × Inoc	**
Rain × Fam × DW	NS
Rain × Inoc × DW	NS
Fam × Inoc × DW	NS
Rain × Fam × Inoc × DW	NS

*: $p < 0.05$, **: $p < 0.01$, ***: $p < 0.001$; NS: not significant; Fam: Family, Inoc: Inoculation.

Tukey pair-wise comparisons are used to denote significant differences during post-hoc analysis. Letters that are the same represent no significant difference, while those that are different represent a significant difference.

3.2. Leaf Chlorophyll/Needle Greeness

Leaf chlorophyll and needle greenness were affected significantly by the rainfall treatment only. Seedlings irrigated 7D (38.2 ± 0.7) were significantly different from those irrigated 4D (35.9 ± 0.3); however, those irrigated 3D (36.8 ± 1.2) were not different from either 7D and 3D seedlings.

3.3. Seedling Volume Increment

Seedling volume increment (mm^3) was affected significantly by rainfall, family and the interaction between rainfall and family. Within the rainfall × family interaction, S1 grew significantly more when watered 7D (Table 4). The remaining families were not significantly affected.

Table 4. Summary of seedling volume increment (mm³) by rainfall and family.

Family	3D			4D			7D		
	Mean (mm³)	95% Confidence Interval	Tukey Pair-Wise	Mean (mm³)	95% Confidence Interval	Tukey Pair-Wise	Mean (mm³)	95% Confidence Interval	Tukey Pair-Wise
S1	1243	±136	A	1377	±115	A	1854	±171	B
S2	1013	±123	A	1166	±90	A	1273	±105	A
T1	3144	±385	A	2729	±220	A	2877	±239	A
T2	3152	±668	A	2424	±184	A	2261	±191	A

Tukey pair-wise comparisons are within families.

3.4. Dry Matter Yield (DMY)

Total DMY was significantly affected by rainfall, family and the interaction between rainfall and family. For the rainfall × family interaction, S1 was the only family to be significantly affected by the rainfall treatment, where it produced more dry matter when watered 7D (Table 5).

Table 5. Summary of seedling total dry matter yield (g) by rainfall and family.

Family	3D			4D			7D		
	Mean (g)	95% Confidence Interval	Tukey Pair-Wise	Mean (g)	95% Confidence Interval	Tukey Pair-Wise	Mean (g)	95% Confidence Interval	Tukey Pair-Wise
S1	8.93	±1.22	A	8.37	±1.14	A	13.43	±1.83	B
S2	8.88	±1.23	A	8.29	±1.13	A	9.17	±1.25	A
T1	17.40	±2.37	A	17.00	±2.32	A	19.95	±2.72	A
T2	23.48	±3.84	A	20.76	±2.83	A	19.20	±2.58	A

Tukey pair-wise comparisons are within families.

Needle DMY was significantly affected by family, inoculation, and rainfall × family. Needle DMY increased significantly when seedlings were inoculated with GH (8.38 ± 0.72) compared to NW and W seedlings (6.73 ± 0.72 and 6.58 ± 0.73, respectively). Seedlings with WM and LT treatment (8.01 ± 0.72 and 7.56 ± 0.72, respectively) were not different from any other treatment. Seedling needle DMY for S1 was significantly affected by the irrigation treatments, with greater needle DMY when watered 7D (Table 6). The remaining families were not significantly affected.

Table 6. Summary of seedling needle dry matter yield (DMY, g) by rainfall and family.

Family	3D			4D			7D		
	Mean (g)	95% Confidence Interval	Tukey Pair-Wise	Mean (g)	95% Confidence Interval	Tukey Pair-Wise	Mean (g)	95% Confidence Interval	Tukey Pair-Wise
S1	4.46	±0.79	A	4.31	±0.77	A	7.02	±1.00	B
S2	4.76	±0.83	A	4.52	±0.79	A	4.72	±0.81	A
T1	8.80	±1.12	A	7.91	±1.06	A	9.50	±1.17	A
T2	10.98	±1.31	A	10.20	±1.21	A	8.59	±1.09	A

Tukey pair-wise comparisons are within families.

Shoot DMY was significantly affected by rainfall, family, and their interaction. Similar to total DMY, family S1 was the only one affected by rainfall treatment, where it produced greater shoot DM when irrigated 7D (Table 7).

Table 7. Summary of seedling shoot dry matter yield (DMY, g) by rainfall and family.

Family	3D			4D			7D		
	Mean (g)	95% Confidence Interval	Tukey Pair-Wise	Mean (g)	95% Confidence Interval	Tukey Pair-Wise	Mean (g)	95% Confidence Interval	Tukey Pair-Wise
S1	2.73	±0.59	A	2.43	±0.55	A	4.46	±0.76	B
S2	2.53	±0.57	A	2.60	±0.57	A	2.51	±0.56	A
T1	4.88	±0.80	A	4.60	±0.77	A	5.65	±0.86	A
T2	6.14	±0.93	A	5.54	±0.85	A	5.83	±0.86	A

Tukey pair-wise comparisons are within families.

Coarse root DMY was significantly affected by rainfall, family, and their interaction. Similar to total DMY, family S1 was the only one affected by rainfall treatment, where it produced more coarse root material when watered 7D (Table 8).

Table 8. Summary of seedling coarse root dry matter yield (DMY, g) by rainfall and family.

Family	3D			4D			7D		
	Mean (g)	95% Confidence Interval	Tukey Pair-Wise	Mean (g)	95% Confidence Interval	Tukey Pair-Wise	Mean (g)	95% Confidence Interval	Tukey Pair-Wise
S1	1.22	±0.27	A	1.32	±0.28	A	2.01	±0.35	B
S2	1.16	±0.27	A	1.05	±0.25	A	1.15	±0.26	A
T1	2.68	±0.41	A	3.43	±0.46	A	3.66	±0.48	A
T2	3.96	±0.52	A	3.52	±0.47	A	3.71	±0.48	A

Tukey pair-wise comparisons are within families.

Fine root DMY was significantly affected by rainfall, family and inoculation; however, no interactions were found to be significant. Seedlings irrigated 3D and 7D (1.12 ± 0.13 and 1.15 ± 0.13, respectively) produced more fine root DM than those watered 4D (0.86 ± 0.13). Susceptible families (S1 and S2) produced less fine root material than tolerant families (T1 and T2) (Table 9). Seedlings inoculated with GH (1.31 ± 0.18) produced significantly more fine root material than control seedlings (NW: 0.94 ± 0.15, W: 0.89 ± 0.15, and WM: 0.98 ± 0.15). Seedlings inoculated with LT (1.07 ± 0.16) were intermediate and not different from any of the inoculated treatments.

Table 9. Summary of seedling fine root dry matter yield (DMY, g) by family.

Family	Mean (g)	95% Confidence Interval	Tukey Pair-Wise
S1	0.64	0.11	A
S2	0.67	0.11	A
T1	1.54	0.17	B
T2	1.47	0.17	B

Tukey pair-wise comparisons are comparing families.

3.5. Lesions and Occlusions

Fungal re-isolation was 75.4% successful for *L. terebrantis* and 73.4% for *G. huntii*. Control seedlings had a 0% re-isolation rate indicating no contamination had occurred. Lesion length for the controls (W and WM) was not significantly different for any treatment or treatment combination, indicating the wounding process was completed with accuracy. This also indicates no effect from the presence of the media (in WM seedlings). The lesion produced from the wounding process was 14.9 ± 2.0 mm. Lesion length/seedling height for all inoculated seedlings was significantly affected by inoculation (Table 10) indicating successful colonization of seedling tissue by root-infecting ophisotomatoid fungi. Only S1 lesion length/seedling height was affected by rainfall; however, this includes the W and WM controls.

Table 10. Summary of lesion length/seedling height ($\times 100$) by family and inoculation treatment.

Inoculation	Family	S1	S2	T1	T2
W	Mean	3.28	3.33	2.59	2.82
	95% Confidence Interval	±0.24	±0.23	±.017	±0.31
	Tukey	A	A	A	A
WM	Mean	3.18	3.40	2.63	2.62
	95% Confidence Interval	±0.23	±0.21	±0.18	±0.24
	Tukey	A	A	A	A
LT	Mean	4.03	5.14	3.17	3.48
	95% Confidence Interval	±0.28	±0.32	±0.19	±0.30
	Tukey	B	B	B	B
GH	Mean	4.37	4.43	3.25	3.39
	95% Confidence Interval	±0.31	±0.30	±0.21	±0.34
	Tukey	B	C	B	B

Tukey pair-wise comparisons are within family.

Occlusion length/seedling height was significantly affected by family, inoculation, rainfall × family, and family × inoculation. The rainfall × family post-hoc analysis yielded results that indicated significant differences between families (not within) and therefore do not warrant further examination. The family × inoculation results show that W and WM seedlings were not different for any family (Table 11). Seedlings inoculated with GH had a lower occlusion length/seedling height than those inoculated with LT for each family with the exception of S1, where they were not significantly different.

Table 11. Summary of occlusion length/seedling height by family and inoculation treatment.

Inoculation	Family	S1	S2	T1	T2
W	Mean	1.52	1.51	1.43	1.45
	95% Confidence Interval	±0.03	±0.03	±0.03	±0.05
	Tukey	A	A	A	A
WM	Mean	1.50	1.53	1.42	1.42
	95% Confidence Interval	±0.03	±0.03	±0.03	±0.04
	Tukey	A	A	A	A
LT	Mean	1.76	1.89	1.66	1.67
	95% Confidence Interval	±0.04	±0.03	±0.03	±0.04
	Tukey	B	B	B	B
GH	Mean	1.69	1.69	1.57	1.55
	95% Confidence Interval	±0.04	±0.03	±0.03	±0.05
	Tukey	B	C	C	C

Tukey pair-wise comparisons are within family.

3.6. Predawn Water Potential

Water potential (megapascals) was affected by many treatments and treatment combinations. Since the rainfall × family × inoculation was significant, the main effects of each of these treatments (and their two-way interactions) cannot be analyzed. The rainfall × family × inoculation post-hoc analysis results showed that significant differences are not comparable (e.g., comparing S1-NW-7D to T2-GH-3D seedlings). There was a significant difference between measurements taken on days following watering (wet: 0.129 ± 0.009) and those taken on days following watering being withheld (dry: 0.146 ± 0.010).

4. Discussion

To our knowledge, no studies have utilized fluctuating moisture availability to mimic predicted changes in rainfall periodicity in tandem with root-pathogen inoculations. The variability in simulated rainfall patterns in the OTCs to simulate precipitation changes due to climate change did not result in an increase in susceptibility of four commonly grown loblolly pine families to the root-infecting ophiostomatoid fungi. We did observe a trend that families chosen for their tolerance to root-infecting ophiostomatoid fungi tended to have greater growth rates and produce more dry matter. Results from this study should be reviewed with caution as the experimental duration (13 weeks) was short, while the impact of changed precipitation patterns could have effects over longer periods. In addition, root-infecting ophiostomatoid fungi associated with SPD affect mature trees; however, the use of seedlings to screen families for tolerance has been useful in predicting the response of mature trees (Eckhardt et al., 2004).

Water stress has been found to result in a decrease in net photosynthesis in loblolly pine [61], which is accompanied by a decrease in transpiration rate [62]. Seiler and Johnson [63] found evidence that water stress conditioning allowed loblolly pine seedlings to photosynthesize at lower water potentials than usual. This may explain why, in our study, seedlings watered 3D were not different from those watered 4D or 7D. The confidence interval for SPAD measurements for seedlings watered 3D was nearly twice that of the other treatments, which could indicate some had begun to become acclimated to the simulated rainfall treatment. Overall, needle greenness was not affected by any other treatment. This could indicate that needle greenness, or more broadly photosynthesis, is unrelated to susceptibility of loblolly pine to root-infecting ophisotomatoid fungi.

Loblolly pine has been shown to have reduced growth when exposed to moisture stress [64,65] and the degree of response is linked to seed source location [63]. In our study, only one susceptible family had reduced growth (volume growth and dry matter yield) when exposed to altered rainfall amounts. In general, tolerant families produce more volume and biomass compared to susceptible families. This could indicate that families that have greater relative growth rates are less susceptible to root-infecting ophiostomatoid fungi. In this study, we found the wounding process to be of particular importance. This could indicate tolerant families allocate photosynthates differently than susceptible families. Future research should investigate resource allocation of photosynthates to structural and chemical properties of wood. We did observe changes in biomass allocation (needles, roots, etc.) with inoculation treatment; however, no strong pattern emerged. Given the short duration of the study, a pattern may have been difficult to detect, and we recommend a longer experimental duration to determine if trends exist regarding biomass allocation and inoculation with fungi associated with SPD. Changes in allocation of biomass can have effects on acquisition (e.g., root length) and storage (e.g., leaf water holding capacity) of water, which in turn can affect plant response to drought. Numerous studies have examined the effects of precipitation magnitude (flooding or drought) on plants and fungal pathogen interactions. These studies usually compare a sufficiently watered control and a reduced water treatment [16,63,64,66,67]. While these findings provide insight into host plant responses to periods of reduced moisture availability, less is known about the impact that variability in moisture availability will have on host-pathogen interactions. Some reports indicate that these alterations result in decreased productivity of loblolly pine [65], but widespread evidence is scarce and lacking with respect to fungal pathogens. In a previous investigation using the same families/genotypes [54], it was observed that inoculation with root-infecting ophiostomatoid fungi increased water stress when compared to non-inoculated control seedlings. In this study, no pattern in water potential emerged with respect to inoculation and the inoculation × rainfall interaction. Overall, the rainfall treatment had a significant effect, indicating the treatments were affecting seedling water relations. Future analysis should utilize a more in-depth analysis of hydraulic features of loblolly pine seedlings.

Gooheen et al. [66] found increased susceptibility of *P. ponderosa* (Laws.) to a root-infecting ophiostomatoid fungi with wetter soil; however, seedlings were inoculated below the root crown. In that study, the success of the fungal pathogen seemed to be driven by increased moisture; however, these results could be caused by the effect of moisture on the pathogen. In our study, inoculation occurred above the soil line [50], which does not directly increase moisture access of the pathogen. Croisé et al. [67] found severe drought stress increased susceptibility of *P. sylvestris* (L.) to *L. wingfieldii* (Morelet) as well as a significant decrease in hydraulic conductivity. Our results do not indicate an increase in susceptibility of loblolly pine or a significant change in plant water status. Working with loblolly pine, Meier et al. [64] found decreases in soil moisture led to decreases in available carbohydrates for both above- and belowground biomass. We observed similar results in that decreased moisture availability decreased total biomass yield. We found that this response was not ubiquitous but was rather specific to family.

5. Conclusions

The results of the study indicate that tolerance to root-infecting ophiostomatoid fungi may be linked to moisture stress sensitivity. One of the two susceptible families used was also increasingly sensitive to moisture stress. The same family that was sensitive to changes in moisture also had a larger lesion or wound when watered less frequently. This was not specific to inoculation with the pathogenic fungi and therefore we reject the hypothesis that altered moisture availability increases loblolly pine susceptibility to root-infecting ophiostomatoid fungi. We can conclude that the strategy to compensate for mechanical stress/wounding is compromised by moisture stress in one family of loblolly pine. We observed that families chosen for their tolerance to root-infecting ophiostomatoid fungi were tolerant to the rainfall treatment itself. Therefore, we fail to reject the hypothesis that families selected for their tolerance would also be more tolerant to changes in water availability. The results from this study indicate that the topic of linkages between tolerance to both drought and root-infecting ophiostomatoid fungi warrants further investigation. The authors recommend utilizing more families of loblolly pine in addition to experiments of longer duration, particularly with trees of varying age.

Acknowledgments: The authors would like to give special thanks to Scott Enebak for his guidance and assistance in preparing and reviewing the manuscript, Ryan Nadel for reviewing the manuscript and offering statistical advice, and Efrem Robbins for assistance with maintaining the field site and data collection. In addition, we would like to acknowledge the critiques of two anonymous reviewers of a previous version of this manuscript. Partial funding was provided by an Alabama Agricultural Experiment Station Internal grant award (AAES-Hatch-Multi-State-04) to Chappelka (PI) and the Forest Health Cooperative at Auburn University.

Author Contributions: All authors conceived and designed the experiments; J.C. performed the experiments; J.C. analyzed the data; L.E. and A.C. contributed reagents/materials/analysis tools; J.C. wrote the paper. A.C. provided the field site and technician to maintain the equipment.

Conflicts of Interest: The authors declare no conflict of interest. The founding sponsors had no role in the design of the study; in the collection, analyses, or interpretation of data; in the writing of the manuscript, and in the decision to publish the results.

References

1. Harrington, T.C.; Cobb, F.W., Jr. Pathogenicity of *Leptographium* and *Verticicladiella* spp. isolated from roots of Western North American conifers. *Phytopathology* **1983**, *73*, 596–599. [CrossRef]
2. Otrosina, W.J.; Hess, N.J.; Zarnoch, S.J.; Perry, T.J.; Jones, J.P. Blue-stain fungi associated with roots of Southern pine trees attacked by the Southern Pine Beetle, *Dendroctonus frontalis*. *Plant Dis.* **1997**, *81*, 942–945. [CrossRef]
3. Eckhardt, L.G.; Weber, A.M.; Menard, R.D.; Jones, J.P.; Hess, N.J. Insect-fungal complex associated with Loblolly Pine Decline in central Alabama. *For. Sci.* **2007**, *53*, 84–92.
4. Eckhardt, L.G.; Sayer, M.A.S.; Imm, D. State of Pine Decline in the Southeastern United States. *South. J. Appl. For.* **2010**, *34*, 138–141.
5. Wingfield, M.J. Association of *Verticicladiella procera* and *Leptographium terebrantis* with Insects in the Lake States. *Can. J. For. Res.* **1983**, *13*, 1238–1245. [CrossRef]
6. Eckhardt, L.G.; Jones, J.P.; Klepzig, K.D. Pathogenicity of *Leptographium* species associated with Loblolly Pine Decline. *Plant Dis.* **2004**, *88*, 1174–1178. [CrossRef]

7. Matusick, G.; Eckhardt, L.G. Variation in virulence among four root-inhabiting ophiostomatoid fungi on *Pinus taeda* L., *P. palustris* Mill, and *P. elliottii* Engelm. seedlings. *Can. J. Plant Pathol.* **2010**, *32*, 361–367. [CrossRef]

8. Brown, H.D.; McDowell, W.E. *Status of Loblolly Pine Die-Off on the Oakmulgee District, Talladega National Forest, Alabama*; USDA Forest Service: Pineville, LA, USA, 1968; Volume 28.

9. Brown, H.D.; Peacher, P.H.; Wallace, H.N. *Status of Loblolly Pine Die-Off on the Oakmulgee District, Talladega National Forest, Alabama*; USDA Forest Service: Pineville, LA, USA, 1969; Volume 9.

10. Roth, E.R.; Peacher, P.H. *Alabama Loblolly Pine Die-Off Evaluation*; USDA Forest Service: Pineville, LA, USA, 1971; Volume 9.

11. Hess, N.J.; Otrosina, W.J.; Jones, J.P.; Goddard, A.J.; Walkinshaw, C.H. Reassessment of Loblolly Pine Decline on the Oakmulgee Ranger District, Talladega National Forest, Alabama. In Proceedings of the Tenth Biennial Southern Silvicultural Research Conference, Shreveport, LA, USA, 16–18 February 1999; pp. 560–564.

12. Hess, N.J.; Otrosina, W.J.; Carter, E.A.; Steinman, J.R.; Jones, J.P.; Eckhardt, L.G.; Weber, A.M.; Walkinshaw, C.H. Assessment of loblolly pine decline in central Alabama. In Proceedings of the Eleventh Southern Silvicultural Research Conference, Asheville, NC, USA, 20–22 March 2002; pp. 558–564.

13. Eckhardt, L.G.; Menard, R.D. Topographic features associated with loblolly pine decline in Central Alabama. *For. Ecol. Manag.* **2008**, *255*, 1735–1739. [CrossRef]

14. Zeng, Y.; Kidd, K.R.; Eckhardt, L.G. The effect of thinning and clear-cut on changes in the relative abundance of root-feeding beetle (Coleoptera: Curculionidae) in *Pinus taeda* plantations in central Alabama and Georgia. *Pest Manag. Sci.* **2017**, *70*, 915–921. [CrossRef] [PubMed]

15. Coyle, D.R.; Klepzig, K.D.; Koch, F.H.; Morris, L.A.; Nowak, J.T.; Oak, S.W.; Otrosina, W.J.; Smith, W.D.; Gandhi, K.J. A review of southern pine decline in North America. *For. Ecol. Manag.* **2015**, *349*, 134–148. [CrossRef]

16. Matusick, G.; Eckhardt, L.G.; Enebak, S.A. Virulence of *Leptographium serpens* on Longleaf Pine Seedlings under Varying Soil Moisture Regimes. *Plant Dis.* **2008**, *92*, 1574–1576. [CrossRef]

17. Matusick, G.G.; Somers, L.; Eckhardt, L.G. Root lesions in large loblolly pine (*Pinus taeda* L.) following inoculation with four root-inhabiting ophiostomatoid fungi. *For. Pathol.* **2012**, *42*, 37–43. [CrossRef]

18. Singh, A.; Anderson, D.; Eckhardt, L.G. Variation in resistance of Loblolly pine (*Pinus taeda* L.) families against *Leptographium* and *Grosmannia* root fungi. *For. Pathol.* **2014**, *44*, 293–298.

19. MacCracken, M.; Barron, E.; Easterling, D.; Felzer, B.; Karl, T. *Scenarios for Climate Variability and Change: The Potential Consequences of Climate Variability and Change for the United States*; US Global Change Research Program, National Science Foundation: Washington, DC, USA, 2000.

20. IPCC. *Climate Change 2013: The Physical Science Basis. Contribution of Working Group I to the Fifth Assessment Report of the Intergovernmental Panel on Climate Change*; Stocker, T.F., Qin, D., Plattner, G.-K., Tignor, M., Allen, S.K., Boschung, J., Nauels, A., Xia, Y., Bex, V., Midgley, P.M., Eds.; Cambridge University Press: Cambridge, UK; New York, NY, USA, 2013; p. 1535.

21. Wang, H.; Fu, R.; Kumar, A.; Li, W. Intensification of summer rainfall variability in the Southeastern United States during recent decades. *J. Hydrometeorol.* **2010**, *11*, 1007–1018. [CrossRef]

22. Li, W.; Li, L.; Fu, R.; Deng, Y.; Wang, H. Changes to the North Atlantic Subtropical High and its role in the intensification of summer rainfall variability in the Southeastern United States. *J. Clim.* **2011**, *24*, 1499–1506. [CrossRef]

23. Seager, R.; Tzanova, A.; Nakamura, J. Drought in the Southeastern United States: Causes, variability over the last millennium, and the potential for future hydroclimate change. *J. Clim.* **2009**, *22*, 5021–5045. [CrossRef]

24. Kunkel, K.E.; Karl, T.R.; Brooks, H.; Kossin, J.; Lawrimore, J.H.; Arndt, D.; Bosart, L. Monitoring and understanding trends in extreme storms: State of knowledge. *Bull. Am. Meteorol. Soc.* **2013**, *94*, 499–514. [CrossRef]

25. Muschinski, T.; Katz, J.I. Trends in hourly rainfall statistics in the United States under a warming climate. *Nat. Clim. Chang.* **2013**, *3*, 577–580. [CrossRef]

26. Hanson, P.J.; Weltzin, J.F. Drought disturbance from climate change: Response of United States forests. *Sci. Total Environ.* **2000**, *262*, 2005–2220. [CrossRef]

27. Neilson, R.P.; Drapek, R.J. Potentially complex biosphere responses to transient global warming. *Glob. Chang. Biol.* **1998**, *4*, 505–521. [CrossRef]

28. Dale, V.H.; Joyce, L.A.; McNulty, S.; Neilson, R.P.; Ayres, M.P.; Flannigan, M.D.; Hanson, P.J. Climate change and forest disturbances: Climate change can affect forests by altering the frequency, intensity, duration, and timing of fire, drought, introduced species, insect and pathogen outbreaks, hurricanes, windstorms, ice storms or landslides. *BioScience* **2001**, *51*, 723–734. [CrossRef]

29. Rouault, G.; Candau, J.; Lieutier, F.; Nageleisen, L.; Martin, J.; Warzée, N. Effects of drought and heat on forest insect populations in relation to the 2003 drought in Western Europe. *Ann. For. Sci.* **2006**, *63*, 613–624. [CrossRef]

30. Sturrock, R.N.; Frankel, S.J.; Brown, A.V.; Hennon, P.E.; Kliejunas, J.T.; Lewis, K.J.; Worrall, J.J.; Woods, A.J. Climate change and forest diseases. *Plant Pathol.* **2011**, *60*, 133–149. [CrossRef]

31. Desprez-Loustau, M.; Marçais, B.; Nageleisen, L.; Piou, D.; Vannini, A. Interactive effects of drought and pathogens in forest trees. *Ann. For. Sci.* **2006**, *63*, 597–612. [CrossRef]

32. Gonthier, P.; Giordana, L.; Nicolotti, G. Further observations of sudden diebacks of Scots pine in the European Alps. *For. Chron.* **2010**, *86*, 110–117. [CrossRef]

33. Mueller, R.; Scrudder, C.; Porter, M.; Trotter, R.; Gehring, C.; Whitman, T. Differential tree mortality in response to severe drought: Evidence from long-term vegetation shifts. *J. Ecol.* **2005**, *93*, 1085–1093. [CrossRef]

34. McDowell, N.; Pockman, W.T.; Allen, C.D.; Breshears, D.D.; Cobb, N.; Kolb, T.; Plaut, J.; Sperry, J.; West, A.; William, D.G.; et al. Mechanisms of plant survival and mortality during drought: Why do some plants survive while other succumb to drought? *New Phytol.* **2008**, *178*, 719–739. [CrossRef] [PubMed]

35. Hanson, J.P.; Todd, D.E.; Amthor, J.S. A six-year study of sapling and large-tree growth and mortality responses to natural and induced variability in precipitation and throughfall. *Tree Physiol.* **2001**, *21*, 345–358. [CrossRef] [PubMed]

36. Huxman, T.E.; Smith, M.D.; Fay, P.A.; Knapp, A.K.; Shaw, M.R.; Loik, M.E.; Smith, S.D.; Tissue, D.T.; Zack, J.C.; Weltzin, J.F.; et al. Convergence across biomes to a common rain-use efficiency. *Nature* **2004**, *429*, 651–654. [CrossRef] [PubMed]

37. Knapp, A.K.; Beier, C.; Briske, D.D.; Classen, A.T.; Luo, Y.; Reichstein, M.; Smith, M.D.; Smith, S.D.; Bell, J.E.; Fay, P.A.; et al. Consequences of more extreme precipitation regimes for terrestrial ecosystems. *BioScience* **2008**, *58*, 811–821. [CrossRef]

38. Vallelian-Bindschedler, L.; Schqeizer, P.; Mosinger, E.; Metraux, J.P. Heat-induced resistance in barley to powdery mildew (*Blumeria graminisf.* sp. *Hordei*) is associated with a burst of active oxygen species. *Physiol. Mol. Plant Pathol.* **1998**, *52*, 185–199.

39. Bowler, C.; Fluhr, R. The role of calcium and activated oxygen as signals for controlling cross-tolerance. *Trends Plant Sci.* **2000**, *5*, 241–246. [CrossRef]

40. Bansal, S.; Hallsby, G.; Löfvenius, M.O.; Nilsson, M.-C. Synergistic, additive and antagonistic impacts of drought and herbivory on *Pinus sylvestrisis*: Leaf, tissue and whole-plant responses and recovery. *Tree Physiol.* **2013**, *33*, 451–463. [CrossRef] [PubMed]

41. Horner, W.E.; Alexander, S.A. Permeability of asymptomatic, resin-soaked and *Verticicladiella procerai*-black-stained pine sapwood. *Phytopathology* **1985**, *75*, 1368.

42. Joseph, G.; Kelsey, R.G.; Thies, W.G. Hydraulic conductivity in roots of ponderosa pine infected with black-stain (*Leptographium wageneri*) or annosus (*Heterobasidion annosum*) root disease. *Tree Physiol.* **1998**, *18*, 333–339. [CrossRef] [PubMed]

43. Contry, J.P.; Smillie, R.M.; Küppers, M.; Bevege, D.I.; Barlow, E.W. Chlorophyll a fluorescence and photosynthetic and growth responses of *Pinus radiate* to phosphorus deficiency, drought stress, and high CO_2. *Plant Physiol.* **1986**, *81*, 423–429.

44. Cregg, B.M.; Zhang, J.W. Physiology and morphology of *Pinus sylvestris* seedlings from diverse sources under cyclic drought stress. *For. Ecol. Manag.* **2001**, *154*, 131–139. [CrossRef]

45. Garrett, K.A.; Dendy, S.P.; Frank, E.E.; Rouse, M.N.; Travers, S.E. Climate change effects on plant disease: Genomes to ecosystems. *Annu. Rev. Phytopathol.* **2006**, *44*, 489–509. [CrossRef] [PubMed]

46. Manning, W.J.; von Tiedemann, A. Climate change: Potential effects of increased atmospheric Carbon dioxide (CO_2), ozone (O_3), and ultraviolet-B (UV-B) radiation on plant diseases. *Environ. Pollut.* **1995**, *88*, 219–245. [CrossRef]

47. Jones, J.; Hatch, U.; Murray, B.; Jagtap, S.; Cruise, J.; Yields, A.C. Potential consequences of climate variability and change for the Southeastern United States. In *Climate Change Impacts on the United States-Foundation Report: The Potential Consequences of Climate Variability and Change*; Cambridge University Press: Cambridge, MA, USA, 2001; Volume 137.

48. Gilliland, N.J.; Chappelka, A.H.; Muntifering, R.B.; Booker, F.L.; Ditchkoff, S.S. Digestive utilization of ozone-exposed forage by rabbits (*Oryctolagus cuniculus*). *Environ. Pollut.* **2012**, *163*, 281–286. [CrossRef] [PubMed]

49. Heagle, A.S.; Philbeck, R.B.; Ferrell, R.E.; Heck, W.W. Design and performance of a large, field exposure chamber to measure effects of air quality on plants. *J. Environ. Qual.* **1989**, *18*, 361–368. [CrossRef]

50. Nevill, R.J.; Kelley, W.D.; Hess, N.J.; Perry, T.J. Pathogenicity to Loblolly pines of fungi recovered from trees attacked by Southern Pine Beetles. *South. J. Appl. For.* **1995**, *19*, 78–83.

51. Chappelka, A.; School of Forestry and Wildlife Sciences, Auburn University. Personal communication, 2014.

52. Hicks, B.R.; Cobb, F.W., Jr.; Gersper, P.L. Isolation of *Ceratocystis wageneri* from forest soil with a selective medium. *Phytopathology* **1980**, *70*, 880–883. [CrossRef]

53. Jacobs, K.; Wingfield, M.J. *Leptographium Species: Tree Pathogens, Insect Associates, and Agents of Blue-Stain*; American Phytopathological Society (APS Press): St. Paul, MN, USA, 2001.

54. Chieppa, J.; Chappelka, A.H.; Eckhardt, L.G. Effects of tropospheric ozone on loblolly pine seedling inoculated with root infecting ophiostomatoid fungi. *Environ. Pollut.* **2015**, *207*, 130–137. [CrossRef] [PubMed]

55. Sinclair, J.B.; Dhingra, O.D. *Basic Plant Pathology Methods*; CRC Press: Boca Raton, FL, USA, 1995.

56. Ruehle, J.L.; Marx, D.H.; Muse, H.D. Calculated nondestructive indices of growth response for young pine seedlings. *For. Sci.* **1984**, *30*, 469–474.

57. Sasek, T.W.; Richardson, C.J.; Fendick, E.A.; Bevington, S.R.; Kress, L.W. Carryover effects of acid rain and ozone on the physiology of multiple flushes of Loblolly pine seedlings. *For. Sci.* **1991**, *37*, 1078–1098.

58. Kaufmann, M.R. Evaluation of the pressure chamber technique for estimating plant water potential of forest tree species. *For. Sci.* **1968**, *14*, 369–374.

59. Box, G.E. Non-normality and tests on variances. *Biometrika* **1953**, *40*, 318–335. [CrossRef]

60. Markowski, C.A.; Markowski, E.P. Conditions for the effectiveness of a preliminary test of variance. *Am. Stat.* **1990**, *44*, 322–326. [CrossRef]

61. Samuelson, L.J.; Pell, C.J.; Stokes, T.A.; Bartkowiak, S.M.; Akers, M.K.; Kane, M.; Markewitz, D.; McGuire, M.A.; Teskey, R.O. Two-year throughfall and fertilization effects on leaf physiology and growth of loblolly pine in the Georgia piedmont. *For. Ecol. Manag.* **2014**, *330*, 29–37. [CrossRef]

62. Groninger, J.W.; Seiler, J.R.; Zedaker, S.M.; Berrang, P.C. Photosynthetic response of Loblolly pine and Sweetgum seedling stands to elevated carbon dioxide, water stress, and nitrogen level. *Can. J. For. Res.* **1996**, *26*, 95–102. [CrossRef]

63. Seiler, J.R.; Johnson, J.D. Physiological and morphological responses of three half-sib families of Loblolly pine to water-stress conditioning. *For. Sci.* **1988**, *34*, 487–495.

64. Meier, S.; Grand, L.F.; Schoeneberger, M.M.; Reinert, R.A.; Bruck, R.I. Growth, ectomycorrhizae and nonstructural carbohydrates of Loblolly pine seedlings exposed to ozone and soil water deficit. *Environ. Pollut.* **1990**, *64*, 11–27. [CrossRef]

65. Tschaplinski, T.J.; Norby, R.J.; Wullschleger, S.D. Responses of Loblolly Pine Seedlings to Elevated CO_2 and Fluctuating Water Supply. *Tree Physiol.* **1993**, *13*, 283–296. [CrossRef] [PubMed]

66. Goheen, D.J.; Cobb, F.W., Jr.; McKibbin, G.N. Influence of soil moisture on infection of Ponderosa pine by *Verticicladiella wagenerii*. *Phytopathology* **1978**, *68*, 913–916. [CrossRef]

67. Croisé, L.; Lieutier, F.; Cochard, H.; Dreyer, E. Effects of drought stress and high density stem inoculations with *Leptographium wingfieldii* on hydraulic properties of young Scots pine trees. *Tree Physiol.* **2001**, *21*, 427–436. [CrossRef] [PubMed]

forests

MDPI

Article

Emerging Needle Blight Diseases in Atlantic Pinus Ecosystems of Spain

Esther Ortíz de Urbina [1], Nebai Mesanza [1,2], Ana Aragonés [1], Rosa Raposo [3,4],
Margarita Elvira-Recuenco [3], Ricard Boqué [5], Cheryl Patten [2,*], Jenny Aitken [6]
and Eugenia Iturritxa [1,*]

[1]   Production and Plant Protection, Neiker Tecnalia, Apartado 46, Vitoria Gasteiz, 01080 Álava, Spain;
      esthertuesta87@gmail.com (E.O.d.U.); nebaimesanzaiturritza@gmail.com (N.M.);
      aaragones@neiker.eus (A.A.)
[2]   Department of Biology, University of New Brunswick, P.O. Box 4400, Fredericton, NB E3B 5A3, Canada
[3]   INIA-CIFOR, Carretera La Coruña Km 7.5, 28040 Madrid, Spain; raposo@inia.es (R.R.);
      elvira@inia.es (M.E.-R.)
[4]   Sustainable Forest Management Research Institute, University of Valladolid-INIA, Avenida Madrid 44,
      34004 Palencia, Spain
[5]   Departamento de Química Analítica y Química Orgánica, Facultad de Química, Universitat Rovira I Virgili,
      Campus Sescelades, 43007 Tarragona, Spain; ricard.boque@urv.cat
[6]   The tree lab, P.O. Box 11236, Palm Beach, Papamoa 3151, New Zealand; jenny_aitken@xtra.co.nz
*   Correspondence: pattenc@unb.ca (C.P.); eiturritxa@neiker.eus (E.I.);
      Tel.: +1-506-458-7599 (C.P.); +34-637-436-343 (E.I.)

Academic Editors: Matteo Garbelotto and Paolo Gonthier
Received: 30 October 2016; Accepted: 23 December 2016; Published: 29 December 2016

Abstract: Red band needle blight caused by *Dothistroma septosporum* and *D. pini*, and brown spot needle blight caused by *Lecanosticta acicola* provoke severe and premature defoliation in *Pinus*, and subsequent reduction of photosynthetic surfaces, vitality, and growth in young and adult trees. The recurrent damage results in branch and tree death. Until recently, pine needle blight diseases have had only minor impacts on native and exotic forest trees in the North of Spain, but in the past five years, these pathogen species have spread widely and caused severe defoliation and mortality in exotic and native plantations of *Pinus* in locations where they were not detected before. In an attempt to understand the main causes of this outbreak and to define the effectiveness of owners' management strategies, four research actions were implemented: a survey of the management activities implemented by the owners to reduce disease impact, the evaluation of specific symptoms and damage associated with infection, and the identification of the causative pathogenic species and their reproductive capacity. Morphological characteristics of the fungus and molecular identification were consistent with those of *Lecanosticta acicola* and *Dothistroma* spp., *D. septosporum*, *D. Pini*, and both mating types were present for the three identified pathogens. The local silvicultural management performed, mainly pruning and thinning, was not resulting in the expected improvement. The results of this study can be applied to establish guidelines for monitoring and controlling the spread of needle blight pathogens.

Keywords: needle blight; *Pinus*; defoliation; *Dothistroma*; *Lecanosticta*

---

1. Introduction

Red band needle blight caused by *Dothistroma* spp. and brown spot needle blight caused by *Lecanosticta acicola* (Thümen) H. Sydow are serious forest diseases in many countries [1–4], particularly when conifers, mainly *Pinus radiata* D. Don, are planted out of their native American forest regions and in European plantations, although recent evidence suggests that in Scandinavia and other

Northern-European countries native *P. silvestris* L. and *P. nigra* Arnold are also suffering severe defoliation from *Dothistroma* [1].

The symptoms of these two diseases are quite similar, including severe defoliation that results in significant growth loss when more than 25 % of the needles are diseased [2–4]. The diseases have caused major epidemics in *Pinus radiata* (Monterey pine) in the Southern Hemisphere, Central Africa, Chile, New Zealand, and Australia [2,5]. In recent decades, they have also been increasing in incidence and severity in the Northern Hemisphere. Currently, serious epidemics are occurring on *Pinus contorta* var. latifolia Dougl. Loud. (lodgepole pine) in British Columbia, Canada [6,7] on *Pinus nigra* Arnold subsp. laricio (Poiret) Maire (Corsican pine) in Britain [8], and *Pinus radiata* in Spain [9,10]. These pathogens are found in most European countries and their spread coincided with importations and plantations of hosts out of their native areas in Europe, Africa, Australasia, and America [1,11–15].

Although red band and brown spot needle blights occur widely on host species in their native area, plantation monocultures are usually regarded as more susceptible to outbreaks than native ecosystems [16]. They frequently cause the most damage as invasive diseases in exotic plantations and have resulted in the abandonment of planting species such as *P. radiata* in East Africa [3]. In addition to the abundance of host material, possible reasons for the disease increasing in severity and incidence are directional climate changes [7], and the occurrence of both mating types, which would enable sexual reproduction and possible increases in the virulence of these pathogens [17].

These pathogens have spread quickly in Central and Northern Europe [15,18] and their control is difficult due to the large size of the infested areas and the successful adaptation of the fungi to climatic and natural conditions in new areas [19]. Both diseases are listed in the EU Plant Directive as quarantine pests, and controls are focused on seedlings, since this is the only recognized pathway for spreading in regulatory terms [20]. *Dothistroma* needle blight is considered endemic in Britain and Finland [21] and eradication is considered non-viable in Britain [22]. Both mating types are already present in some European countries. Sexual reproduction increases genetic diversity that can promote fungal proliferation in new environments, virulence on native and exotic hosts, and fungicide resistance [15,17,22,23]. It is not yet known if isolates of the pathogen from different countries differ in virulence. Until this is known it is recommended to restrict the transfer of isolates from countries in which disease presence has already been confirmed [24].

The main aim of this study was to evaluate strategies to prevent the spread of these pathogens that are employed in the region of Northern Spain. The incidence and severity of defoliations were evaluated in *Pinus* plantations, the main causal agents of needle blight and their corresponding mating types were identified, and the effectiveness of the disease management activities performed by local owners was analyzed.

2. Materials and Methods

2.1. Study Area, Field Observations, and Sampling

This study covers forest ecosystems (natural forest and plantations) in the Spanish Atlantic climate region. Field observations and sampling were conducted from spring to late summer in 2015. Surveys were conducted in 311 plots across 1650 km^2 of primarily radiata pine plantations. The plots were randomly located along accessible tracks and averaged 2.25 ha in size. Needle samples from 5 to 10 trees per plot were collected and transported in a cooling box to the laboratory. These were used to identify the causal agents of needle blight. Fungal and foliar samples were maintained in a collection at the technological institute Neiker in Arkaute (Spain).

For each plot, the percentage of trees (scale = 0–100) affected by the disease, disease incidence (S inc), and the severity (scale of 0–1) of the disease (A sev) in affected trees was estimated using the 5% step method. The stand level product of these measurements (S inc × A sev) was used to determine the severity of needle blight within each plot (S sev, scale = 0–100) [25].

Evaluation of damages in the stands was conducted by checking perimeters and interior areas to determine the mean level of damage. For stands with up to 15% overall damage, a general estimation after a tour through the stand was carried out. If the level of infection was 15% or over, an evaluation of 100–200 trees was implemented, following a transect through the longest axis of the area.

Additional information was compiled about age of the trees and management activities implemented in the sampled plantations aimed to reduce the disease impact (high pruning, low pruning, no pruning, thinning, and disposal of pruning waste).

2.2. Laboratory Analyses

Reliable identification of pathogens was performed through evidence of the characteristic conidia in the anamorphic state and by using molecular methods. Needles with brown spot and red band symptoms were sampled and immersed in NaOCl (commercial bleach, 2% active chlorine) for 60 s, and then rinsed in sterile water. Fruiting bodies and spores were observed by optical microscopy of typical conidiospores, which are produced in the conidiomata developed on symptomatic needles that were sampled directly or after incubation in a humid chamber.

In addition, molecular methods were employed to confirm morphological identification of the fungi. DNA samples were obtained from symptomatic and asymptomatic needles using the extraction Kit DNeasy Plant Mini Kit (QIAGEN Gmb, Hilden, Germany).

Blight species were identified using specific primers (LAtef.F, LAtef.R, DStub2-F, DStub2-R, DPtef-F, and DPtef-R) [26]. PCR conditions consisted of PCR buffer (500 mM KCl, 100 mM Tris-HCL pH 8.8, 0.1% Tween-20, 15 mM MgCl$_2$), 200 µM dNTP, 8 pmol of each specific primer, 0.5 U *Taq* DNA Polymerase (BIORON GmbH, Ludwigshafen am Rhein, Germany), and 10–20 ng DNA template in a total volume of 20 µL. Cycling conditions consisted of 10 min denaturation at 94 °C, 35 cycles of 30 s at 94 °C, 30 s at 60 °C, 45 s, at 72 °C, and a last extension at 72 °C for 10 min.

Mating types for each detected species were identified using specific primers. Primers specific for *L. acicola* were Md MAT1-1F, Md MAT1-1R, Md MAT1-2F, and Md MAT1-2R [27]. PCR conditions consisted of PCR buffer (500 mM KCl, 100 mM Tris-HCL pH 8.8, 0.1% Tween-20, 15 mM MgCl$_2$), 200 µM dNTP, 6.4 pmol of each specific primer, 0.5 U *Taq* DNA Polymerase (BIORON GmbH, Ludwigshafen am Rhein, Germany), and 10–20 ng DNA template in a total volume of 20 µL. Cycling conditions consisted of 5 min denaturation at 94 °C, 35 cycles of 30 s at 94 °C, 30 s at 58 °C, 45 s, at 72 °C, and a last extension at 72 °C for 7 min.

For *D. septosporum* (G. Doroguine) Morelet (as '*septospora*') and *D. pini* Hulbary mating type identification the following specific primers were used: DseptoMat1f *Dothistroma septosporum MAT1-1-1*-specific primer. DpiniMat1f2 *D. pini MAT1-1-1MAT1*-specific primer, DotMat1r *Dothistroma MAT1-1-1*-specific primer, DseptoMat2f *D. septosporum MAT1-2*-specific primer, DpiniMat2f *D. pini MAT1-2*-specific primer, DotMat2r *Dothistroma MAT1-2*-specific primer, for *D. pini*, and *D. septosporum* [17]. The same PCR procedure described for the characterization of *MAT1-1* and *MAT1-2* sequences in *L. acicola* DNA samples were applied. The cycling profile used was: denaturation at 94 °C for 5 min followed by 40 cycles at 94 °C for 30 s, 65 °C for 30 s, and 72 °C for 45 s, and a final extension at 72 °C for 7 min.

PCR products obtained were separated by electrophoresis at 100 V for 30 min on a 1.5% (w/v) agarose gel in 1× Tris-acetate-EDTA buffer (0.4 M Tris, 0.05 M sodium acetate, and 0.01 M EDTA, pH 7.8) and visualized under UV light.

2.3. Statistical Methods

As a preliminary exploratory analysis, multiple correspondence analysis (MCA) was applied on the categorized variables to represent the relationships between the variables. MCA is analogous to principal component analysis (PCA) for categorical (qualitative) variables and allows the projection of samples and variables in a reduced space, facilitating visual interpretation for large datasets. This analysis converts a matrix of data into a graphical display known as factor planes. The rows and columns of the matrix are plotted (or represented) as points in the factor planes and allow a geometrical representation of the information [28].

In addition, a binary logistic regression model was used to complement the MCA findings. Binary logistic regression is a special type of regression where one dependent binary variable (presence/absence of the pathogen) is related to a set of explanatory variables, listed in Table 1.

Table 1. Characteristics of the plantations ($n = 311$) included in this study.

Characteristic	Category Codes	Description of Characteristic	Detection of Needle Blight Species		Number of Plantations
			Positive	Negative	
Age (years)	1	<15	28	30	58
	2	15–<20	43	22	65
	3	20–<30	55	39	94
	4	≥30	39	55	94
Pruning	0	no pruning	78	89	167
	1	low	72	39	111
	2	high	17	16	33
Severity (%)	1	<30	28	68	96
	2	30–<60	33	26	59
	3	60–<90	51	18	69
	4	≥90	64	23	87
Defoliation at the base (defb %)	1	<20	16	99	115
	2	20–<40	37	28	65
	3	40–<60	37	6	43
	4	60–<80	46	2	48
	5	≥80	36	4	40
Defoliation at the middle (defi %)	1	<20	24	106	130
	2	20–<40	49	15	74
	3	40–<60	51	8	59
	4	60–<80	18	3	21
	5	≥80	27	0	27
Defoliation at the top (deft %)	1	<20	106	125	231
	2	20–<40	36	7	43
	3	40–<60	14	0	14
	4	60–<80	17	1	18
	5	≥80	5	0	5

3. Results

3.1. Fungal Species, Host Distribution and Mating Type Detection in the Studied Area

Morphological characteristics of the lesions and molecular identification of the fungi obtained from trees displaying symptoms of brown spot and red band needle blights were consistent with those of *L. acicola*, and *D. septosporum* or *D. pini*, respectively (Figure 1); the expected PCR product sizes is indicated for each fungal species.

The frequency of species of pine blight pathogens identified in various host trees is shown in Table 2. *L. acicola* was most common (in 44.7% of samples) and was mainly detected on *P. radiata*. *D. septosporum* was present in 10% of samples and was found predominantly on *P. nigra*. Occasionally two fungal species were detected in the same sample (*D. septosporum* and *L. acicola*, or *D. septosporum* and *D. pini*; in 1.6% of the samples). Detection of *D. pini* was rare (in less than 1% of samples), and it was always detected with *D. septosporum*. In terms of stand age, 81% of the plantations in which needle blight species were found were less than 25 years old.

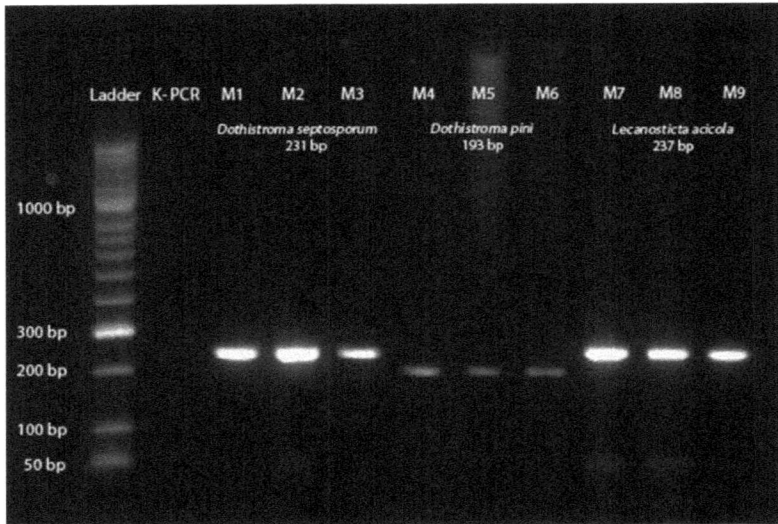

Figure 1. Polymerase chain reaction products using specific primers to detect *Lecanosticta acicola* (LAtef.F, LAtef.R), *Dothistroma septosporum* (DStub2-F, DStub2-R), and *D. pini* (DPtef-F, DPtef-R). The names M1–M9 at the top of the gel refer to sample numbers.

A high diversity of mating types was detected in the study area. Different mating types of *Dothistroma* spp. and *L. acicola* were present in samples from trees exhibiting symptoms of red band and brown spot needle blight (Figure 2). All samples showed either Mat 1, Mat 2, or both Mat 1 and Mat 2. *L. acicola* Mat 1 was detected in 75% of the samples, Mat 2 in 16%, and both Mat 1 and Mat 2 in 8% of the cases. For *D. septosporum*, Mat 2 was most frequent in 63% of samples, Mat 1 appeared in 13% of the samples, and both Mat 1 and Mat 2 were present in 24% of the plantations where this species was detected. *D. pini* was only present in three samples; Mat 2 in two samples, and Mat 1 and Mat 2 together in the third.

Table 2. Frequency of detection of species of *Dothistroma* and *Lecanosticta acicola* in relation to species of hosts.

Fungal spp./Hosts	nig [1]	hal [2]	pin [3]	pine [4]	rad [5]	syl [6]	men [7]	Total
None detected	44	3	15	1	70	9	1	143
D. septosporum	21	0	2	0	0	2	1	26
D. septosporum and *D. pini*	3	0	0	0	0	0	0	3
D. septosporum and *L. acicola*	0	0	0	0	1	1	0	2
Lecanosticta acicola	3	0	0	0	134	0	0	137
Number of plots	72	3	17	1	205	12	2	311

[1] *Pinus nigra*, [2] *P. halepnesis* Mill., [3] *P. pinaster* Aiton, [4] *P. pinea* L., [5] *P. radiata*, [6] *P. sylvestris*, [7] *Pseudotsuga menziesii* (Mirb.) Franco.

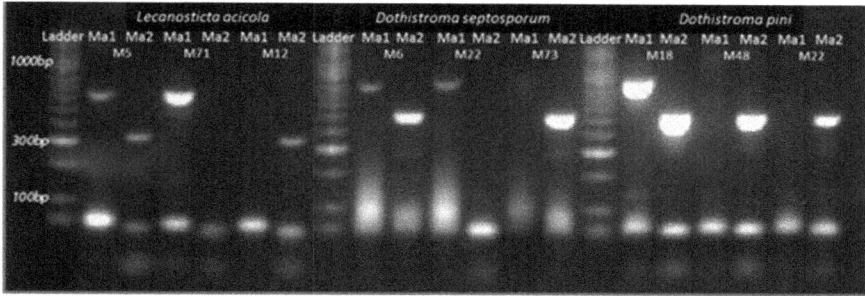

Figure 2. PCR amplicons of *L. acicola*, *D. septosporum*, and *D. pini* obtained using the MAT primers in PCR. The numbers at the top of the gel refer to sample number; Ma 1 Ma 2 refer to mating type MAT 1 and MAT 2, respectively; M5, M71, M12, M6, M22, M73, M18, M48, and M22 refer to sample numbers.

3.2. Disease Impact and Its Connection with Implemented Silvicultural Management

Field characteristics of the studied plantations were recorded including the age and disease status of the plots as well as silvicultural practices implemented by the owners (Table 1). In most plantations trees were not pruned. Where they were pruned the pruning was directed to remove live or dead branches for further improvement of crops to produce knot free wood.

The estimated severity of needle blight in plantations ranged from 5% to 95%. Defoliation was greater at the base of the diseased trees (55.6% ± 27.8%), and less in the middle (43.5% ± 26.1%) and at the top (18.2% ± 21.7%) of the diseased trees. In comparison, in plantations where the disease was not present 16.6% ± 11.6%, 12.4% ± 7.8%, and 5.8% ± 12% of the trees were defoliated at the base, middle, and top, respectively. Defoliation at the base was three times greater in infested plantations. Defoliation at the base in healthy plantations can be associated with self-pruning due to the lack of light.

In MCA (Figure 3), the spread of the category quantifications for every variable is represented, and reflects the relationships between variables in each dimension. MCA revealed that the first horizontal dimension explained 45.6% of the total inertia (variance), as the first factor plane represents the largest inertia, while the second vertical dimension explained 26.3%. Additional dimensions explained less than 1.5% each and hence had no practical significance (Table 3). With respect to the damage caused by these diseases, the first dimension is related to detection and disease symptoms (mainly defoliation at the base and in the middle part of the trees) and the second dimension is related also to defoliation at the base and intermediate part of the trees. Figure 4 shows, for each variable, a measure of its importance, which can be regarded as a squared component loading that is computed for each dimension. This measure is also the variance of the quantified variable in that dimension. Variables such plantation age and pruning, located very close to the origin, do not highlight correspondence in any dimension. With regard to management practices in the plantations, statistically significant differences were not observed in the disease severity of unhealthy plots subjected to different management activities such as extent of thinning or removal of diseased branches after pruning ($p > 0.05$; χ^2 test).

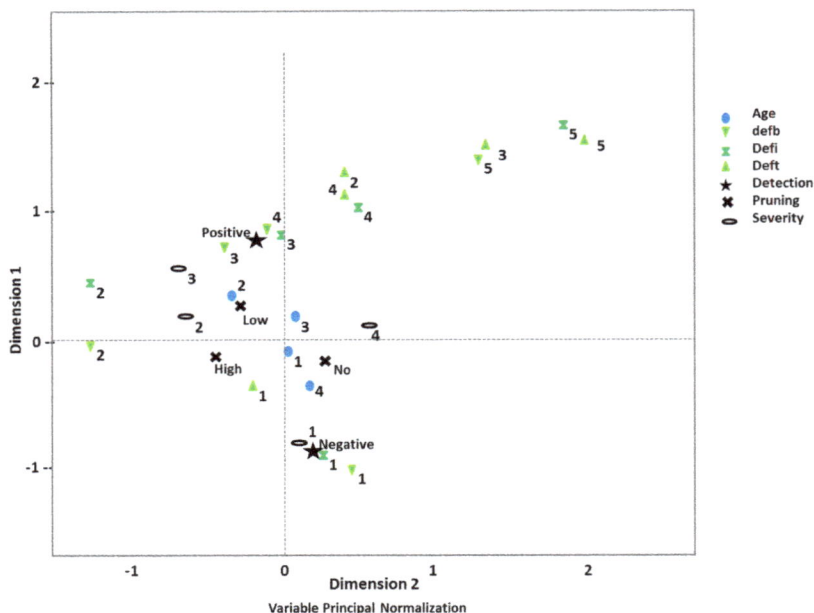

Figure 3. Relationship among fungal infection severity, symptoms, and tree characteristics in the surveyed plantations. Variety and dimensions of dichotomized presence absence of needle pathogenic species, healthy vs. unhealthy, visualized with multiple correspondence analysis and the rest of the variables: Age, defb = defoliation of the tree at the base; defi = defoliation at the middle of the tree; deft = defoliation at the top of the tree; Pruning, severity = severity of the disease.

Table 3. Logistic Regression Analysis of disease detection: Factors explaining disease development in plantations of *Pinus* (*n* = 311).

Variables	B (S.E.)	Wald	ODDS (C.I. 95%)
Age		5.37	
Age (1)	1.27 (0.70) *	3.24	3.54 (0.89–14.08)
Age (2)	1.50 (0.71) **	4.35	4.47 (1.09–18.23)
Age (3)	0.64 (0.60)	0.98	1.9 (0.53–6.75)
Pruning	ns	0.91	
Waste	ns	0.001	
Thining	ns	0.002	
Severity		15.98	
Severity (1)	−1.69 (0.76) **	4.96	0.18 (0.04–0.81)
Severity (2)	−0.81 (0.61)	1.74	0.44 (0.13–1.48)
Severity (3)	1.11 (0.69)	2.52	3.03 (0.77–11.89)
defb		16.07	
defb (1)	−2.89 (1.33) **	4.73	0.06 (0.00–0.75)
defb (2)	−1.57 (1.14)	1.89	0.21 (0.02–1.95)
defb (3)	0.53 (0.97)	0.3	1.71 (0.25–11.53)
defb (4)	3.28 (1.42) **	5.28	26.56 (1.62–43.53)
defi	ns	5.71	
deft	ns	0.82	

p values: ns, ≥0.10; *, <0.10; **, <0.05. Model Chi^2 (1) = 168.50, *p* < 0.001; Nagelkerke R^2 = 0.70; B, Coefficient for the constant or "intercept"; S.E., standard error around the coefficient for the constant; Wald, Wald criterion; ODDS and 95.0% C.I., Exponentiation of the B coefficient, which is an odds ratio (this value is given by default because odds ratios can be easier to interpret than the coefficient, which is in log-odds units); C.I., Confidence Interval.

The logistical analysis was conducted to predict detection of blight needle disease using as predictors the coded variable of age, defoliation at the base, in the middle, and at the top of the trees, plot severity, and management activities. A test of the full model against a constant only was statistically significant, indicating that the predictors as a set reliably distinguished between positive and negative detection of needle blight (chi square = 168.551, $p < 001$ with $df = 21$). Nagelkerke's R^2 of 0.70 indicated a moderately strong relationship between prediction and grouping. Prediction success overall was 87.4% (81.2% for negative detection and 90.8% for positive detection of the pathogens). The Wald criterion demonstrated the significant contribution of predictors to the model ($p < 0.05$). The most significant factor was the highest defoliation on the base defb (4), $p < 0.05$ and ODD = 26.56, followed by Age (1) and Age (2) ($p < 0.05$; ODDs 3.54 and 4.46, respectively) and Severity (1) ($p < 0.05$; ODD = 0.18). Management activities such as pruning and removal of pruning debris "waste" did not have a significant effect on the disease severity ($p > 0.05$) (Table 3).

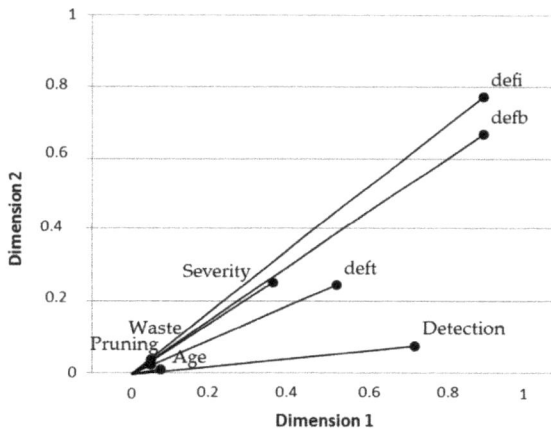

Figure 4. Multiple correspondence analysis (MCA) dimensions contain variances, indicating which variables are related along which dimension.

This study revealed two distinctive and significant dimensions of disease presence. There was a clear relationship among age (≤2), disease severity (1), high defoliation at the tree base (4), and damage to trees from the pathogens. On the other hand, the management activities carried out in infested plantations in an attempt to reduce the negative effect of the diseases had a low or null effect on the extent of damage. There were no differences in detection or severity of the disease associated with traditional management activities that are normally recommended to reduce the disease impact.

4. Discussion

The purpose of this work was to determine the extent of needle blight diseases caused by pathogenic fungi in susceptible forest ecosystems in the North of Spain and the impact of management activities in disease development. The fungal species causing needle blight and their corresponding mating types were identified, their severity was evaluated, and the effect of the main management activities performed by local owners in infested plantations was analyzed. Assessment methods of the disease were selected to identify factors correlated with disease incidence in areas where management practices are possible. The study was not aimed at comparing presence vs. absence of disease, but rather it was designed to discriminate between factors associated with high or low disease levels where disease was confirmed.

In this study we discovered that three needle blight pathogens, *L. acicola*, *D. septosporum*, and *D. pini* are present in Spain and, moreover, that both mating types for each are present, in some cases in the same plantation. The presence of both mating types of the three species in the same plantations could indicate the presence of sexual crossing and the potential to increase the development and spread of the diseases. In this context of increased genetic diversity of the pathogens, the implementation of successful control measures becomes even more complicated. It has been reported that the introduction of the second mating type of a pathogen can aggravate the disease severity and it may increase the resistance of the pathogens to chemical or biological treatments due to a rapid increase in virulence as a consequence of genetic exchange [29].

The symptoms of both red band and brown spot needle blights are most severe at the lower and middle part of the crowns and in trees less than 25 years old. This is the standard rotation age for *P. radiata* in New Zealand and Europe and it could be the reason why the symptoms are more prevalent in these plantations because in contrast to other *Pinus* species, there is some evidence that *P. radiata* trees develop resistance to the fungus gradually as they mature [11,30–32].

Two very different forest management strategies are currently in place in the plantations in the study area. Although some owners are still undertaking major pruning, a large part of the collective is managed using low cost methods that do not involve pruning and thinning regimes. On plantations where pruning and thinning are implemented, only low pruning (to about 2.2 m when the trees are 8–10 years old) or high pruning (to about 5.5 m at the age of 13–15 years) are employed, and usually only a single thinning. The pruning height does not exceed half of the tree height [33].

Silvicultural practices such as thinning, pruning, and removal of pruning debris are conducted in these plantations mainly for two objectives. Firstly, they increase the value of the wood products. These silvicultural practices encourage trees to develop a strong structure and produce knot-free wood. Knots are the primary reason for reduction in lumber value. Secondly, they improve tree health by increasing the airflow through the stands, making the microclimate conditions less favorable to disease development. Removing broken or damaged branches encourages wound closure and prevents diseases from entering the tree [34,35].

Our study revealed that silvicultural practices did not significantly reduce needle blight disease severity in infested plantations, in contrast to expectations. Although they did not eliminate the pathogens, these practices have previously been reported to reduce the inoculum and the disease level. Bulmen et al., Gadgil, and Mullet et al. [4,36,37] showed evidence for a reduction in disease levels from thinning and pruning of the lower branches in at least one season. However, other authors [38–40] did not report positive effects of low pruning on disease reduction, which is consistent with our observations. In two cases [39,40] the lack of an observed effect was attributed to the size of the blocks used in the trials. On the other hand, Gibson et al. [41] reported some evidence that pruning may accelerate the onset of mortality in affected stands. In one highly infested plantation, unpruned plots had 2.8% mortality compared with 8.8% mortality in pruned plots one year after treatment.

The plantations evaluated in this study implemented management practices at different times; some plantations were recently pruned and thinned and others were pruned and thinned several years ago (>5 years). This may explain the observed effect on needle blight disease severity of these practices. In addition, the strong influence of climate on the incidence of needle blight diseases may mask the effect of thinning and pruning since the infested plantations in this study are located for the most part (84%) in a region with the highest climate risk factors which may influence the development of needle blight disease [7,42–44]. Climate change may impact tree health and place managed plantations at high risk by altering the disturbance dynamics of native forest insect pests and microbial pathogens, as well as facilitating the establishment and spread of nonindigenous species [45]. Changes in the patterns of disturbance by forest pests are expected under a changing climate as a result of warmer temperatures, changes in precipitation, increased drought frequency, and higher carbon dioxide concentrations [46].

In the case of the studied area in the North of Spain, climate projections under greenhouse gas emission scenarios indicate that this area will experience changes in climate throughout the 21st century, including warming of surface air (especially heat wave episodes) and intensification of extreme daily rainfall (10%). Observations made in the studied area throughout the 20th century indicate increases (albeit slight) in air temperature and mean sea level that are in agreement with these projections. The result may be changes in the regime of flood events and the torrential character of the draining rivers [47].

In recent years there has been a drastic intensification in the severity of red band needle blight caused by *D. septosporum* and *D. pini* and brown spot needle blight caused by *L. acicola* in western Canada, the United States, and Europe [6,7,19,48]. The decline in forest health in these countries could be explained by a combination of factors including the presence and high density of the host, the cosmopolitan nature of the fungal species present in the regions with tropical, subtropical, temperate, Mediterranean, Atlantic, continental, and subarctic climates, climate conditions suitable for pathogen growth, and directional climate change that improves growth conditions [7,43,49].

To minimize the environmental and economic impact of needle blight disease, and encourage the sustainability of the most susceptible forest ecosystems, cultural practices and control strategies may require the combination of several methods [4,10,50]. Cultivation of alternative tolerant, resistant, or non-host forest species adapted to the local growing conditions is recommended in areas with the highest disease risk. However, since disease resistance is believed to be associated with tree maturation [32], prevalence of natural biological control agents, and genetic diversity of fungal populations [17], application of biological and copper-containing fungicides, and the use of fertilizers could encourage optimal growth of the trees and contribute to the recovery from fungal damage and other stressful factors, especially in seedlings and young plantations [51]. These measures do not eradicate the causal pathogens, and the application of chemical treatments in these ecosystems is not common, especially in adult plantations. In addition, there is a reluctance to apply aerial fungicides in forest ecosystems in EU and in New Zealand [44]. Exploration of new host species and provenances, breeding for increased resistance, and forest diversification have been reported as the key options to improve disease management in the future [10].

5. Conclusions

Needle blight is an economically important disease in many parts of the world and estimates of the economic injury level made by different researchers are consistent. Growth losses have been demonstrated when disease levels reach about 20% of the crown. The disease impact and fungal diversity of the pathogens revealed in this study suggest that implementation of successful control measures is not an easy task. The wide spread of the disease, its severity in plantations in which silvicultural management practices were implemented, the presence of three different pathogens identified as species causing severe defoliation, and the report of two mating types in the tree species complicate efficient control strategies.

Despite the fact that there are several options for disease control, each option has to be weighed against the availability of resources, cost of implementation, and political, environmental, and market restrictions. Substitution of a non-susceptible tree species for one that is highly susceptible may seem a reasonable measure but, in the studied area, widespread planting of a new species may not be feasible, especially when processing plants and markets are set up for the susceptible species. In addition, it is difficult to guarantee the future health of substitute forest species in the context of a changing climate and the global trade in wood products that is conducive to the introduction of pathogens.

Acknowledgments: To perform this study the authors were funded by their respective institutions and by the Project RTA 2013-00048-C03-03 INIA, and Project: Healthy Forest: LIFE14 ENV/ES/000179.

Author Contributions: Ana Aragonés, Margarita Elvira-Recuenco, Cheryl Patten, Jenny Aitken, Nebai Mesanza, Rosa Raposo and Eugenia Iturritxa developed conceptual ideas, designed the study, conducted measurements and data analysis in field experiments, and wrote the paper. Ricard Boqué designed the study, conducted data analysis for the field experiments, and wrote the paper. Esther Ortíz de Urbina developed conceptual ideas and conducted measurements and analyses of samples in the lab.

Conflicts of Interest: The authors declare no conflict of interest.

References

1. Drenkhan, R.; Tomešová-Haataja, V.; Fraser, S.; Bradshaw, R.E.; Vahalík, P.; Mullett, M.S.; Martín-García, J.; Bulman, L.S.; Wingfield, M.J.; Kirisits, T.; et al. Global geographic distribution and host range of *Dothistroma* species: A comprehensive review. *For. Pathol.* **2016**, *46*, 408–442. [CrossRef]
2. Gibson, I.A.S. Pests and diseases of pines in the tropics. *For. Pathol.* **1979**, *9*, 126–127.
3. Gibson, I.A.S. Impact and control of Dothistroma blight of pines. *For. Pathol.* **1974**, *4*, 89–100. [CrossRef]
4. Bulman, L.; Ganley, R.J.; Dick, M. *Needle Diseases of Radiata Pine in New Zealand*; Scion Client Report No. 13010; Forest Biosecurity Research Council: Rotorua, New Zealand, 2008.
5. Bradshaw, R.E. Dothistroma (red-band) needle blight of pines and the Dothistromin toxin: A review. *For. Pathol.* **2004**, *34*, 163–185. [CrossRef]
6. Woods, A.J. Species diversity and forest health in northwest British Columbia. *For. Chron.* **2003**, *79*, 892–897. [CrossRef]
7. Woods, A.J.; Coates, K.D.; Hamann, A. Is an unprecedented Dothistroma needle blight epidemic related to climate change? *Bioscience* **2005**, *55*, 761–769. [CrossRef]
8. Brown, A.; Webber, J. *Red band needle blight of conifers in Britain*; Forestry Commission: Edinburgh, UK, 2008; pp. 1–8.
9. Piou, D.; Ioos, R. First report of *Dothistroma pini*, a recent agent of the Dothistroma needle blight (DNB), on Pinus radiata in France. *Plant Dis.* **2014**, *98*, 841. [CrossRef]
10. Bulman, L.S.; Bradshaw, R.E.; Fraser, S.; Martín-García, J.; Barnes, I.; Musolin, D.L.; La Porta, N.; Woods, A.J.; Diez, J.; Koltay, A.; et al. A worldwide perspective on the management and control of Dothistroma needle blight. *For. Pathol.* **2016**, *46*, 472–488. [CrossRef]
11. Gibson, I.A.S. Dothistroma blight of *Pinus radiata*. *Annu. Rev. Phytopathol.* **1972**, *10*, 51–72. [CrossRef]
12. Ana Magan, F.J.F. Red band disease of *Pinus radiata*. *Comun. I.N.I.A. Prot. Veg.* **1975**, *3*, 1–16.
13. Van der Pas, J.B. Reduced early growth rate of *Pinus radiata* caused by *Dothistroma pini*. *N. Zeal. J. For. Pathol.* **1981**, *11*, 210–220.
14. Evans, H.C. *The Genus Mycosphaerella and Its Anamorphs Cercoseptoria, Dothistroma and Lecanosticta on Pines*; Mycology Paper No. 153; Commonwealth Mycological Institute: Kew, Surrey, UK, 1984; pp. 1–102.
15. Drenkhan, R.; Hantula, J.; Vuorinen, M.; Jankovsky, L.; Müller, M.M. Genetic diversity of *Dothistroma septosporum* in Estonia, Finland and Czech Republic. *For. Pathol.* **2012**, *136*, 71–85. [CrossRef]
16. Jactel, H.; Brockerhoff, E.; Duelli, P. A test of the biodiversity-stability theory: Meta-analysis of tree species diversity effects on insect pest infestations, and re-examination of responsible factors. In *Forest Diversity and Function—Temperate and Boreal Systems*; Scherer-Lorenzen, M., Körner, C., Schulze, E.-D., Eds.; Springer: Heidelberg/Berlin, Germany, 2005; Volume 176, pp. 235–262.
17. Groenewald, M.; Barnes, I.; Bradshaw, R.E.; Brown, A.V.; Dale, A.; Groenewald, J.Z.; Lewis, K.J.; Wingfield, B.D.; Wingfield, M.J.; Crous, P.W. Characterization and distribution of mating type genes in the Dothistroma needle blight pathogens. *Phytopathology* **2007**, *97*, 825–834. [CrossRef] [PubMed]
18. Woods, A.J.; Martín-García, J.; Bulman, L.; Vasconcelos, M.W.; Boberg, J.; La Porta, N.; Peredo, H.; Vergara, G.; Ahumada, R.; Brown, A.; et al. Dothistroma needle blight, weather and possible climatic triggers for the disease's recent emergence. *For. Pathol.* **2016**, *46*, 443–452. [CrossRef]
19. Jankovsky, L.; Palovcikova, D.; Dvorak, M.; Tomsovsky, M. Records of brown spot needle blight related to Lecanosticta acicola in the Czech Republic. *Plant Prot. Sci.* **2009**, *45*, 16–18.
20. European and Mediterranean Plant Protection Organization (EPP/EPPO). *Exigences Spécifiques de Quarantaine*; EPPO Technical Documents; EPPO: Paris, France, 1990.
21. European and Mediterranean Plant Protection Organization (OEPP/EPPO). *Mycosphaerella Dearnessii and Mycosphaerella Pini.*; Bulletin 38; EPPO: Paris, France, 2008; pp. 349–362.
22. Brown, A.; Clayden, H. Time for action: Dothistroma (red band) needle blight in Scotland. *Forestry* **2012**, *18*, 16–17.
23. Barnes, I.; Walla, J.A.; Bergdahl, A.; Wingfield, M.J. Four new host and three new state records of Dothistroma Needle Blight caused by *Dothistroma pini* in the United States. *Plant Dis.* **2014**, *98*, 1443. [CrossRef]

24. Bradshaw, R.E.; Bhatnagar, D.; Ganley, R.J.; Gillman, C.J.; Monahan, B.J.; Seconi, J.M. *Dothistroma pini*, a forest pathogen, contains homologs of aflatoxin biosynthetic pathway genes. *Appl. Environ. Microb.* **2002**, *68*, 2885–2892. [CrossRef]

25. Bulman, L.S.; Gadgil, P.D.; Kershaw, D.J.; Ray, J.W. *Assessment and Control of Dothistroma Needle-Blight*; Forest Research Bulletin No. 229; Forest Research: Rotorua, New Zealand, 2004; pp. 1–48.

26. Ioos, R.; Fabre, B.; Saurat, C.; Fourrier, C.; Frey, P.; Marcais, B. Development, comparison, and validation of real-time and conventional PCR tools for the detection of the fungal pathogens causing brown spot and red band needle blights of pine. *Phytopathology* **2010**, *100*, 105–114. [CrossRef] [PubMed]

27. Janoušek, J.; Krumböck, S.; Kirisits, T.; Bradshaw, R.E.; Barnes, I.; Jankovský, L.; Stauffer, C. Development of microsatellite and mating type markers for the pine needle pathogen *Lecanosticta acicola*. *Australas. Plant Pathol.* **2014**, *43*, 161–165. [CrossRef]

28. Greenacre, M.; Hastie, T. The Geometric Interpretation of Correspondence Analysis. *J. Am. Stat. Assoc.* **1987**, *82*, 437–447. [CrossRef]

29. Paoletti, M.; Buck, K.W.; Brasier, C.M. Selective acquisition of novel mating type and vegetative incompatibility genes via interspecies gene transfer in the globally invading eukaryote Ophiostoma novo-ulmi. *Mol. Ecol.* **2006**, *15*, 249–262. [CrossRef] [PubMed]

30. Ivory, M.H. Resistance to Dothistroma needle blight induced in *Pinus radiata* by maturity and shade. *Br. Mycol. Soc.* **1972**, *59*, 205–212. [CrossRef]

31. Garcia, J.; Kummerow, J. Infection of Monterey Pine graftings with *Dothistroma pini*. *Plant Dis. Rep.* **1970**, *54*, 403–404.

32. Power, A.B.; Dodd, R.S. Early differential susceptibility of juvenile seedlings and more mature stecklings of *Pinus radiata* to *Dothistroma pini*. *N. Z. J. For. Sci.* **1984**, *14*, 223–228.

33. Mead, D.J. *Sustainable Management of Pinus Radiata Plantations*; FAO Forestry Paper No. 170; FAO: Rome, Italy, 2013.

34. Emmingham, W.; Fitzgerald, S. *Pruning to Enhance Tree and Stand Value*; Extension Service. Publication number EC 1457; Oregon State University: Orvallis, OR, USA, 1995; pp. 1–12.

35. O'Hara, K. Pruning Wounds and Occlusion: A Long-Standing Conundrum. *J. For.* **2007**, *105*, 131–138.

36. Gadgil, P.D. *Dothistroma Needle Blight*; Forest Pathology in New Zealand No. 5; Forest Research Institute: Rotorua, New Zealand, 1984; pp. 1–8.

37. Mullett, M.S.; Tubby, K.V.; Webber, J.F.; Brown, A.V. A reconsideration of natural dispersal of the pine pathogen *Dothistroma septosporum*. *Plant Pathol.* **1984**. [CrossRef]

38. Scott, C.A. The Influence of Low Pruning on *Dothistroma Pini* Infection in *Pinus radiata* in Kaingaroa Forest. Forest Pathology Report No.39; Forest Research Institute: Rotorua, New Zealand, 1973; pp. 1–21. [CrossRef]

39. Hood, I.A.; Ramsden, M. *Dothistroma Needle Blight on Pinus Radiata at Gambubal Forest, QFRI Disease Management Research Trials*; Interim Report; Queensland Forest Research Institute: Indooroopilly, Austarlia, 1996.

40. Bulman, L.S.; Dick, M.A.; Ganley, R.J.; McDougal, R.L.; Schwelm, A.; Bradshaw, R.E. *Dothistroma Needle Blight*; Gonthier, P., Nicolotti, G., Eds.; CABI: Boston, MA, USA, 2013; pp. 436–457.

41. Gibson, I.A.S.; Christensen, P.S.; Munga, F.M. First observations in Kenya of a foliage disease of Pines caused by *Dothistroma pini* Hulbary. *Commonw. For. Rev.* **1964**, *43*, 31–48.

42. Sutherst, R.W.; Maywald, G.F.; Kriticos, D.J. CLIMEX Version 3: User's Guide. Available online: http://www.hearne.software/getattachment/0343c9d5-999f-4880-b9b2-1c3eea908f08/Climex-User-Guide.aspx (accessed on 27 December 2016).

43. Watt, M.S.; Kriticos, D.J.; Alcaraz, S.; Brown, A.V.; Leriche, A. The hosts and potential geographic range of Dothistroma needle blight. *For. Ecol. Manag.* **2009**, *257*, 1505–1519. [CrossRef]

44. EFSA Panel on Plant Health (PLH). Scientific Opinion on the risk to plant health posed by *Dothistroma septosporum* (Dorog.) M. Morelet (*Mycosphaerella pini* E. Rostrup, syn. *Scirrhia pini*) and *Dothistroma pini* Hulbary to the EU territory with the identification and evaluation of risk. *EFSA J.* **2013**, *11*, 173. [CrossRef]

45. Hepting, G.H. Climate and forest diseases. *Annu. Rev. Phytopathol.* **1963**, *1*, 31–50. [CrossRef]

46. Dale, V.H.; Joyce, L.A.; McNulty, S.; Neilson, R.P.; Ayres, M.P.; Flannigan, M.D.; Hanson, P.J.; Irland, L.C.; Lugo, A.E.; Peterson, C.J.; et al. Climate change and forest disturbances. *BioScience* **2001**, *51*, 723–734. [CrossRef]

47. Moncho, R.; Chust, G.; Caselles, V. Análisis de la precipitación del País Vasco en el período 1961–2000 mediante reconstrucción espacial. *Nimbus* **2009**, *23*, 149–170.

48. Brown, A.; Green, S.; Hendry, S. Needle diseases of pine. Forestry Commission: Edinburgh, UK, 2005; pp. 1–12.

49. Guernier, V.; Hochberg, M.E.; Guegan, J.F.O. Ecology drives the worldwide distribution of human diseases. *PLoS Biol.* **2004**, *2*, 740–746. [CrossRef] [PubMed]

50. Koltay, A. Incidence of *Dothistroma septospora* (Dorog.) Morlet in the Austrian pine (*Pinus nigra* Arn.) stands in Hungary and results of chemical control trials. *Novenytermeles* **2001**, *37*, 231–235.

51. Forestry Commission. Forests and biodiversity, UK Forestry Standard Guidelines. Forestry Commission: Edinburgh, UK, 2011. Available online: http://www.forestry.gov.uk/PDF/FCGL001.pdf/$FILE/FCGL001.PDF (accessed on 27 December 2016).

forests

MDPI

Article

Invasive Everywhere? Phylogeographic Analysis of the Globally Distributed Tree Pathogen *Lasiodiplodia theobromae*

James Mehl [1], Michael J. Wingfield [1], Jolanda Roux [2] and Bernard Slippers [3,*]

[1] Department of Microbiology and Plant Pathology, DST-NRF Centre of Excellence in Tree Health Biotechnology (CTHB), Forestry and Agricultural Biotechnology Institute (FABI), University of Pretoria, Private Bag X20, Hatfield, Pretoria 0028, South Africa; james.mehl@fabi.up.ac.za (J.M.); mike.wingfield@fabi.up.ac.za (M.J.W.)
[2] Department of Plant and Soil Sciences, DST-NRF CTHB, FABI, University of Pretoria, Private Bag X20, Hatfield, Pretoria 0028, South Africa; jolanda.roux@fabi.up.ac.za
[3] Department of Genetics, DST-NRF CTHB, FABI, University of Pretoria, Private Bag X20, Hatfield, Pretoria 0028, South Africa
* Correspondence: bernard.slippers@fabi.up.ac.za; Tel.: +27-12-420-3938

Academic Editors: Matteo Garbelotto and Paolo Gonthier
Received: 19 March 2017; Accepted: 22 April 2017; Published: 27 April 2017

Abstract: Fungi in the Botryosphaeriaceae are important plant pathogens that persist endophytically in infected plant hosts. *Lasiodiplodia theobromae* is a prominent species in this family that infects numerous plants in tropical and subtropical areas. We characterized a collection of 255 isolates of *L. theobromae* from 52 plants and from many parts of the world to determine the global genetic structure and a possible origin of the fungus using sequence data from four nuclear loci. One to two dominant haplotypes emerged across all loci, none of which could be associated with geography or host; and no other population structure or subdivision was observed. The data also did not reveal a clear region of origin of the fungus. This global collection of *L. theobromae* thus appears to constitute a highly connected population. The most likely explanation for this is the human-mediated movement of plant material infected by this fungus over a long period of time. These data, together with related studies on other Botryosphaeriaceae, highlight the inability of quarantine systems to reduce the spread of pathogens with a prolonged latent phase.

Keywords: Botryosphaeriaceae; latent pathogen; endophyte; fungal ecology; fungal invasion; quarantine

---

1. Introduction

The health of both native and planted forests is under increasing pressure from rapid changes in the environment (many related to the growing impacts of human society) or the introduction of non-native, invasive pathogens and pests [1,2]. The rise in the number of invasive pathogens and pests is thought to be driven primarily by increasing international movement and trade in plants and plant products [2,3]. This problem might be even more severe than previously realized, because quarantine mechanisms designed to reduce such movement are oblivious of the multitude of cryptic and endophytic microbes that occur asymptomatically within plants [3,4]. A prominent group of fungi that reflect this threat is the Botryosphaeriaceae.

The Botryosphaeriaceae includes many important plant pathogens such as well-known species in *Botryosphaeria*, *Diplodia*, *Dothiorella*, *Lasiodiplodia*, *Macrophomina*, and *Neofusicoccum* [5]. These fungi can persist endophytically within apparently asymptomatic plant material, from where they can cause disease when the host is stressed [4,6]. Many Botryosphaeriaceae species infect multiple plant hosts and commonly occur on both native and non-native hosts in a region [7–11]. Consequently, they can easily be spread when plants or plant material are moved between regions [3,4].

The majority of the Botryosphaeriaceae have relatively limited distributions [12–15]. This is perhaps not surprising given that their spread is closely linked with rainfall and associated wind dispersal, and is consequently expected to be relatively local [6,16]. While stepwise, long-distance spread would be possible, a continuous distribution of available hosts would be needed, making spread across oceans or other major physical barriers unlikely. A few species, however, have very broad global distributions, including *Botryosphaeria dothidea*, *Diplodia sapinea*, *D. seriata*, *Dothiorella sarmentorum*, *Neofusicoccum parvum*, and *Lasiodiplodia theobromae* [4,11,17–20]. These species are commonly associated with agriculture, forestry, or urban environments, and it is thought that human-assisted dispersal has played a significant role in their distributions [15,18,19].

A number of previous studies have suggested that human-assisted dispersal of the Botryosphaeriaceae might in some cases occur on a large scale. For example, *D. sapinea* has been introduced to all areas where *Pinus* species have been planted in the southern hemisphere [21]. Population genetic studies on this fungus suggest that, in most areas, these introductions have been so extensive that the diversity of the non-native populations exceeds that of some local native populations of the fungus [22,23]. Another example is *N. parvum*, which is also highly genetically diverse, with 12 lineages identified using microsatellite markers, many of which are shared between different countries and on different continents [18]. In the case of *Macrophomina phaseolina*, Sarr et al. [24] identified three lineages using DNA sequence data for six loci, also with shared geographic ranges. Analyses of a global collection of isolates of *B. dothidea* using two DNA sequence markers, showed that isolates grouped into two main haplotypes, with no structure based on either host genus or country of origin [19].

Lasiodiplodia theobromae is one of the most commonly reported species in the Botryosphaeriaceae. This fungus has been associated with at least 500 plant hosts from many tropical and subtropical regions globally [17,25]. However, many of these host associations and disease reports for *L. theobromae* predate the use of DNA sequencing for species identification, and at least some could be attributed to cryptic species related to *L. theobromae* [12,17]. In recent years, many cryptic species have been described for isolates previously treated as *L. theobromae* due to their morphological similarity, but that are distinct based mainly on DNA sequence data from two loci, the internal transcribed spacer ribosomal DNA (ITS) and translation elongation factor 1α (*tef1α*) [26–28]. At present, the genus *Lasiodiplodia* comprises 31 species [20], mostly distinguished using sequence data. Furthermore, Cruywagen et al. [27] recently showed that four species of *Lasiodiplodia* represent hybrid species, based on more complete isolate collections or sequence data of more loci than originally used. In view of all of these studies, there is no overall clarity on the host or geographic distribution of what can be considered *L. theobromae sensu stricto*, based on current DNA-based definitions of this taxon. It is also not clear where the fungus might have originated, where it is invasive, or to what extent humans have facilitated the dispersal of this fungus globally.

The first aim of this study was to screen a global collection of isolates putatively identified as *L. theobromae* and thus to identify a collection that represented *L. theobromae sensu stricto* based on DNA sequence data. Sequence data from four nuclear loci were then used to determine whether there was genetic structure amongst this global collection of *L. theobromae* isolates. Finally, we considered whether the data revealed a possible area of origin for the fungus.

2. Materials and Methods

2.1. Isolate Collections and DNA Extractions

A total of 426 fungal isolates designated as *Botryosphaeria* sp. or *L. theobromae* were obtained from the culture collection (CMW) of the Forestry and Agricultural Biotechnology Institute (FABI) at the University of Pretoria, Pretoria, South Africa. These isolates originated from collections made in Australia, Benin, Brazil, Cameroon, China, Colombia, Ecuador, Indonesia, Madagascar, Mexico, Oman, Peru, South Africa, Thailand, Uganda, the United States of America (USA), Venezuela, and Zambia (Figure 1). Several isolates identified as *L. theobromae* were also sourced from the culture collection of the Westerdijk Fungal Biodiversity Institute (previously known as the Centraalbureau voor Schimmelcultures), Utrecht, the Netherlands. In addition, sequences were sourced from GenBank for taxa labeled as "*Botryosphaeria rhodina*" or "*Lasiodiplodia theobromae*" and were included in datasets for analyses (Table 1).

Isolates assembled for this study were purified by transferring single hyphal tips to clean culture plates following the method described in Mehl et al. [30]. DNA was extracted from isolates using the method described by Wright et al. [31] with pellets suspended in 50 µL Tris Ethylenediaminetetraacetic acid (TE) buffer. DNA concentrations were determined using a NanoDrop® ND-1000 and accompanying software (NanoDrop Technologies, DuPont Agricultural Genomics Laboratories, Wilmington, DE, USA).

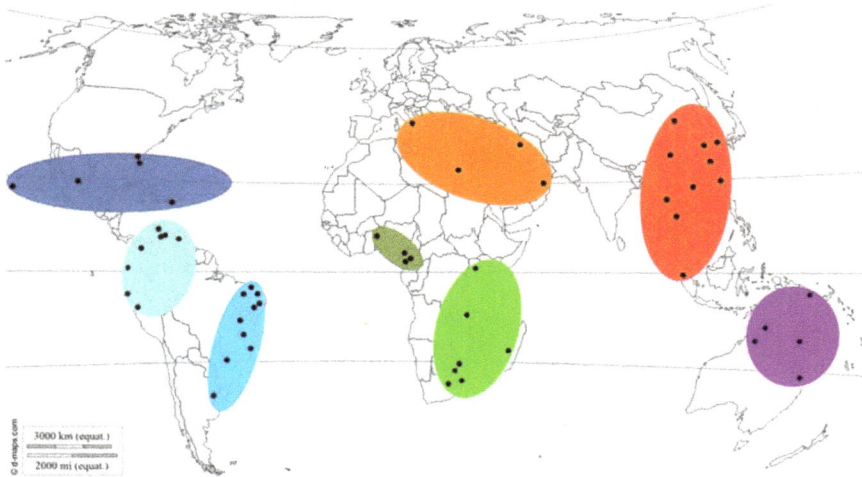

Figure 1. Sites (black circles) and biogeographic regions (shaded) where isolates originated from. Map source: [29].

Table 1. List of isolates used for genetic analyses. Isolates are ordered geographically, moving from North America eastwards to Australia. Countries in each region are arranged alphabetically. Sequences from GenBank are italicized.

Region	Country, Locality	Isolate	Host	Plant Family	ITS	tef1α	tub2	rpb2
North America	Hawaii	CBS111530	*Leucospermum* sp.	Proteaceae	FJ150695	EF622064	KU887531	KU696382
	Mexico	BOM230	*Carica papaya*	Caricaceae	KR001856	KT075154		
	Mexico	BOS104	*Car. papaya*	Caricaceae	KR001857	KT075158		
	Mexico	BOT112	*Car. papaya*	Caricaceae	KT075139	KT075155		
	Mexico	BOT359	*Car. papaya*	Caricaceae	KR001859	KT075159		
	Mexico	LAM118	*Car. papaya*	Caricaceae	KT075141	KT075156		
	Puerto Rico	K286	*Mangifera indica*	Anacardiaceae	KC631660	KC631656	KC631652	
	Puerto Rico	K8	*Man. indica*	Anacardiaceae	KC631659	KC631655	KC631651	
	Puerto Rico	PHLO10	*Dimocarpus longan*	Sapindaceae	KC964547	KC964554	KC964550	
	Puerto Rico	PHLO9	*Dim. longan*	Sapindaceae	KC964546	KC964553	KC964549	
	USA	CBS124.13	Unknown	Sapindaceae	DQ458890	DQ458875	DQ458858	KY472887
	USA, Florida	CMW34107	*Eucalyptus amplifolia*	Myrtaceae	KY473070	KY473018		
	USA, Florida, Apopka	SEFL3	*Vaccinium* sp.	Ericaceae	JN607091	JN607114	JN607138	
	USA, Florida, Alaucha	UF05161	*Vacc. corymbosum*	Ericaceae	GQ845096	GQ850468		
	Country	WFP92	*Vacc. corymbosum*	Ericaceae	GQ845095	GQ850467		
Western South America	Colombia, Andes	CMW34303	Unknown	Fabaceae	KY473031	KY472979	KY472913	KY472842
	Ecuador	CMW4694	*Schizolobium parahyba*	Fabaceae	KY473033	KY472981	KY472914	KY472843
	Ecuador	CMW4695	*Sch. parahyba*	Fabaceae	KY473034	KF886730	KY472915	
	Ecuador	CMW4696	*Sch. parahyba*	Fabaceae	KY473035	KY472982	KY472916	KY472844
	Ecuador	CMW9273	*Sch. parahyba*	Fabaceae	KF886709	KY472983	KY472911	KY472840
	Ecuador, Esmeraldas	CMW2924	*Sch. parahyba*	Fabaceae	KY473032	KF886732	KY472912	KY472841
	Ecuador, Esmeraldas	CMW22926	*Sch. parahyba*	Fabaceae	KY473048	KY472996	KY472935	
	Peru	CMW31861	*Theobroma cacao*	Malvaceae	KY473049	KY472997	KY472936	KY472862
	Peru	CMW31867	*Th. cacao*	Malvaceae	KY473050	KY472998	KY472937	KY472863
	Peru	CMW31899	*Th. cacao*	Malvaceae				
	Peru, Cienneguillo Norte, Piura	LA-SJ1	*Vitis vinifera*	Vitaceae	KM401976	KM401973		
	Peru, Sol-Sol, Piura	LA-SOL1	*Vis. vinifera*	Vitaceae	KM401974	KM401971		
	Peru, San Vicente, Piura	LA-SV1	*Vis. vinifera*	Vitaceae	KM401975	KM401972		
	Venezuela, Guayana	A10	*Acacia mangium*	Fabaceae	JX545093	JX545113	JX545133	
	Venezuela, Guayana	A13	*Ac. mangium*	Fabaceae	JX545094	JX545114	JX545134	
	Venezuela, Acarigua	CMW13490	*Euc. urophylla*	Myrtaceae	KY473071	KY473019	KY472962	KY472888
	Venezuela, Cojedes	CMW13501	*Ac. mangium*	Fabaceae	KY473072	KY473020	KY472963	KY472889
	Venezuela, Falcon State	CMW13519	*Pinus caribaea* var. *hondurensis*	Pinaceae	KY473073	KY473021	KY472964	KY472890
	Venezuela, Falcon State	CMW13527	*Pin. caribaea* var. *hondurensis*	Pinaceae	KY473074	KY473022	KY472965	KY472891
Eastern South America	Brazil	ARM122	*Jatropha curcas*	Euphorbiaceae	KF553895	KF553896		
	Brazil, Vicosa, MG	CDA425	*Cocos nucifera*	Arecaceae	KP244697	KP308475	KP308531	
	Brazil, Vicosa, MG	CDA444	*Coc. nucifera*	Arecaceae	KP244699	KP308477	KP308532	
	Brazil, Vicosa, MG	CDA450	*Coc. nucifera*	Arecaceae	KP244688	KP308478	KP308533	
	Brazil, Vicosa, MG	CDA455	*Coc. nucifera*	Arecaceae	KP244689	KP308463	KP308534	
	Brazil, Juazeiro, BA	CDA465	*Coc. nucifera*	Arecaceae	KP244701	KP308465	KP308535	
	Brazil, Juazeiro, BA	CDA467	*Coc. nucifera*	Arecaceae	KP244702	KP308473	KP308536	
	Brazil, Juazeiro, BA	CDA469	*Coc. nucifera*	Arecaceae	KP244691	KP308466	KP308537	
	Brazil, Juazeiro, BA	CDA472	*Coc. nucifera*	Arecaceae	KP244692	KP308467	KP308538	

Table 1. Cont.

Region	Country, Locality	Isolate	Host	Plant Family	ITS	tef1α	tub2	rpb2
	Brazil, Sao Francisco Valley	CMM 0307	Vts. vinifera	Vitaceae	KJ450879	KJ417879		
	Brazil, Sao Francisco Valley	CMM 0310	Vts. vinifera	Vitaceae	KJ450880	KJ417880		
	Brazil, Sao Francisco Valley	CMM 0384	Vts. vinifera	Vitaceae	KJ450876	KJ417876		
	Brazil, Sao Francisco Valley	CMM 0455	Vts. vinifera	Vitaceae	KJ450878	KJ417878		
	Brazil, Sao Francisco Valley	CMM 0820	Vts. vinifera	Vitaceae	KJ450877	KJ417877		
	Brazil	CMM1476	Man. indica	Anacardiaceae	JX464083	JX464057		
	Brazil	CMM1481	Man. indica	Anacardiaceae	JX464095	JX464021		
	Brazil	CMM1517	Man. indica	Anacardiaceae	JX464060	JX464054		
	Brazil	CMM2168	Car. papaya	Caricaceae	KC484817	KC481572		
	Brazil	CMM2179	Car. papaya	Caricaceae	KC484787	KC481569		
	Brazil	CMM2183	Car. papaya	Caricaceae	KC484824	KC481573		
	Brazil	CMM2190	Car. papaya	Caricaceae	KC484780	KC481518		
	Brazil	CMM2193	Car. papaya	Caricaceae	KC484826	KC481550		
	Brazil	CMM2208	Car. papaya	Caricaceae	KC484776	KC481575		
	Brazil	CMM2209	Car. papaya	Caricaceae	KC484784	KC481578		
	Brazil	CMM2210	Car. papaya	Caricaceae	KC484783	KC481577		
	Brazil	CMM2231	Car. papaya	Caricaceae	KC484775	KC481515		
	Brazil	CMM2232	Car. papaya	Caricaceae	KC484785	KC481521		
	Brazil	CMM2235	Car. papaya	Caricaceae	KC484779	KC481517		
	Brazil	CMM2237	Car. papaya	Caricaceae	KC484819	KC481547		
	Brazil	CMM2238	Car. papaya	Caricaceae	KC484771	KC481512		
	Brazil	CMM2239	Car. papaya	Caricaceae	KC484786	KC481522		
	Brazil	CMM2241	Car. papaya	Caricaceae	KC484790	KC481571		
	Brazil	CMM2261	Car. papaya	Caricaceae	KC484789	KC481579		
	Brazil	CMM2262	Car. papaya	Caricaceae	KC484822	KC481581		
	Brazil	CMM2265	Car. papaya	Caricaceae	KC484772	KC481574		
	Brazil	CMM2267	Car. papaya	Caricaceae	KC484777	KC481576		
	Brazil	CMM2268	Car. papaya	Caricaceae	KC484818	KC481580		
	Brazil	CMM2269	Car. papaya	Caricaceae	KC484821	KC481585		
	Brazil	CMM2276	Car. papaya	Caricaceae	KC484820	KC481548		
	Brazil	CMM2278	Car. papaya	Caricaceae	KC484781	KC481519		
	Brazil	CMM2280	Car. papaya	Caricaceae	KC484773	KC481513		
	Brazil	CMM2282	Car. papaya	Caricaceae	KC484827	KC481551		
	Brazil	CMM2294	Car. papaya	Caricaceae	KC484828	KC481552		
	Brazil	CMM2295	Car. papaya	Caricaceae	KC484774	KC481514		
	Brazil	CMM2297	Car. papaya	Caricaceae	KC484823	KC481582		
	Brazil	CMM2303	Car. papaya	Caricaceae	KC484816	KC481546		
	Brazil	CMM2306	Car. papaya	Caricaceae	KC484788	KC481570		
	Brazil	CMM2310	Car. papaya	Caricaceae	KC484782	KC481520		
	Brazil	CMM2327	Car. papaya	Caricaceae	KC484778	KC481516		
	Brazil	CMM2328	Car. papaya	Caricaceae	KC484825	KC481549		
	Brazil	CMM3612	Jat. curcas	Euphorbiaceae	KF234546	KF226692	KF254929	
	Brazil	CMM3647	Jat. curcas	Euphorbiaceae	KF234548	KF226704	KF254932	
	Brazil	CMM3654	Jat. curcas	Euphorbiaceae	KF234555	KF226716	KF254939	
	Brazil	CMM3831	Jat. curcas	Euphorbiaceae	KF234556	KF226717	KF254940	

Table 1. *Cont.*

Region	Country, Locality	Isolate	Host	Plant Family	ITS	tef1α	tub2	rpb2
	Brazil	CMM4019	*Mangifera indica*	Anacardiaceae	JX464096	JX464026		
	Brazil	CMM4021	*Man. indica*	Anacardiaceae	JX464064	JX464047		
	Brazil	CMM4033	*Man. indica*	Anacardiaceae	JX464081	JX464032		
	Brazil	CMM4039	*Man. indica*	Anacardiaceae	JX464065	JX464041		
	Brazil	CMM4041	*Man. indica*	Anacardiaceae	KC184891	JX464042		
	Brazil	CMM4042	*Man. indica*	Anacardiaceae	JX464070	JX464017		
	Brazil	CMM4043	*Man. indica*	Anacardiaceae	JX464087	JX464056		
	Brazil	CMM4046	*Man. indica*	Anacardiaceae	JX464091	JX464027		
	Brazil	CMM4047	*Man. indica*	Anacardiaceae	JX464082	JX464025		
	Brazil	CMM4048	*Man. indica*	Anacardiaceae	JX464093	JX464048		
	Brazil	CMM4050	*Man. indica*	Anacardiaceae	JX464062	JX464024		
	Brazil	CMM4499	*Anacardium occidentale*	Anacardiaceae	KT325578	KT325587		
	Brazil	CMM4508	*Ana. occidentale*	Anacardiaceae	KT325576	KT325588		
	Brazil	CMM4513	*Ana. occidentale*	Anacardiaceae	KT325577	KT325589		
	Brazil	CMW32099	Unknown	Anacardiaceae	KY473028	KY472971	KY472897	
	Brazil, Vicosa, MG	COAD 1788	*Coc. nucifera*	Arecaceae	KP244698	KP308476	KP308528	
	Brazil, Vicosa, MG	COAD 1789	*Coc. nucifera*	Arecaceae	KP244700	KP308474	KP308529	
	Brazil, Juazeiro, BA	COAD 1790	*Coc. nucifera*	Arecaceae	KP244703	KP308468	KP308530	
	Brazil, Cattuana, Ceará	IBL340	*Spondias purpurea*	Anacardiaceae	KT247466	KT247472	KT247475	
	Brazil, Itapipoca, Ceara	IBL375	*Talisia esculenta*	Sapindaceae	KT247467	KT247473	KT247474	
	Brazil, Buíque, Piauí	IBL404	*Ana. occidentale*	Anacardiaceae	KT247468	KT247470	KT247476	
	Brazil, Buíque, Piauí	IBL405	*Ana. occidentale*	Anacardiaceae	KT247469	KT247471	KT247477	
	Uruguay, Paysandú	Fi2359	*Malus domestica*	Rosaceae	KR071127	KT191041		
Western Africa	Benin	CMW33290	*Adansonia digitata*	Bombacaceae	KY473027	KY472970	KY472896	KY472828
	Cameroon, Mbalmayo-Bilink	CMW28311	*Terminalia ivorensis*	Combretaceae	GQ469932	GQ469898	KY472898	KY472829
	Cameroon, Kribi	CMW28317	*Ter. catappa*	Combretaceae	FJ900602	FJ900648	KY472899	KY472830
	Cameroon, Kribi	CMW28319	*Ter. catappa*	Combretaceae	FJ900603	FJ900650		
	Cameroon, Kribi	CMW28547	*Ter. mentaly*	Combretaceae	GQ469919	KY472972	KY472900	KY472831
	Cameroon, Kribi	CMW28548	*Ter. mentaly*	Combretaceae	GQ469920	KY472973	KY472901	KY472832
	Cameroon, Kribi	CMW28550	*Ter. mentaly*	Combretaceae	GQ469921	KY472974	KY472902	KY472833
	Cameroon, Mbalmayo-Ebogo	CMW28570	*Ter. ivorensis*	Combretaceae	GQ469923	GQ469896	KY472903	KY472834
	Cameroon, Mbalmayo-Ebogo	CMW28571	*Ter. ivorensis*	Combretaceae	GQ469924	GQ469897	KY472904	KY472835
	Cameroon, Mbalmayo-Ebogo	CMW28573	*Ter. ivorensis*	Combretaceae	GQ469925	KY472975	KY472905	KY472836
	Cameroon, Mbalmayo-Ekombitie	CMW28625	*Ter. ivorensis*	Combretaceae	GQ469933	KY472976	KY472906	KY472837
	Cameroon, Lombel	CMW36127	*Ad. digitata*	Bombacaceae	KY473029	KY472977	KY472907	
Southern and Eastern Africa	Madagascar, Madamo	CMW27810	*Ter. catappa*	Combretaceae	FJ900605	FJ900651	KY472923	KY472851
	South Africa, Mpumalanga	CMW18422	*Pin. patula*	Pinaceae	DQ103544	DQ103562		
	South Africa, Mpumalanga	CMW18423	*Pin. patula*	Pinaceae	DQ103545	DQ103563		
	South Africa, Mpumalanga	CMW18425	*Pin. patula*	Pinaceae	DQ103546	DQ103561		
	South Africa, Mpumalanga	CMW22663	*Pterocarpus angolensis*	Fabaceae	FJ888468	FJ888450		KY472864
	South Africa, Mpumalanga	CMW22664	*Pt. angolensis*	Fabaceae	FJ888469	FJ888451		KY472865
	South Africa, Kwazulu-Natal	CMW24125	*Sclerocarya birrea*	Anacardiaceae	KU997372	KU997111		KY472866
	South Africa, Mpumalanga	CMW25212	*Man. indica*	Anacardiaceae	KU997392	KU997128	KU997566	
	South Africa, Limpopo	CMW26616	*Euphorbia ingens*	Euphorbiaceae	KY473051	KY472999	KY472941	KY472867
	South Africa, Limpopo	CMW26630	*Euph. ingens*	Euphorbiaceae	KY473052	KY473000	KY472942	KY472868

Table 1. *Cont.*

Region	Country, Locality	Isolate	Host	Plant Family	ITS	tef1α	tub2	rpb2
	South Africa, Kwazulu-Natal	CMW26715	*Ter. catappa*	Combretaceae	FJ900604	FJ900649	KY472943	KY472869
	South Africa, Kwazulu-Natal	CMW32018	*Pin. elliottii*	Pinaceae	KY473053	KY473001	KY472944	KY472870
	South Africa, Mpumalanga	CMW32498	*Pin. patula*	Pinaceae	KY473054	KY473002	KY472945	KY472871
	South Africa, Mpumalanga	CMW32536	*Pin. elliottii*	Pinaceae	KY473055	KY473003	KY472946	KY472872
	South Africa, Kwazulu-Natal	CMW32544	*Pin. elliottii*	Pinaceae	KY473056	KY473004	KY472947	KY472873
	South Africa, Kwazulu-Natal	CMW32549	*Pin. elliottii*	Pinaceae	KY473057	KY473005	KY472948	KY472874
	South Africa, Kwazulu-Natal	CMW32571	*Pin. elliottii*	Pinaceae	KY473058	KY473006	KY472949	KY472875
	South Africa, Kwazulu-Natal	CMW32603	*Pin. elliottii*	Pinaceae	KY473059	KY473007	KY472950	KY472876
	South Africa, Kwazulu-Natal	CMW32604	*Pin. elliottii*	Pinaceae	KY473060	KY473008	KY472951	KY472877
	South Africa, Kwazulu-Natal	CMW32606	*Pin. elliottii*	Pinaceae	KY473061	KY473009	KY472952	KY472878
	South Africa, Kwazulu-Natal	CMW32651	*Pin. elliottii*	Pinaceae	KY473062	KY473010	KY472953	KY472879
	South Africa, Kwazulu-Natal	CMW32666	*Pin. elliottii*	Pinaceae	KY473063	KY473011	KY472954	
	South Africa, Kwazulu-Natal	CMW32669	*Pin. elliottii*	Pinaceae	KY473064	KY473012	KY472955	KY472880
	South Africa, Mpumalanga	CMW33658	*Man. indica*	Anacardiaceae	KY473065	KY473013	KY472956	
	South Africa, Gauteng	CMW38120	*Vachellia karroo*	Fabaceae	KC769935	KC769843	KC769887	
	South Africa, Gauteng	CMW38121	*Vac. karroo*	Fabaceae	KC769936	KC769844	KC769888	
	South Africa, Gauteng	CMW38122	*Vac. karroo*	Fabaceae	KC769937	KC769845	KC769889	
	South Africa, Gauteng	CMW39290	*Vac. karroo*	Fabaceae	KF270061	KF270021		
	South Africa, Gauteng	CMW39291	*Vac. karroo*	Fabaceae	KF270062	KF270022		
	South Africa, Kwazulu-Natal	CMW41214	*Barringtonia racemosa*	Lecythidaceae	KP860842	KU666547	KP860765	KU587889
	South Africa, Kwazulu-Natal	CMW41222	*Bar. racemosa*	Lecythidaceae	KP860836	KU666549	KP860759	KU587881
	South Africa, Kwazulu-Natal	CMW41223	*Bar. racemosa*	Lecythidaceae	KP860837	KU666548	KP860760	KU587882
	South Africa, Kwazulu-Natal	CMW41360	*Bar. racemosa*	Lecythidaceae	KP860841	KP860686	KP860764	KU587888
	South Africa, Kwazulu-Natal	CMW42341	*Bar. racemosa*	Lecythidaceae	KP860843	KU587945	KU587866	
	South Africa, Kwazulu-Natal	MTU53	*Syzygium cordatum*	Myrtaceae	KY052943	KY024622	KY000125	
	Uganda	CMW10130	*Vitex donniana*	Lamiaceae	AY236951	AY236900	AY236929	KY472883
	Uganda, Mbale	CMW18420	*Casuarina cunninghamii*	Casuarinaceae	DQ103534	DQ103564	KY472959	KY472884
	Uganda, Mbale	CMW32245	*Cas. cunninghamii*	Casuarinaceae	KY473068	KY473016	KY472960	KY472885
	Uganda, Mbale	CMW32246	*Cas. cunninghamii*	Casuarinaceae	KY473069	KY473017	KY472961	KY472886
	Zambia, Samfya	CMW30103	*Syg. cordatum*	Myrtaceae	FJ747640	FJ871114		
	Zambia, Samfya	CMW30104	*Syg. cordatum*	Myrtaceae	FJ747641	FJ871115		
	Zambia, Samfya	CMW30105	*Syg. cordatum*	Myrtaceae	FJ747642	FJ871116		

Table 1. Cont.

Region	Country, Locality	Isolate	Host	Plant Family	ITS	tef1α	tub2	rpb2
Middle East and Europe	Egypt	BOT23	*Man. indica*	Anacardiaceae	JN814400	JN814427		
	Egypt	BOT4	*Man. indica*	Anacardiaceae	JN814395	JN814422		
	Egypt	BOT5	*Man. indica*	Anacardiaceae	JN814376	JN814403		
	Egypt	BOT6	*Man. indica*	Anacardiaceae	JN814399	JN814426		
	Egypt	BOT7	*Man. indica*	Anacardiaceae	JN814396	JN814423		
	Egypt	BOT9	*Man. indica*	Anacardiaceae	JN814392	JN814419		
	Iran	CJA198	Unknown		GU973871	GU973863		
	Iran	CJA199	Unknown		GU973872	GU973864		
	Iran	IRAN1233C	Unknown	Anacardiaceae	GU973868	GU973860		
	Iran	IRAN1496C	*Man. indica*	Anacardiaceae	GU973869	GU973861		
	Iran	IRAN1499C	*Man. indica*	Anacardiaceae	GU973870	GU973862		
	Italy, Foggia	B159	*Vis. vinifera*	Vitaceae	KM675760	KM822731		
	Italy, Cerignola	B202	*Vis. vinifera*	Vitaceae	KM675761	KM822732		
	Italy, Cerignola	B215	*Vis. vinifera*	Vitaceae	KM675762	KM822733		
	Italy, Cerignola	B342	*Vis. vinifera*	Vitaceae	KM675763	KM822734		
	Italy, Cerignola	B85	*Vis. vinifera*	Vitaceae	KM675759	KM822730		
	Oman, Barka	CMW20506	*Man. indica*	Anacardiaceae	KY473037	KY472985	KY472924	KY472852
	Oman, Barka	CMW20508	*Man. indica*	Anacardiaceae	KY473038	KY472986	KY472925	KY472853
	Oman, Barka	CMW20511	*Man. indica*	Anacardiaceae	KY473039	KY472987	KY472926	KY472854
	Oman, Barka	CMW20512	*Man. indica*	Anacardiaceae	KY473040	KY472988	KY472927	KY472855
	Oman	CMW20537	Unknown		KY473041	KY472989	KY472928	KY472856
	Oman	CMW20542	Unknown		KY473042	KY472990	KY472929	
	Oman	CMW20543	Unknown		KY473043	KY472991	KY472930	KY472857
	Oman	CMW20546	Unknown		KY473044	KY472992	KY472931	KY472858
	Oman	CMW20560	Unknown		KY473045	KY472993	KY472932	KY472859
	Oman	CMW20573	Unknown		KY473046	KY472994	KY472933	KY472860
	Oman	CMW20579	Unknown		KY473047	KY472995	KY472934	KY472861
Asia	China, Fangshan, Pingtung	B838	*Man. indica*	Anacardiaceae	GQ502456	GQ980001	GU056852	
	China, Guantian, Tainan	B852	*Man. indica*	Anacardiaceae	GQ502457	GQ980002	GU056851	
	China, Chiayi	B886	*Man. indica*	Anacardiaceae	GQ502452	GQ980005	GU056847	
	China, Guantian, Tainan	B902	*Man. indica*	Anacardiaceae	GQ502459	GQ980004	GU056849	
	China, Guantian, Tainan	B918	*Man. indica*	Anacardiaceae	GQ502458	GQ980003	GU056850	
	China, Guantian, Tainan	B961	*Man. indica*	Anacardiaceae	GQ502453	GQ979999	GU056845	
	China, Guantian, Tainan	B965	*Man. indica*	Anacardiaceae	GQ502454	GQ980000	GU056854	
	China	BL1331	*Albizia falcataria*	Fabaceae	KU712499	KU712500	KU712501	
	China	CBS122127	*Homo sapiens*		EF622017	EF622018		
	China, GuangDong Province	CERC1983	*Polyscias balfouriana*	Araliaceae	KP822979	KP822997	KP823012	
	China, GuangDong Province	CERC1985	*Pol. balfouriana*	Araliaceae	KP822980	KP822998	KP823013	
	China, GuangDong Province	CERC1988	*Pol. balfouriana*	Araliaceae	KP822981	KP822999	KP823014	
	China, GuangDong Province	CERC1989	*Euc. GU hybrid*	Myrtaceae	KP822982	KP823000	KP823015	
	China, GuangDong Province	CERC1991	*Euc. GU hybrid*	Myrtaceae	KP822983	KP823001	KP823016	
	China, GuangDong Province	CERC1996	*Euc. GU hybrid*	Myrtaceae	KP822984	KP823002	KP823017	
	China, GuangDong Province	CERC2049	*Bougainvillea spectabilis*	Nyctaginaceae	KP822985	KP823003	KP823018	
	China, GuangDong Province	CERC3820	*Rosa rugosa*	Rosaceae	KR816831	KR816837	KR816843	

Table 1. *Cont.*

Region	Country, Locality	Isolate	Host	Plant Family	ITS	*tef1α*	*tub2*	*rpb2*
	China, GuangDong Province	CERC3821	*R. rugosa*	Rosaceae	*KR816832*	*KR816838*	*KR816844*	
	China, GuangDong Province	CERC3822	*R. rugosa*	Rosaceae	*KR816833*	*KR816839*	*KR816845*	
	China, GuangDong Province	CERC3823	*R. rugosa*	Rosaceae	*KR816834*	*KR816840*	*KR816846*	
	China, GuangDong Province	CERC3824	*R. rugosa*	Rosaceae	*KR816835*	*KR816841*	*KR816847*	
	China, GuangDong Province	CERC3825	*R. rugosa*	Rosaceae	*KR816836*	*KR816842*	*KR816848*	
	China, Dong Men Forest Farm	CMW24701	Euc. GU hybrid	Myrtaceae	*HQ332193*	*HQ332209*	*KY472908*	*KY472838*
	China, Dong Men Forest Farm	CMW24702	Euc. GU hybrid	Myrtaceae	*HQ332194*	*HQ332210*	*KY472909*	*KY472839*
	China	CMW33957	*Eucalyptus* sp.	Myrtaceae	*KY473030*	*KY472978*	*KY472910*	
	China	FXPZ	*Vis. vinifera*	Vitaceae	*KR232666*	*KR232660*	*KR232674*	
	China	HD1332	*Alb. falcataria*	Fabaceae	*KU712502*	*KU712503*	*KU712504*	
	China	HN74	*Hevea brasiliensis*	Euphorbiaceae	*KT947466*	*KU925617*	*KU925616*	
	China, Guangxi Province	L1	*Man. indica*	Anacardiaceae	*KR260791*	*KR260808*	*KR260820*	
	China, Guangxi Province	L2	*Man. indica*	Anacardiaceae	*KR260792*	*KR260809*	*KR260821*	
	China, Guangxi Province	L3	*Man. indica*	Anacardiaceae	*KR260793*	*KR260810*	*KR260822*	
	China, Guangxi Province	L4	*Man. indica*	Anacardiaceae	*KR260794*	*KR260811*	*KR260823*	
	China, Guangxi Province	L5	*Man. indica*	Anacardiaceae	*KR260795*	*KR260812*	*KR260824*	
	China, Guangxi Province	L6	*Man. indica*	Anacardiaceae	*KR260796*	*KR260813*	*KR260825*	
	China, Guangxi Province	L7	*Man. indica*	Anacardiaceae	*KR260797*	*KR260814*	*KR260826*	
	China, Guangxi Province	L8	*Man. indica*	Anacardiaceae	*KR260798*	*KR260815*	*KR260827*	
	China, Guangxi Province	L9	*Man. indica*	Anacardiaceae	*KR260799*	*KR260816*	*KR260828*	
	China, Guangxi Province	L10	*Man. indica*	Anacardiaceae	*KR260800*	*KR260817*	*KR260829*	
	China, Guangxi Province	L11	*Man. indica*	Anacardiaceae	*KR260801*	*KR260818*	*KR260830*	
	China, Guangxi Province	L15	*Man. indica*	Anacardiaceae	*KR260802*	*KR260819*	*KR260831*	
	China, Sichuan	Mht-5	*Actinidia deliciosa*	Actinidiaceae	*JQ658976*	*JQ658977*	*JQ658978*	
	China, Shanghai	SHYAG	*Vitis vinifera*	Vitaceae	*JX275794*	*JX462302*	*JX462276*	
	China, Zhejiang	ZHn411	*Pyrus pyrifolia*	Rosaceae	*KC960899*	*KC961038*	*KC960992*	
	Indonesia, Sumatra	CMW22881	*Euc. grandis*	Myrtaceae	*KY473036*	*KY472984*	*KY472917*	*KY472845*
	Indonesia, Logas	CMW23003	*Ac. mangium*	Fabaceae	*EU588629*	*EU588609*	*KY472918*	*KY472846*
	Indonesia, Logas	CMW23008	*Ac. mangium*	Fabaceae	*EU588630*	*EU588610*	*KY472919*	*KY472847*
	Indonesia, Logas	CMW23018	*Ac. mangium*	Fabaceae	*EU588631*	*EU588611*	*KY472920*	*KY472848*
	Indonesia, Teso	CMW23031	*Ac. mangium*	Fabaceae	*EU588632*	*EU588612*	*KY472921*	*KY472849*
	Indonesia, Logas	CMW23073	*Ac. mangium*	Fabaceae	*EU588633*	*EU588613*	*KY472922*	*KY472850*
	Korea	ML1001	*Man. indica*	Anacardiaceae	*JN542561*	*JN542563*		
	Korea	ML1005	*Man. indica*	Anacardiaceae	*JN542562*	*JN542564*		
	Thailand, Prajinburi	CMW15680	*Euc. camaldulensis*	Myrtaceae	*KY473066*	*KY473014*	*KY472957*	*KY472881*
	Thailand, Prajinburi	CMW15682	*Euc. camaldulensis*	Myrtaceae	*KY473067*	*KY473015*	*KY472958*	*KY472882*
	Thailand, Chiang Mai	CPC 22766	*Pin. kesiya*	Pinaceae	*KM006436*	*KM006467*		
	Thailand, Chiang Mai	CPC 22780	*Manilkara zapota*	Sapotaceae	*KM006442*	*KM006473*		
	Thailand, Chiang Mai	CPC 22798	*Syz. samarangense*	Myrtaceae	*KM006454*	*KM006485*		
	Thailand, Chiang Mai	MFLUCC12 0293	*Tectona grandis*	Lamiaceae	*KM396896*	*KM409634*	*KM510354*	
Australasia	Australia	CMW40630	*Syzygium* sp.	Myrtaceae	*KY473023*	*KY472966*	*KY472892*	*KY472825*
	Australia	CMW40635	*Syz. neorosum*	Myrtaceae	*KY473024*	*KY472967*	*KY472893*	
	Australia	CMW40636	*Syz. neorosum*	Myrtaceae	*KY473025*	*KY472968*	*KY472894*	*KY472826*
	Australia	CMW40637	*Syz. neorosum*	Myrtaceae	*KY473026*	*KY472969*	*KY472895*	*KY472827*
	Darwin, Australia	MUCC737	*Ad. gregorii*	Bombacaceae	*GU199387*	*GU199407*		
	Papua New Guinea, Madang	CBS164.96	Fruit along coral reef coast		*AY640255*	*AY640258*	*KU887532*	*KU696383*

2.2. PCR Amplifications, DNA Sequencing, and Confirmation of Species Identity

Isolate identities were confirmed as *L. theobromae* using data from four loci; the ITS rDNA (including the ITS1, 5.8S nuclear ribosomal RNA (nrRNA) and ITS2), *tef1α*, β-tubulin-2 (*tub2*) and RNA polymerase II (*rpb2*) loci. Preliminary identification was done for all isolates using maximum likelihood phylogenetic analysis of sequence data from the *tef1α* locus, which was then supported by data for the other three loci. The dataset for *tef1α* included all other *Lasiodiplodia* species known at the time of the analyses.

For PCR amplifications, the primer sets ITS1 and ITS4 [32], EF1F and EF2R [33], EF688F and EF1251R [34], Bt-2a and Bt-2b [35], and RPB2-LasF and RPB2-LasR [27] were used to amplify the ITS, *tef1α*, *tub2*, and *rpb2* loci, respectively. PCR mixes were the same as those that included KAPA Taq and MyTaq DNA polymerases as described by Mehl et al. [36] and PCR cycling conditions and product visualization were the same as those used by Mehl et al. [37]. PCR product purification and sequencing were done as described by Mehl et al. [30] and sequences were examined and edited using MEGA 6 [38].

Sequence datasets were aligned using MAFFT 6 [39] with the G-INS-I algorithm selected and alignment errors corrected visually. For the *tef1α* dataset that included isolates of species other than *L. theobromae*, the best nucleotide substitution model was determined using JMODELTEST 2.1.3 [40] with the corrected Akaike Information Criterion selected. The dataset was analyzed with PHYML 3.0.1 [41] using the same model parameters as determined by JMODELTEST and the robustness of the generated tree was evaluated using 1000 bootstrap replicates. Sequences generated in this study were deposited in GenBank (Table 1).

2.3. Haplotype Assignment and Networks

To ascertain the number of haplotypes for each dataset and to identify where haplotypes occurred, sequence datasets were generated for each locus separately, along with one combined dataset for the ITS and *tef1α* regions. The combined dataset was generated because it included the majority of isolates and provided a better representation of the diversity inherent in the populations and regions. For each dataset, isolates were assigned to different haplotypes using the map program in Mobyle SNAP Workbench [42]. Sites that violated the infinite sites model, as well as indels, were removed prior to assigning haplotypes. Median joining haplotype networks were then constructed for each dataset, as well as for the combined dataset using NETWORK 4.6.1.3 [43,44].

2.4. Population and Regional Structure and Diversity

To determine whether there was genetic structure present in the datasets and to test for potential population subdivision, haplotype assignments for all four loci, as determined by Mobyle SNAP Workbench, were analyzed using the program STRUCTURE 2.3.4 [45,46]. STRUCTURE uses a Bayesian clustering algorithm to evaluate the possibility of multiple lineages being present. Two sets of analyses were made, the first of which evaluated whether there was genetic structure in the dataset for all isolates. The second set of analyses involved grouping isolates into five populations based on the continent of origin (North America, South America, Africa, Eurasia, and Australasia) and then running STRUCTURE analyses on pairs of populations to determine whether there was genetic structure between any of the populations (10 pairs including every possible combination).

For all analyses, burnin was set at 300,000 and the number of Markov Chain Monte Carlo (MCMC) repeats was set at 900,000, so that more than 1,000,000 repeats were done to generate robust results. Initially lambda was computed based on five runs at K = 1. The model selected entailed admixture with independent allele frequencies and the lambda value computed. Twenty iterations were done for each value of K = 1 to K = 10. Results were parsed through STRUCTURE HARVESTER [47] and the DeltaK [48] output used to identify possible subpopulations.

Population statistics, including gene and nucleotide diversities, were inferred using ARLEQUIN 3.5.1.2 [49] on the ITS, *tef1α*, combined ITS and *tef1α*, and *tub2* sequence datasets for every geographic country and region assigned. Pairwise population differentiation (Φ_{ST}) comparisons were computed for all populations and regions using ARLEQUIN on the same dataset.

2.5. Putative Geographic Origin of Lasiodiplodia theobromae

To determine the possible centre of origin for *L. theobromae*, scenarios of how populations could have arisen were simulated and the summary statistics of these compared to those of the observed dataset using DIYABC 2.0.4 [50]. For these analyses, the sequence datasets of isolates (with data from all four loci) were grouped according to continent of origin, similar to the arrangements for the second set of analyses using STRUCTURE. To determine whether any of the populations could be ancestral, pairs of populations were evaluated using three possible scenarios (Figure 2): scenario 1—the first population is ancestral to both, scenario 2—the second population is ancestral to both, scenario 3—both populations diverged from an unknown ancestral population. For each scenario, 1,000,000 datasets were simulated.

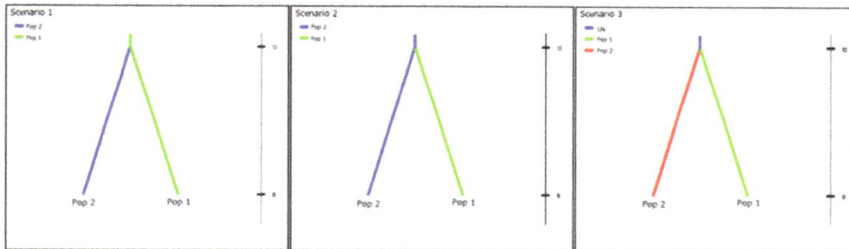

Figure 2. Scenarios evaluated to determine possible ancestry between any of the pairs of populations tested. In scenario 1, population 1 is ancestral to both. In scenario 2, population 2 is ancestral to both. In scenario 3, both populations diverged from an unknown source population.

Posterior probabilities of scenarios for each analysis step were computed using polychotomous logistic regression on 1% of the simulated datasets closest to the dataset provided. The best scenario was the one having the highest probability and with 95% confidence intervals that did not overlap with those of the other scenarios tested.

3. Results

3.1. Isolate Collections and Confirmation of Species Identity

The *tef1α* sequence dataset that included all isolates, as well as representatives of other *Lasiodiplodia* species, consisted of 340 characters (151 parsimony informative, 22 parsimony uninformative, 167 constant). The model selected by JMODELTEST was HKY (transitions:transversions (ti/tv) = 1.719, $\gamma = 0.407$). The resulting tree contained a clade of 255 isolates, from 26 countries, that was considered to represent *L. theobromae sensu stricto* as it included authentic isolates of this species (Figure S1). Of these, 95 isolates represented a global collection assembled over many years and stored in the CMW culture collection. The other isolates sampled from this collection grouped with *Botryosphaeria dothidea*, *D. pseudoseriata*, *L. brasiliense*, *L. crassispora*, *L. gilanensis*, *L. gonubiensis*, *L. hormozganensis*, *L. iraniensis*, *L. laeliocattleyae*, *L. mahajangana*, *L. margaritacea*, *L. parva*, *L. pseudotheobromae*, *L. viticola*, *Neofusicoccum parvum*, and *N. vitifusiforme* (data not shown) and were thus excluded. Four isolates were from the collection of the Westerdijk Fungal Biodiveristy Institute. The remaining sequences for 156 additional isolates were sourced from GenBank (Table 1, Figure 1). Thus, all subsequent analyses were based on data for this core group of 255 isolates from 52 plant hosts.

Countries considered in the analyses were grouped into eight geographic regions, including north America (Hawaii, Mexico, Puerto Rico, United States of America—USA), western south America (Colombia, Ecuador, Peru, Venezuela), eastern south America (Brazil, Uruguay), western Africa (Benin, Cameroon), southern and eastern Africa (Madagascar, South Africa, Uganda, Zambia), Middle East and Europe (Egypt, Iran, Italy, Oman), Asia (China, Indonesia, Korea, Thailand), and Australasia (Australia, Papua New Guinea) (Tables 1 and 2).

3.2. Haplotype Assignment and Networks

The ITS dataset (252 isolates) consisted of 333 characters (two parsimony informative, 23 parsimony uninformative, 308 constant) and yielded 11 haplotypes with 17 fixed single nucleotide polymorphisms (SNPs) (Table S1, Figure 3a). The *tef1α* dataset (255 isolates) consisted of 216 characters (five parsimony informative, 11 parsimony uninformative, 200 constant) and yielded eight haplotypes with 14 SNPs (Table S1, Figure 3b). The *tub2* dataset (153 isolates) consisted of 309 characters (six parsimony informative, nine parsimony uninformative, 294 constant) and yielded 12 haplotypes with 15 SNPs (Table S1, Figure 3c). The *rpb2* dataset (73 isolates) consisted of 535 characters (zero parsimony informative, zero parsimony uninformative, 535 constant) and yielded a single haplotype. The combined ITS and *tef1α* dataset consisted of 549 characters (seven parsimony informative, 34 parsimony uninformative, 508 constant) and yielded 17 haplotypes (Figure 4).

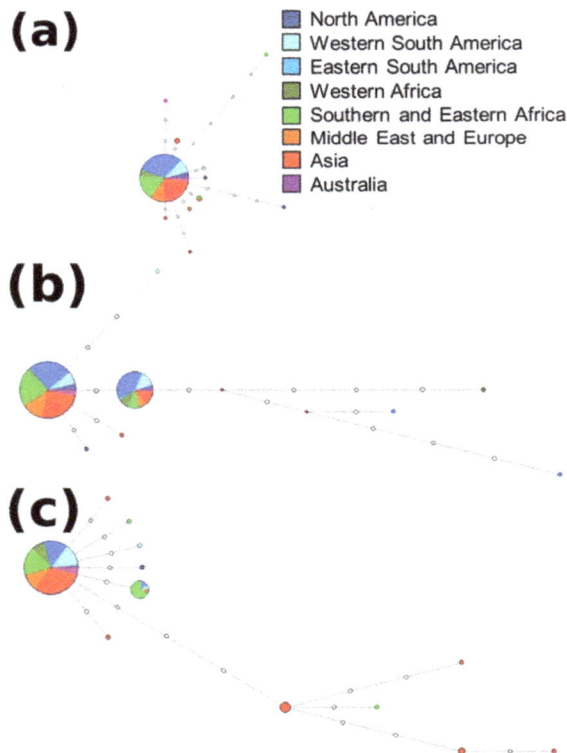

Figure 3. Haplotype networks generated for the (**a**) internal transcribed spacer rDNA (ITS), (**b**) translation elongation factor 1α (*tef1α*), and (**c**) β-tubulin-2 (*tub2*) loci. Only one haplotype resulted from analysis of the RNA polymerase II (*rpb2*) locus and is not included. Colours represent the different regions isolates were obtained from.

Table 2. Standard genetic and nucleotide diversity measures for isolates collected in each country and region, for the ITS, tef1α, combined ITS and tef1α, and tub2 sequence datasets. Included are sample size (N), number of haplotypes found (H), gene diversity (H_E) and nucleotide diversity (π). Sample sizes are also recorded for the tub2 dataset as sequence data for this locus was not available for all isolates. Totals for each region are also listed.

Region	Country	ITS				tef1α			ITS + tef1α			tub2			
		N	H	H_E	π ($\times10^{-3}$)	H	H_E	π ($\times10^{-3}$)	H	H_E	π ($\times10^{-3}$)	N	H	H_E	π ($\times10^{-3}$)
North America	Hawaii	1	1	0	0	1	0	0	1	0	0	1	1	0	0
	Mexico	5	1	0	0	1	0	0	1	0	0	0	0	0	0
	Puerto Rico	4	1	0	0	1	0	0	1	0	0	4	1	0	0
	USA	5	3	0.356	4.271	2	0.356	1.646	4	0.385	4.210	2	2	0.667	2.157
	Total	15	3	0.129	1.546	3	0.405	3.746	4	0.193	2.110	7	2	0.264	0.854
Western South America	Colombia	1	1	0	0	1	0	0	1	0	0	0	0	0	0
	Ecuador	6	1	0	0	3	0.384	5.331	3	0.384	2.097	6	1	0	0
	Peru	6	1	0	0	2	0.303	1.403	2	0.303	0.552	3	2	0.533	1.726
	Venezuela	6	2	0.303	0.910	2	0.303	1.403	3	0.303	1.104	6	2	0.303	0.981
	Total	19	2	0.102	0.308	3	0.201	2.792	4	0.176	1.285	15	3	0.129	0.833
Eastern South America	Brazil	76	1	0	0	4	0.097	3.131	4	0.097	1.232	19	2	0.185	0.597
	Uruguay	1	1	0	0	1	0	0	1	0	0	0	0	0	0
	Total	77	1	0	0	4	0.096	3.106	4	0.096	1.222	19	2	0.185	0.597
Western Africa	Benin	1	1	0	0	1	0	0	1	0	0	1	1	0	0
	Cameroon	11	1	0	0	3	0.222	5.131	3	0.222	2.019	10	1	0	0
	Total	12	1	0	0	3	0.220	5.099	3	0.220	2.006	11	1	0	0
Southern and Eastern Africa	Madagascar	1	1	0	0	1	0	0	1	0	0	1	1	0	0
	South Africa	32	3	0.064	0.953	3	0.112	0.520	5	0.062	0.560	29	4	0.119	2.302
	Uganda	4	1	0	0	2	0.429	1.984	2	0.429	0.781	4	2	0.429	1.387
	Zambia	3	1	0	0	1	0	0	1	0	0	0	0	0	0
	Total	40	3	0.051	0.760	4	0.248	1.147	6	0.072	0.782	34	4	0.119	2.301
Middle East and Europe	Egypt	6	1	0	0	1	0	0	1	0	0	0	0	0	0
	Iran	5	1	0	0	1	0	0	1	0	0	0	0	0	0
	Italy	5	1	0	0	1	0	0	1	0	0	0	0	0	0
	Oman	11	3	0.173	1.040	3	0.308	1.424	3	0.173	0.631	11	2	0.173	0.560
	Total	27	3	0.073	0.436	4	0.308	1.424	5	0.151	0.825	11	2	0.173	0.560
Asia	China	43	3	0.606	0.546	3	0.108	1.003	5	0.080	0.726	42	7	0.153	4.939
	Indonesia	6	1	0	0	2	0.485	2.245	2	0.485	0.883	6	1	0	0
	Korea	2	1	0	0	1	0	0	1	0	0	1	0	0	0
	Thailand	6	2	0.303	0.910	3	0.485	2.245	3	0.394	1.435	3	1	0	0
	Total	57	4	0.043	0.518	6	0.139	1.288	6	0.075	0.821	51	7	0.130	4.202
Australasia	Australia	5	2	0.356	2.135	2	0.356		2	0.356	1.295	4	1	0	0
	Papua New Guinea	1	1	0	0	1	0	0	1	0	0	1	1	0	0
	Total	6	2	0.303	1.820	3	0.303	1.403	2	0.303	1.656	5	1	0	0
All		255	11	0.001		8	0.003		17	0.001		153	12	0.002	

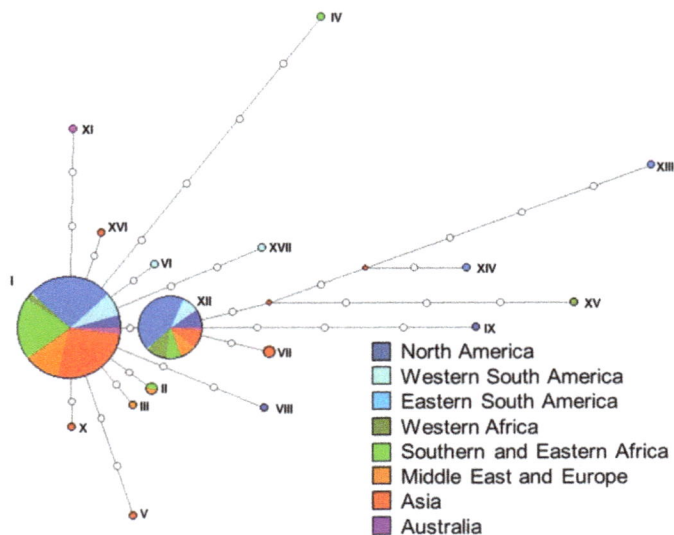

Figure 4. Haplotype network generated for the combined ITS and *tef1α* dataset. Colours represent the different regions isolates were obtained from. Haplotypes designated by Roman numerals (I–XVII). Open circles represent inferred haplotypes.

There was no clear grouping of isolates based on region of origin. Analyses of the ITS and *tub2* loci (Figure 3) showed that one haplotype was most common. The *rpb2* dataset was not analyzed further as it constituted only one haplotype. For the *tef1α* dataset and the combined dataset of ITS and *tef1α*, two closely related (separated by a single mutation) haplotypes were most common. These common haplotypes represented isolates sourced from all eight regions sampled (Figures 3 and 4, Table S2).

An analysis of haplotypes (Table S3) showed that Asia and North America had the greatest number of unique haplotypes (10 and four, respectively) across all three loci (ITS, *tef1α*, and *tub2*). For the remaining regions, one to three unique haplotypes were detected. When considering the individual loci, three unique ITS haplotypes and six unique *tub2* haplotypes were observed amongst isolates from Asia. For all other regions, two or fewer unique haplotypes were found. Upon closer examination, these unique haplotypes were confined to specific countries. Two of the five isolates collected from the USA (North America) had unique haplotypes, while 15 isolates collected from three locations in China over a period of four years had unique haplotypes.

3.3. Population and Regional Structure and Diversity

There was no evidence of sub-populations present in either set of the STRUCTURE analyses. In the first set of analyses that considered all isolates, the significantly highest DeltaK value was at K = 8 populations, but the corresponding barplot showed that no structure was present (Figure 5). Similarly, in the second set of analyses that evaluated genetic structure between the pairs of populations, the highest DeltaK values obtained differed for each population pair tested and varied from K = 2 to K = 8. However, the corresponding barplots for these values of K all showed that no structure was present in the data (Figure S2a–j).

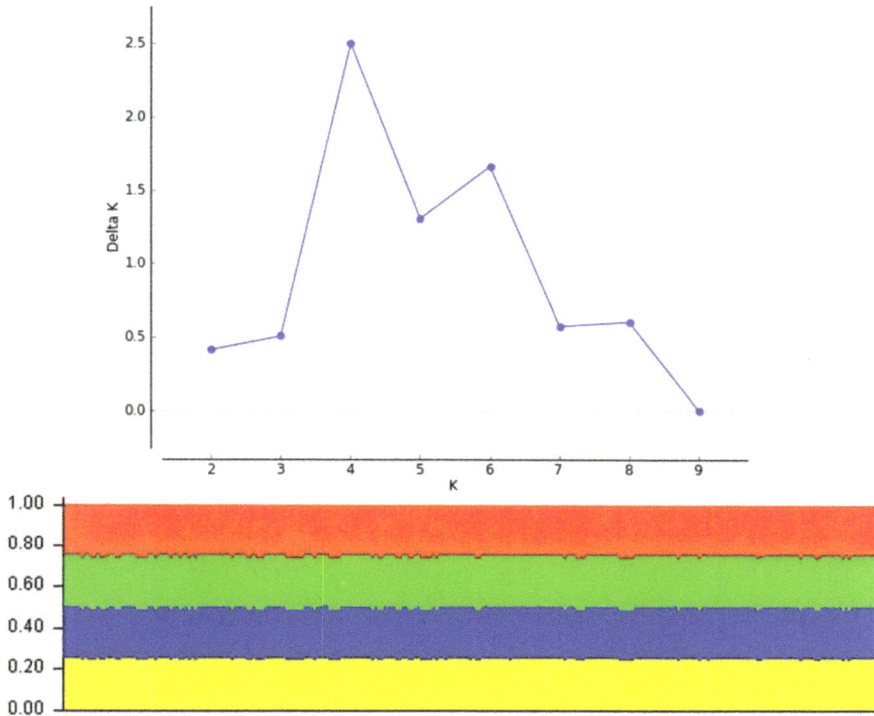

Figure 5. Structure output on the combined dataset of all four loci. The output from the DeltaK analysis from STRUCTURE HARVESTER (top) resulted in the highest peak at K = 8 populations, but the corresponding barplot (bottom) showed no structure.

Gene diversity was low for most countries and regions sampled. High gene diversity (>0.4) was detected for individual loci in countries including USA, Peru, Uganda, China, Indonesia, and Thailand, and in North America (Table 2). High nucleotide diversity was detected in the above-mentioned countries, as well as in Ecuador, Venezuela, Brazil, Cameroon, South Africa, Oman and Australia, and in several regions including western and eastern South America, western Africa, and Australasia (Table 2).

When combining the gene and nucleotide diversities across the three individual loci (ITS, *tef1α*, *tub2*) (Table 2), the greatest diversity overall was recorded for North America (H_E = 0.798, $\pi \times 10^{-3}$ = 6.146). High gene diversity (H_E > 0.4) was also detected for Australasia (H_E = 0.606, $\pi \times 10^{-3}$ = 6.146), Middle East and Europe (H_E = 0.554, $\pi \times 10^{-3}$ = 2.420), western South America (emphH_E = 0.432, $\pi \times 10^{-3}$ = 3.933), and southern and eastern Africa (H_E = 0.418, $\pi \times 10^{-3}$ = 4.208). Asia and western Africa had low levels of gene diversity, but high levels of nucleotide diversity (Asia: H_E = 0.312, $\pi \times 10^{-3}$ = 6.008; western Africa: H_E = 0.220, $\pi \times 10^{-3}$ = 5.099). Eastern South America had the lowest diversity overall (H_E = 0.281, $\pi \times 10^{-3}$ = 3.703).

Most populations were not highly genetically differentiated, based on Φ_{ST} values. The greatest genetic differentiation was seen in the north American and western African populations, with moderate to very high levels of genetic differentiation [51] compared to the other populations assessed (Table 3).

Table 3. Pairwise population differentiation (Φ_{ST}) comparisons between the regions that isolates were obtained from, based on the combined ITS and *tef1α* dataset.

Region	N	North America	Western South America	Eastern South America	Western Africa	Southern and Eastern Africa	Middle East and Europe	Asia	Australasia
North America	15								
Western South America	19	0.047							
Eastern South America	77	0.026	0.014						
Western Africa	12	0.040	0.165	0.121	-				
Southern and Eastern Africa	40	0.189	0.051	0.105	0.367				
Middle East and Europe	27	0.109	0.008	0.045	0.272	0.01			
Asia	57	0.166	0.032	0.080	0.343	0.006	0.002		
Australasia	6	0.087	0.041	0.087	0.205	0.075	0.056	0.068	

3.4. Putative Geographic Origin of Lasiodiplodia theobromae

Posterior probabilities for all of the scenarios tested for the pairs of populations were low (Table S3) when a posterior probability of 0.7 or more was considered high. Ninety-five percent (95%) confidence intervals for different scenarios for the same pairwise comparison often overlapped (Table S4), indicating a lack of resolution in choosing one specific scenario over the others. These results are likely due to the lack of variation in the markers. However, they support the conclusions of other analyses reported above that did not identify any specific region as an evolutionary origin of the fungus over others.

4. Discussion

Results of this study suggest that isolates associated with *L. theobromae* collected from many different hosts and countries of the world represent a single globally distributed species, with no obvious phylogeographic structure. This was evident from various analyses on sequence datasets for four loci (only three of which were variable) in 255 isolates from 52 hosts from all continents other than Antarctica. We thus contend that the only likely explanation for this result is the large-scale human dispersal of this fungal species.

The lack of population structure in *L. theobromae* on a global scale is in contrast to studies on other broadly distributed fungi that infect commercially cultivated plants or are medically important (e.g., [52–54]). These previous studies have typically revealed phylogeographic structure within species, with multiple cryptic lineages linked to geographic regions, leading to the conclusion that, for fungi, "nothing is generally everywhere" [54,55]. Subsequent studies have shown that lineages in some of these fungi (e.g., *Fusarium graminearum* and *Histoplasma capsulatum*) represent cryptic species [56,57]. An exception to this rule is *Aspergillus fumigatus*, which has very small (2–3 µm), wind-dispersed conidia. This special case is hypothesized to possibly arise from human influence, especially through environmental impact, which has created ideal habitats for the fungus [58,59].

Amongst the Botryosphaeriaceae, the shared genetic diversity across continents is not unique to *L. theobromae*. *Neofusicoccum parvum* also appears to have a similar global distribution of diversity [18]. Recently, Marsberg et al. [19] reported a similar lack of structure amongst a global collection of *B. dothidea* isolates. All three of these species have exceptionally broad host ranges across many plant families, and this has no doubt facilitated their broad distribution. Furthermore, *N. parvum* was reported to be more common in human-associated and disturbed environments, such as plantations, orchards, and urban environments [15], which could facilitate invasion (similar to *A. fumigatus*).

Lasiodiplodia theobromae, *B. dothidea*, and *N. parvum* are ideal systems in which to further test these hypotheses regarding the role of host and human association in facilitating invasions.

The absence of phylogeographic structure amongst global collections of Botryosphaeriaceae such as *L. theobromae* is surprising in the light of their spore dispersal mechanism. Spores of the Botryosphaeriaceae, including those of *L. theobromae*, emerge in a sticky matrix and are relatively large (the most common spores, conidia, range between $10–35 \times 8–15$ μm; [12]) and are naturally dispersed by wind and rain splash [6,16,60–62]. Consequently spores are not expected to be spread over large distances or across geographic barriers and certainly not between continents. The limited ability of these fungi to disperse over long distances would be expected to result in a vicariant population structure with differences at a regional level between populations. The lack of population structure and dominance of identical multilocus haplotypes on distant continents can only be explained by assisted dispersal. In this case, human-mediated movement of plant material [1,3,63] has most likely facilitated this global dispersal.

A large number of the plant hosts from which isolates of *L. theobromae* were obtained for the present study are commercially important and traded globally as part of the nursery trade, or cultivated either for agriculture (e.g., *Carica papaya*, *Mangifera indica*, and *Vitis vinifera*) or forestry (e.g., *Acacia mangium*, *Eucalyptus* species). The Botryosphaeriaceae, including *L. theobromae*, are common endophytes in such plants and plant products, including fruits [4,64]. Endophytic infections by these fungi are typically invisible and are thus not detected by quarantine systems [3,19,65]. The present study highlights how widely species of the Botryosphaeriaceae, specifically *L. theobromae*, can be spread as a consequence of such human-assisted movement.

Results of this study were consistent with those of previous studies that used microsatellite markers to study populations of *L. theobromae* [66–68]. These previous studies considered populations of isolates from Mexico, South Africa, Venezuela, India, and Cameroon, and detected extensive gene flow and shared genotypes from different hosts [66–68] and from different countries [66]. Our analyses provide a broader representation with consistent results, including publicly available data combined with data from our own collection of *L. theobromae* isolates.

No clear centre of origin for *L. theobromae* emerged from this study based on gene diversity. The greatest cumulative diversity obtained by combining the diversities for the individual loci was detected for the North American collections. Population differentiation tests highlighted the North American and west African populations as being moderately to fairly distinct from the rest. The North American and Asian regions had higher numbers of unique haplotypes (four and ten respectively), but these haplotypes were present only in some countries (USA and China, respectively).

The diversity of *L. theobromae* in the USA was especially noticeable given that only a few isolates were available for that country. Further sampling would be needed to confirm whether this reflects a possible native population or is the result of introductions through trade with various other regions [55]. It has been shown for other organisms, for example lizards, that the invasive populations could be more diverse than native populations if introduced multiple times and from various isolated native populations [69]. This has also been observed in fungi such as *D. sapinea* in parts of its invasive range (e.g., in South Africa; [22,23]).

This study provides a valuable foundation for future studies that will investigate the genetic structure, movement, and origins in *L. theobromae* and other important species of the Botryosphaeriaceae. The loci used were chosen to allow for the inclusion of publicly available sequence data so as to obtain a more comprehensive global perspective. We excluded cryptic lineages based on previous studies that have resolved the taxonomy of *Lasiodiplodia* spp. and have defined these lineages as distinct species, including hybrid species [12,27]. As such, the current collection represents a valuable resource to represent a *sensu stricto* definition of the species. This information can now serve as a basis for further collections targeted at more isolated areas that could reveal the potential origin of the fungus. Other markers, such as microsatellite markers, would also provide further insights into origins and patterns of spread of this fungus. However, this will require greater numbers of isolates and ideally a more structured sampling regime than was possible for this study [18].

5. Conclusions

The results of this study, together with other recent investigations on diversity amongst global populations of Botryosphaeriaceae, have highlighted the fact that human-mediated movement of plant material infected by these fungi can facilitate their movement globally. The extent of movement of this serious pathogen around the world suggests a major shortcoming in the ability of quarantine systems to inhibit or stop its movement. These fungi, and their hosts, are also likely to increasingly be influenced by global climate change. Because the earth is subjected to more extreme weather events, plants are likely to become increasingly stressed and more susceptible to disease by pathogens [70], including opportunistic and generalist pathogens such as the Botryosphaeriaceae. Consequently, the Botryosphaeriaceae, including *L. theobromae*, will become increasingly prominent and important for the management of health in both native and commercially cultivated woody plants. Serious attention should be given to strategies that could reduce the extent of such movement. Such management strategies are likely to also be relevant to the numerous other endophytes and potential latent pathogens that inhabit plants and plant material traded around the world.

Supplementary Materials: The following are available online at www.mdpi.com/1999-4907/8/5/145/s1, Figure S1: Maximum likelihood tree of the *tef1α* sequence dataset for the initial identification of isolates for inclusion in this study. Included were type and paratype strains of other *Lasiodiplodia* species, Figure S2: STRUCTURE output from pairwise comparisons of populations. Each plot includes the DeltaK analysis from STRUCTURE HARVESTER (top) and the corresponding barplot for the highest value of K. Pairwise comparisons as follows: (a) north America and south America, (b) north America and Africa, (c) north America and Eurasia, (d) north America and Australasia, (e) south America and Africa, (f) south America and Eurasia, (g) south America and Australasia, (h) Africa and Eurasia, (i) Africa and Australasia and (j) Eurasia and Australasia, Table S1: Polymorphic sites for the respective haplotypes for the ITS, *tef1α* and *tub2* datasets, Table S2: Haplotype assignments for every isolate used in this study, based on the sequence datasets, Table S3: Summary of haplotypes obtained and unique haplotypes (listed in brackets) found for each locus, Table S4: Posterior probabilities (with 95% confidence intervals) of pairwise comparisons for three scenarios to test for possible ancestry between populations done in DIYABC. In scenario 1, population 1 is ancestral to both. In scenario 2, population 2 is ancestral to both. In scenario 3, both populations diverged from an unknown source population.

Acknowledgments: We thank the Department of Science and Technology (DST)-National Research Foundation (NRF) Centre of Excellence in Tree Health Biotechnology (CTHB) and members of the Tree Protection Co-operative Programme (TPCP), South Africa, for financial support. Mr. Victor Kalbskopf and Ms. Elmien Slabbert assisted the lead author with some of the laboratory work required for this study and their assistance is gratefully acknowledged.

Author Contributions: James Mehl conducted the laboratory work, analyzed the data, and drafted the manuscript. Bernard Slippers, Jolanda Roux, and Michael J. Wingfield conceived the study, assembled collections of isolates, assisted with the analyses, contributed to and assisted in writing the manuscript.

Conflicts of Interest: The authors declare no conflict of interest.

References

1. Wingfield, M.J.; Brockerhoff, E.G.; Wingfield, B.D.; Slippers, B. Planted forest health: The need for a global strategy. *Science* **2015**, *349*, 832–836. [CrossRef] [PubMed]
2. Ghelardini, L.; Pepori, A.L.; Luchi, N.; Capretti, P.; Santini, A. Drivers of emerging fungal diseases of forest trees. *For. Ecol. Manag.* **2016**, *381*, 235–246. [CrossRef]
3. Burgess, T.I.; Crous, C.J.; Slippers, B.; Hantula, J.; Wingfield, M.J. Tree invasions and biosecurity: Eco-evolutionary dynamics of hitchhiking fungi. *AoB Plants* **2016**, *8*, plw076. [CrossRef] [PubMed]
4. Slippers, B.; Wingfield, M.J. Botryosphaeriaceae as endophytes and latent pathogens of woody plants: Diversity, ecology and impact. *Fungal Biol. Rev.* **2007**, *21*, 90–106. [CrossRef]
5. Slippers, B.; Crous, P.W.; Jami, F.; Groenewald, J.Z.; Wingfield, M.J. Diversity in the Botryosphaeriales: Looking back, looking forward. *Fungal Biol.* **2017**, *121*, 307–321. [CrossRef] [PubMed]

6. Mehl, J.W.M.; Slippers, B.; Roux, J.; Wingfield, M.J. Cankers and other diseases caused by the Botryosphaeriaceae. In *Infectious Forest Diseases*; Gonthier, P., Nicolotti, G., Eds.; CAB International: Boston, MN, USA, 2013; pp. 298–317.
7. Taylor, K.; Barber, P.A.; Hardy, G.E.S.J.; Burgess, T.I. Botryosphaeriaceae from tuart (*Eucalyptus gomphocephala*) woodland, including descriptions of four new species. *Mycol. Res.* **2009**, *113*, 337–353. [CrossRef] [PubMed]
8. Perez, C.A.; Wingfield, M.J.; Slippers, B.; Altier, N.A.; Blanchette, R.A. Endophytic and canker-associated Botryosphaeriaceae occurring on non-native *Eucalyptus* and native Myrtaceae trees in Uruguay. *Fungal Divers.* **2010**, *41*, 53–69. [CrossRef]
9. Bihon, W.; Burgess, T.I.; Slippers, B.; Wingfield, M.J.; Wingfield, B.D. Distribution of *Diplodia pinea* and its genotypic diversity within asymptomatic *Pinus patula* trees. *Australas. Plant Pathol.* **2011**, *40*, 540–548. [CrossRef]
10. Sakalidis, M.L.; Hardy, G.E.S.J.; Burgess, T.I. Class III endophytes, clandestine movement amongst hosts and habitats and their potential for disease; a focus on *Neofusicoccum australe*. *Australas. Plant Pathol.* **2011**, *40*, 510–521. [CrossRef]
11. Jami, F.; Slippers, B.; Wingfield, M.J.; Gryzenhout, M. Botryosphaeriaceae species overlap on four unrelated, native South African hosts. *Fungal Biol.* **2014**, *118*, 168–179. [CrossRef] [PubMed]
12. Phillips, A.J.L.; Alves, A.; Abdollahzadeh, J.; Slippers, B.; Wingfield, M.J.; Groenewald, J.Z.; Crous, P.W. The Botryosphaeriaceae: Genera and species known from culture. *Stud. Mycol.* **2013**, *76*, 51–167. [CrossRef] [PubMed]
13. Slippers, B.; Roux, J.; Wingfield, M.J.; Van der Walt, F.J.J.; Jami, F.; Mehl, J.W.M.; Marais, G.J. Confronting the constraints of morphological taxonomy in the Botryosphaeriales. *Persoonia* **2014**, *33*, 155–168. [CrossRef] [PubMed]
14. Jami, F.; Slippers, B.; Wingfield, M.J.; Loots, M.T.; Gryzenhout, M. Temporal and spatial variation of Botryosphaeriaceae associated with *Acacia karroo* in South Africa. *Fungal Ecol.* **2015**, *15*, 51–62. [CrossRef]
15. Pavlic-Zupanc, D.; Wingfield, M.J.; Boissin, E.; Slippers, B. The distribution of genetic diversity in the *Neofusicoccum parvum/N. ribis* complex suggests structure correlated with level of disturbance. *Fungal Ecol.* **2015**, *13*, 93–102. [CrossRef]
16. Úrbez-Torres, J.R.; Battany, M.; Bettiga, L.J.; Gispert, C.; McGourty, G.; Roncoroni, J.; Smith, R.J.; Verdegaal, P.; Gubler, W.D. Botryosphaeriaceae species spore-trapping studies in California vineyards. *Plant Dis.* **2010**, *94*, 717–724. [CrossRef]
17. Punithalingam, E. *Botryodiplodia theobromae*. *C.M.I. Descript. Fungi Bact.* **1976**, *519*, 1–2.
18. Sakalidis, M.L.; Slippers, B.; Wingfield, B.D.; Hardy, G.E.S.J.; Burgess, T.I. The challenge of understanding the origin, pathways and extent of fungal invasions: Global populations of the *Neofusicoccum parvum-N. ribis* species complex. *Divers. Distrib.* **2013**, *19*, 873–883. [CrossRef]
19. Marsberg, A.; Kemler, M.; Jami, F.; Nagel, J.H.; Postma-Smidt, A.; Naidoo, S.; Wingfield, M.J.; Crous, P.W.; Spatafora, J.W.; Hesse, C.N.; et al. *Botryosphaeria dothidea*: A latent pathogen of global importance to woody plant health. *Mol. Plant Pathol.* **2017**, in press.
20. Dissanayake, A.J.; Phillips, A.J.L.; Li, X.H.; Hyde, K.D. Botryosphaeriaceae: Current status of genera and species. *Mycosphere* **2016**, *7*, 1001–1073.
21. Burgess, T.; Wingfield, M.J. Quarantine is important in restricting the spread of exotic seed-borne tree pathogens in the southern hemisphere. *Int. For. Rev.* **2002**, *4*, 56–65.
22. Burgess, T.I.; Wingfield, M.J.; Wingfield, B.D. Global distribution of *Diplodia pinea* genotypes revealed using simple sequence repeat (SSR) markers. *Australas. Plant Pathol.* **2004**, *33*, 513–519. [CrossRef]
23. Bihon, W.; Burgess, T.; Slippers, B.; Wingfield, M.J.; Wingfield, B.D. High levels of genetic diversity and cryptic recombination is widespread in introduced *Diplodia pinea* populations. *Australas. Plant Pathol.* **2012**, *41*, 41–46. [CrossRef]
24. Sarr, M.P.; Ndiaye, M.; Groenewald, J.Z.; Crous, P.W. Genetic diversity in *Macrophomina phaseolina*, the causal agent of charcoal rot. *Phytopathol. Mediterr.* **2014**, *53*, 250–268.
25. Farr, D.F.; Rossman, A.Y. Fungal Databases. Systematic Mycology and Microbiology Laboratory, ARS, USDA. Available online: http://nt.ars-grin.gov/fungaldatabases/ (accessed on 28 February 2017).
26. Coutinho, I.B.L.; Freire, F.C.O.; Lima, C.S.; Lima, J.S.; Gonçalves, F.J.T.; Machado, A.R.; Silva, A.M.S.; Cardoso, J.E. Diversity of genus *Lasiodiplodia* associated with perennial tropical fruit plants in northeastern Brazil. *Plant Pathol.* **2017**, in press. [CrossRef]

27. Cruywagen, E.M.; Slippers, B.; Roux, J.; Wingfield, M.J. Phylogenetic Species Recognition and hybridisation in *Lasiodiplodia*: A case study on species from baobabs. *Fungal Biol.* **2017**, *121*, 420–436. [CrossRef] [PubMed]

28. Netto, M.S.B.; Lima, W.G.; Correia, K.C.; da Silva, C.F.B.; Thon, M.; Martins, R.B.; Miller, R.N.G.; Michereff, S.J.; Câmara, M.P.S. Analysis of phylogeny, distribution and pathogenicity of Botryosphaeriaceae species associated with gummosis of *Anacardium* in Brazil, with a new species of *Lasiodiplodia*. *Fungal Biol.* **2017**, *121*, 437–451. [CrossRef] [PubMed]

29. World Europe and Africa Centered. Available online: http://www.d-maps.com/carte.php?num_car= 126805&lang=en (accessed on 28 February 2017).

30. Mehl, J.W.M.; Slippers, B.; Roux, J.; Wingfield, M.J. Botryosphaeriaceae associated with *Pterocarpus angolensis* (kiaat) in South Africa. *Mycologia* **2011**, *103*, 534–553. [CrossRef] [PubMed]

31. Wright, L.P.; Davis, A.J.; Wingfield, B.D.; Crous, P.W.; Brenneman, T.; Wingfield, M.J. Population structure of *Cylindrocladium parasiticum* infecting peanuts (*Arachis hypogaea*) in Georgia, USA. *Eur. J. Plant Pathol.* **2010**, *127*, 199–206. [CrossRef]

32. White, T.J.; Bruns, T.; Lee, S.; Taylor, J. Amplification and direct sequencing of fungal ribosomal RNA genes for phylogenetics. In *PCR Protocols: A Guide to Methods and Applications*; Innis, M.A., Gelfand, D.H., Sninsky, J.J., White, T.J., Eds.; Academic Press: San Diego, CA, USA, 1990; pp. 315–322.

33. Jacobs, K.; Bergdahl, D.R.; Wingfield, M.J.; Halik, S.; Seifert, K.A.; Bright, D.E.; Wingfield, B.D. *Leptographium wingfieldii* introduced into North America and found associated with exotic *Tomicus piniperda* and native bark beetles. *Mycol. Res.* **2004**, *108*, 411–418. [CrossRef] [PubMed]

34. Alves, A.; Crous, P.W.; Correia, A.; Phillips, A.J.L. Morphological and molecular data reveal cryptic speciation in *Lasiodiplodia theobromae*. *Fungal Divers.* **2008**, *28*, 1–13.

35. Glass, N.L.; Donaldson, G.C. Development of primer sets designed for use with the PCR to amplify conserved genes from filamentous ascomycetes. *Appl. Environ. Microb.* **1995**, *61*, 1323–1330.

36. Mehl, J.W.M.; Slippers, B.; Roux, J.; Wingfield, M.J. Overlap of latent pathogens in the Botryosphaeriaceae on a native and agricultural host. *Fungal Biol.* **2017**, *121*, 405–419. [CrossRef] [PubMed]

37. Mehl, J.W.M.; Slippers, B.; Roux, J.; Wingfield, M.J. Botryosphaeriaceae associated with die-back of *Schizolobium parahyba* trees in South Africa and Ecuador. *For. Pathol.* **2014**, *44*, 396–408.

38. Tamura, K.; Stecher, G.; Peterson, D.; Filipski, A.; Kumar, S. MEGA6: Molecular evolutionary genetics analysis version 6.0. *Mol. Biol. Evol.* **2013**, *30*, 2725–2729. [CrossRef] [PubMed]

39. Katoh, K.; Toh, H. Recent developments in the MAFFT multiple sequence alignment program. *Brief. Bioinform.* **2008**, *9*, 286–298. [CrossRef] [PubMed]

40. Darriba, D.; Taboada, G.L.; Doallo, R.; Posada, D. JMODELTEST 2: More models, new heuristics and parallel computing. *Nat. Methods* **2012**, *9*, 772. [CrossRef] [PubMed]

41. Guindon, S.; Dufayard, J.-F.; Lefort, V.; Anisimova, M.; Hordijk, W.; Gascuel, O. New algorithms and methods to estimate maximum-likelihood phylogenies: Assessing the performance of PHYML 3.0. *Syst. Biol.* **2010**, *59*, 307–321. [CrossRef] [PubMed]

42. Monacell, J.T.; Carbone, I. Mobyle SNAP Workbench: A web-based analysis portal for population genetics and evolutionary genomics. *Bioinformatics* **2014**, *30*, 1488–1490. [CrossRef] [PubMed]

43. Bandelt, H.-J.; Forster, P.; Röhl, A. Median-joining networks for inferring intraspecific phylogenies. *Mol. Biol. Evol.* **1999**, *16*, 37–48. [CrossRef] [PubMed]

44. Fluxus Technology Ltd. NETWORK Version 4.6.1.3. Available online: http://www.fluxus-engineering.com/ sharenet.htm (accessed on 28 February 2017).

45. Pritchard, J.K.; Stephens, M.; Donnelly, P. Inference of population structure using multilocus genotype data. *Genetics* **2000**, *155*, 945–959. [PubMed]

46. Hubisz, M.J.; Falush, D.; Stephens, M.; Pritchard, J.K. Inferring weak population structure with the assistance of sample group information. *Mol. Ecol. Resour.* **2009**, *9*, 1322–1332. [CrossRef] [PubMed]

47. Earl, D.A.; von Holdt, B.M. STRUCTURE HARVESTER: A website and program for visualizing STRUCTURE output and implementing the Evanno method. *Conserv. Genet. Resour.* **2012**, *4*, 359–361. [CrossRef]

48. Evanno, G.; Regnaut, S.; Goudet, J. Detecting the number of clusters of individuals using the software STRUCTURE: A simulation study. *Mol. Ecol.* **2005**, *14*, 2611–2620. [CrossRef] [PubMed]

49. Excoffier, L.; Lischer, H.E. ARLEQUIN suite ver 3.5: A new series of programs to perform population genetics analyses under Linux and Windows. *Mol. Ecol. Resour.* **2010**, *10*, 564–567. [CrossRef] [PubMed]

50. Cornuet, J.M.; Pudlo, P.; Veyssier, J.; Dehne-Garcia, A.; Gautier, M.; Leblois, R.; Marin, J.M.; Estoup, A. DIYABC v2.0: A software to make approximate Bayesian computation inferences about population history using single nucleotide polymorphism, DNA sequence and microsatellite data. *Bioinformatics* **2014**, *30*, 1187–1189. [CrossRef] [PubMed]

51. Wright, S. *Evolution and the Genetics of Populations: A Treatise in Four Volumes: Vol. 4: Variability within and among Natural Populations*; University of Chicago Press: Chicago, IL, USA, 1978.

52. O'Donnell, K.; Kistler, H.; Tacke, B.; Casper, H. Gene genealogies reveal global phylogeographic structure and reproductive isolation among lineages of *Fusarium graminearum*, the fungus causing wheat scab. *Proc. Natl. Acad. Sci. USA* **2000**, *97*, 7905–7910. [CrossRef] [PubMed]

53. Kasuga, T.; White, T.J.; Koenig, G.; Mcewen, J.; Restrepo, A.; Castaneda, E.; Da Silva Lacaz, C.; Heins-Vaccari, E.M.; De Freitas, R.S.; Zancopé-Oliveira, R.M.; et al. Phylogeography of the fungal pathogen *Histoplasma capsulatum*. *Mol. Ecol.* **2003**, *12*, 3383–3401. [CrossRef] [PubMed]

54. Taylor, J.W.; Turner, E.; Townsend, J.P.; Dettman, J.R.; Jacobson, D. Eukaryotic microbes, species recognition and the geographic limits of species: Examples from the kingdom Fungi. *Philos. Trans. R. Soc. B* **2006**, *361*, 1947–1963. [CrossRef] [PubMed]

55. Gladieux, P.; Feurtey, A.; Hood, M.E.; Snirc, A.; Clavel, T.J.; Dutech, C.; Roy, M.; Giraud, T. The population biology of fungal invasions. *Mol. Ecol.* **2015**, *24*, 1969–1986. [CrossRef] [PubMed]

56. O'Donnell, K.; Ward, T.J.; Geiser, D.M.; Kistler, H.C.; Aoki, T. Genealogical concordance between the mating type locus and seven other nuclear genes supports formal recognition of nine phylogenetically distinct species within the *Fusarium graminearum* clade. *Fungal Genet. Biol.* **2004**, *41*, 600–623. [CrossRef] [PubMed]

57. Teixeira, M.D.M.; Patané, J.S.; Taylor, M.L.; Gómez, B.L.; Theodoro, R.C.; de Hoog, S.; Engelthaler, D.M.; Zancopé-Oliveira, R.M.; Felipe, M.S.; Barker, B.M. Worldwide phylogenetic distributions and population dynamics of the genus *Histoplasma*. *PLoS Negl. Trop. Dis.* **2016**, *10*, e0004732. [CrossRef] [PubMed]

58. Pringle, A.; Baker, D.M.; Platt, J.L.; Wares, J.P.; Latge, J.P.; Taylor, J.W. Cryptic speciation in the cosmopolitan and clonal human pathogenic fungus *Aspergillus fumigatus*. *Evolution* **2005**, *59*, 1886–1899. [CrossRef] [PubMed]

59. Ramírez-Camejo, L.A.; Zuluaga-Montero, A.; Lázaro-Escudero, M.; Hernández-Kendall, V.; Bayman, P. Phylogeography of the cosmopolitan fungus *Aspergillus flavus*: Is everything everywhere? *Fungal Biol.* **2012**, *116*, 452–463. [CrossRef] [PubMed]

60. Swart, W.J.; Wingfield, M.J.; Knox-Davies, P.S. Conidial dispersal of *Sphaeropsis sapinea* in three climatic regions of South Africa. *Plant Dis.* **1987**, *71*, 1038–1040. [CrossRef]

61. Pusey, P.L. Availability and dispersal of ascospores and conidia of *Botryosphaeria* in peach orchards. *Phytopathology* **1989**, *79*, 635–639. [CrossRef]

62. Amponsah, N.T.; Jones, E.E.; Ridgway, H.J.; Jaspers, M.V. Rainwater dispersal of *Botryosphaeria* conidia from infected grapevines. *N. Z. Plant Prot.* **2009**, *62*, 228–233.

63. Santini, A.; Ghelardini, L.; Pace, C.D.; Desprez-Loustau, M.L.; Capretti, P.; Chandelier, A.; Cech, T.; Chira, D.; Diamandis, S.; Gaitniekis, T.; et al. Biogeographical patterns and determinants of invasion by forest pathogens in Europe. *New Phytol.* **2013**, *197*, 238–250. [CrossRef] [PubMed]

64. Slippers, B.; Smit, W.A.; Crous, P.W.; Coutinho, T.A.; Wingfield, B.D.; Wingfield, M.J. Taxonomy, phylogeny and identification of Botryosphaeriaceae associated with pome and stone fruit trees in South Africa and other regions of the world. *Plant Pathol.* **2007**, *56*, 128–139. [CrossRef]

65. Crous, P.W.; Groenewald, J.Z.; Slippers, B.; Wingfield, M.J. Global food and fibre security threatened by current inefficiencies in fungal identification. *Philos. Trans. R. Soc. B* **2016**, *371*. [CrossRef] [PubMed]

66. Mohali, S.; Burgess, T.I.; Wingfield, M.J. Diversity and host association of the tropical tree endophyte *Lasiodiplodia theobromae* revealed using simple sequence repeat markers. *For. Pathol.* **2005**, *35*, 385–396. [CrossRef]

67. Shah, M.D.; Verma, K.S.; Singh, K.; Kaur, R. Morphological, pathological and molecular variability in *Botryodiplodia theobromae* (Botryosphaeriaceae) isolates associated with die-back and bark canker of pear trees in Punjab, India. *Genet. Mol. Res.* **2010**, *9*, 1217–1228. [CrossRef] [PubMed]

68. Begoude, A.D.B.; Slippers, B.; Perez, G.; Wingfield, M.J.; Roux, J. High gene flow and outcrossing within populations of two cryptic fungal pathogens on a native and non-native host in Cameroon. *Fungal Biol.* **2012**, *116*, 343–353. [CrossRef] [PubMed]

69. Kolbe, J.J.; Glor, R.E.; Schettino, L.R.; Lara, A.C.; Larson, A.; Losos, J.B. Genetic variation increases during biological invasion by a Cuban lizard. *Nature* 2004, *431*, 177–181. [CrossRef] [PubMed]
70. Sturrock, R.N.; Frankel, S.J.; Brown, A.V.; Hennon, P.E.; Kliejunas, J.T.; Lewis, K.J.; Worrall, J.J.; Woods, A.J. Climate change and forest diseases. *Plant Pathol.* 2011, *60*, 133–149. [CrossRef]

forests

Review

Epidemiological History of Cypress Canker Disease in Source and Invasion Sites

Roberto Danti and Gianni Della Rocca *

Institute for Sustainable Plant Protection, National Research Council, Via Madonna del Piano 10,
Sesto Fiorentino, Florence 50019, Italy; roberto.danti@ipsp.cnr.it
* Correspondence: gianni.dellarocca@ipsp.cnr.it; Tel.: +39-055-522-5663

Academic Editors: Matteo Garbelotto and Paolo Gonthier
Received: 27 February 2017; Accepted: 12 April 2017; Published: 15 April 2017

Abstract: *Seiridium cardinale* is a fungal pathogen responsible for pandemic cypress canker disease (CCD). The fungus has shown the ability to infect different hosts in many areas throughout the globe, but its spread and impact were favored by conducive environmental conditions. The most severe epidemics were reported in California and the Mediterranean, the former considered the source area of the pathogen from which the Mediterranean infestation have originated. Here we reconstruct the epidemiological history of the disease in California and the Mediterranean. Evolution of the disease in the two contrasting areas was weighed in relation to differences between the two environments in terms of climate, landscape properties, and adopted management practices. In addition, differences emerged among the source and invasive populations in terms of genetic and phenotypic variability, structure, and mode of reproduction allow a few comments to be made about the environmental implications and related quarantine of new introductory events.

Keywords: *Seiridium cardinale*; fungal pathogen; *Cupressus*; alien species; landscape properties; population genetic; phenotypic traits; disease management; resistance

---

1. Introduction

The introduction of alien species ranks second only to habitat destruction as one of the major threats to native ecosystems [1,2]. In the case of pathogenic species, lack of co-evolution between hosts and invasive pathogens can result in devastating impacts on the host species [3–5]. As a result of the introduction of alien species, native species can be decimated or displaced, local biodiversity can be dramatically reduced, landscape can be spoiled, and when plants of economic importance are affected, the damage can amount to millions or billions of dollars per year [6–9].

Seiridium cardinale (W.W. Wagener) B. Sutton & I.A.S. Gibson is a pathogenic fungus responsible for cypress canker—a pandemic disease which has caused significant mortality worldwide in many species of *Cupressaceae* [10–12]. Cypress canker is also caused by other sister species—*S. cupressi* (Guba) Boesew. (anamorph of *Lepteutypa cupressi* (Nattrass, C. Booth & B. Sutton) H.J. Swart) and *S. unicorne* (Cooke & Ellis) B. Sutton—but they are less widespread and aggressive, and their epidemiology has been substantially less studied compared to *S. cardinale*. Therefore, this paper will focus on the most known of the three fungi.

The first epidemic of cypress canker disease (CCD) was reported in California in 1928 on Monterey cypress (*Cupressus macrocarpa* Hartw. ex Gordon), which over a period of only a few years was wiped out in the plantations located in the inland districts [13]. Mediated by the movement of infected plant material over long distances, in the course of the following decades, the disease spread progressively over the five continents: to New Zealand, France, and Chile; to Italy, Argentina, and Greece [14–19], and subsequently to the entire Mediterranean basin and to other countries in central-northern Europe (UK, Ireland, and Germany), to Canada, North and South Africa, and Australia [20–27]. Cypress canker

181

disease has been reported to affect species of *Cupressus, Chamaecyparis, Cryptomeria, Juniperus, Thuja,* and ×*Cupressocyparis* [11,12,28].

Although the pathogen has shown the ability to infect different host species and to establish in many countries in which it was introduced, development, impact, and invasiveness of the disease varied in the different countries. Epidemics were certainly favored by climatic conditions suited to the fungal pathogen and by the density and continuity of susceptible hosts [11]. Generally, the Mediterranean climate—which is characterized by a marked seasonality, with a substantial thermal range between summer and winter and by mild rainy and humid springs and autumns—is highly conducive to the development of an epidemic of cypress canker disease [10]. Microclimatic conditions may play a major role in the development of the disease, and also explain the presence of nuclei of the host that have escaped the disease in close proximity to areas that have been severely affected [12]. The occurrence of *S. cardinale* infections is known to be favored by the presence of small wounds in the periderm through which the conidia or mycelium enter the inner bark. Young trunks and branches are particularly exposed to infections, because they are more greatly subjected to bark injuries due to cold, frost, hail, or to forced growth by fertilizers, insects, rodents, etc. [10].

Though high levels of mortality were experienced in many areas around the globe, spread and impact of the disease were determined by the presence of conducive factors. The disease has even been reported in north-central Europe [21–29] and in SE Alaska and Canada [23–30], where it maintained a sporadic character as the climate is too cold for the rapid spread of the pathogen. On the other hand, the disease was reported in isolated spots in some countries of North Africa [27–31], where the hot and dry climate also slows the spread of the disease. In other countries, such as South Africa and New Zealand [26–32], where climatic conditions are quite similar to the Mediterranean climate, bark canker showed an invasive character, and represented a limiting factor for cypress plantations. After the introduction of the pathogen, CCD assumed the character of a dreadful epidemic, with a high degree of invasiveness and destructiveness in the Mediterranean basin, where both favorable climatic conditions and the continuity and density of the susceptible host (*Cupressus sempervirens* L.) occurred over a vast territory [11,12].

Studies on population genetics and phenotypic traits enabled comparison of the *S. cardinale* populations of California and the Mediterranean, the two areas where the pathogen gave rise to the most severe outbreaks [33–35]. It appeared that the Mediterranean cypress canker epidemics stemmed from the introduction of the fungus from California, probably following a single introductory event.

Here we reconstruct the epidemiological history of the disease in California and the Mediterranean. The evolution of the disease in the two contrasting areas is weighed in relation to differences between the two environments in terms of climate, landscape properties, and adopted management practices. In addition, differences emerged among the source and invasive populations in terms of genetic and phenotypic variability, structure, and mode of reproduction, raising a few points about the environmental implications and the related quarantine of new introductory events.

2. Epidemiology of CCD in California (*S. cardinale* Source Population)

Cypress canker disease caused by *S. cardinale* was identified and described for the first time in Palo Alto around San Francisco Bay (western-central California) by Wagener [13] on Monterey cypress (*C. macrocarpa*), although evidence has indicated that it had been present there since at least 1885 (Jepson field book). In the following years, the disease was detected in Alameda, Santa Clara, and San Mateo (Sacramento county, report uncertain) counties [13]. During the initial years after these reports, the pathogen epidemic was so violent that more than 30,000 trees of Monterey and Common cypress (*C. sempervirens*) were killed [36]. In California, San Francisco and Los Angeles—the two main ports along the Californian pacific coast—were identified as the two main foci of the disease [37].

Since its first report, the disease assumed major epidemic proportions in many districts through the western half of California and practically wiped out most inland plantations of Monterey cypress [37]. As reported by Wagener in 1928, based on the natural spread of infections, in California the main host species of *S. cardinale* were: *C. macrocarpa, C. sempervirens, C. pygmaea* Sarg., *C. forbesii* Jeps., *C. lusitanica* Mill., *Thuja orientalis* L., *Ch. Lawsoniana* Murray Parl., *Calocedrus Decurrens* Florin, *Juniperus chinensis* L. var. *femina*, and *J. sabina* L. var. *tamariscifolia*. No traces of the pathogen were found in the natural range of some Californian *Cupressus* species, including *C. macrocarpa* and *C. bakeri* Jeps., *C. macnabiana* Murray, *C. pygmaea, C. forbesii*, as well as in natural stands of others *Cupressaceae*, such as *Cal. decurrens, Ch. lawsoniana, J. occidentalis* Hook., and *J. californica* Carr. [37].

In spite of the efforts pursued between 1936 and 1939 to eradicate all cases of the disease in numerous plantings [37], losses from cypress mortality in California were extensive. Eight years after the identification of the disease, three-fourths of all planted Monterey cypress and a considerable number of other susceptible species (e.g., Common cypress) were estimated to be lost as a result of *S. cardinale* attacks [37,38]. These heavy wholesale losses resulted in the substitution of cypress windbreak around citrus groves with other tree species, such as *Eucalyptus* sp. [38].

No more studies dealt with the epidemiology of the cypress canker in California after 1939, when the limits of its geographic spread limits were Glenn County in the Sacramento Valley, as isolated infection centre, Sonoma county as area generally infected (North), San Diego county (South), Sacramento, and San Joaquin county (inland/East) [37] (Figure 1). After this first outbreak, cypresses have not been used for extensive plantings in California, but they were still planted for hedges or as shade trees or, in some cases, for protective purposes, e.g., *C. macrocarpa* along the coastal area.

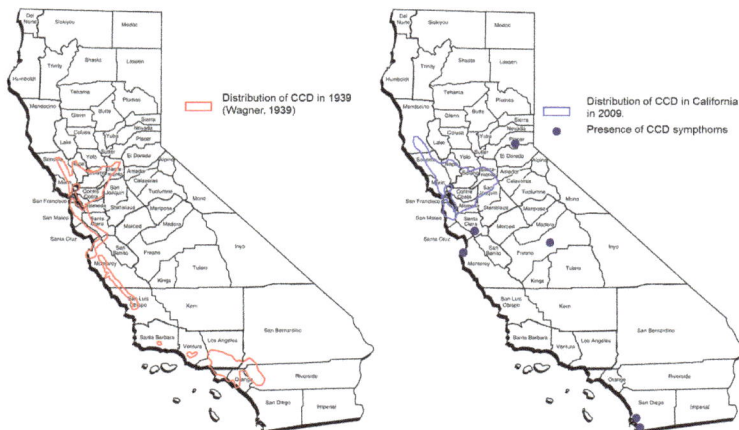

Figure 1. Spread of *Seiridium cardinale* in California. Maps of distribution of cypress canker disease (CCD) in 1939 and in 2009.

Today, both *S. cardinale* and *S. unicorne* are known to be present in California [33–39], threatening ornamental plantations of *Cupressus* sp.—mainly *C. sempervirens, C. macrocarpa*, ×*Cup. leylandii*, and *Juniperus* spp.

The current spread of CCD in California has been determined by surveying locations where its main hosts are present, such as *C. sempervirens, C. macrocarpa*, and ×*Cup. leylandii* during early summer 2009 (Figure 1). The survey included ground determination of infected trees along freeways and national roads, particularly focusing beyond the *S. cardinale* distribution sites reported by Wagener in the 1930s. Many locations were also taken into consideration following reports of local advisors, extension specialists in plant pathology from UC Berkeley, and homeowners. The following counties were covered during the survey: Mendocino, Sonoma, Napa, Marin, Contra Costa, Alameda, Sacramento, El Dorado, San Joaquin, San Francisco, San Mateo, Santa Clara, Stanislaus, Mercedes, Madera, Fresno, Monterey, Tulare, Kern, San Luis Obispo, Orange, Los Angeles, and San Diego. A total of 2385 trees of *C. sempervirens* (790), *C. macrocarpa* (611) (Figure 2), and ×*Cup. leylandii* (984) were surveyed for the presence of bark canker infection in 24 counties of California (Table 1). An incidence rate of 11.7% of *S. cardinale* was recorded.

Figure 2. *Cupressus macrocarpa* affected by CCD in Marin county (California, CA, USA) (2009). Monterey cypress is the major host of CCD in California.

Table 1. CCD incidence per county and host species resulting from the survey conducted in California in 2009 (Della Rocca and Garbelotto, unpublished).

County	Host	% Infected Trees Per Species	% of Diseased Trees in the County
Mendocino	×*Cup. leylandii*	11.4	11.4
Sonoma	×*Cup. leylandii* *C. sempervirens* *C. macrocarpa*	19 12.1 7.7	- 12.9 -
Napa	×*Cup. leylandii* *C. sempervirens*	13.5 17.2	- 15.3
Marin	×*Cup. leylandii* *C. sempervirens* *C. macrocarpa*	23.2 3.2 17.9	- - 14.8
Contra Costa	×*Cup. leylandii* *C. sempervirens* *C. macrocarpa*	29.4 1.2 26.2	- - 18.9
Alameda	*C. sempervirens* *C. macrocarpa*	1.6 25	- 13.3
San Joaquin	×*Cup. leylandii* *C. sempervirens*	34.6 1.7	- 18.1
San Francisco	*C. sempervirens* *C. macrocarpa*	3.6 7.1	- 5.4
San Mateo	×*Cup. leylandii* *C. macrocarpa*	32.1 21.6	- 26.9
San Diego	×*Cup. leylandii*	3.9	3.9
Monterey	*C. sempervirens* *C. macrocarpa*	3.3 2.8	- 3.1
Orange	*C. sempervirens*	1.4	-
Los Angeles	×*Cup. leylandii* *C. sempervirens*	3.3 3.8	- 3.6
San Luis Obispo	*C. sempervirens* *C. macrocarpa*	4.3 12.7	- 8.5
Tulare	×*Cup. leylandii* *C. sempervirens*	4.4 3.1	- 3.8

The highest disease incidence was scored on Leyland cypress, ranging from 34.6% to 3.3% in San Joaquin and Los Angeles counties, respectively. In Monterey cypress, incidence of the disease ranged from 26.2% in Contra Costa to 2.8% in Monterey; and in Common cypress, incidence ranged from 17.2% in Napa to 1.2% in Contra Costa. The incidence rate was highly variable among counties, depending on ecology and host density at the local level (Table 1). The relationship between the mean percentage of diseased trees and the density of cypresses in an area surveyed in a given county resulted significant (Pearson correlation coefficient $r = 0.807$; $n = 15$; $p = 0.01$). In general, disease incidence decreased with the distance (both northward and southward) from the area surrounding San Francisco Bay, which seemed to represent the centre of gravity of the disease.

Today, common cypress is widely used in inland central California, strictly for ornamental purposes: single or in small groups in parks, gardens, flowerbeds, or in rows along some country drives to form hedges, and accompanying catholic sacred places. In this species, the highest incidence of cypress canker was found where it had been planted at higher frequencies, such as in some rural areas of the Sacramento valley and in the Napa and Sonoma valleys. *Cupressus sempervirens* is less common near to the coast due to the colder climate and sea spray. In northern counties (Humboldt, Trinity, Mendocino), *C. sempervirens* is rare due to the cold winter, whereas in southern counties such as Los Angeles and San Diego it is quite uncommon and replaced with palms in the landscape, because of the hot and dry desert climate.

The highest disease incidence in Monterey cypress was around the San Francisco bay area, where it reached 25% and 26.2% in Alameda and Contra Costa counties, respectively, while lower incidence was recorded in the southern county except for San Luis Obispo county, where it was still relatively high (12.7%) (Table 1). No CCD symptoms were observed in the native area of Monterey cypress in Monterey and Santa Cruz coastland, where few *S. cardinale* symptoms were observed only in common cypress. Monterey cypress is broadly spread along the central-northern Pacific coast of the State, where it has been successfully planted (sometimes naturalized) for protective purposes thanks to its tolerance to the strong wind and sea spray (Table 1).

Leyland cypress is a fast-growing evergreen hybrid used extensively in horticulture for hedges and screens to enforce privacy. It is very frequently planted in private gardens around cottages, pruned at 2–3 m height to form a "green barrier" in coastal and inland areas of California. Infected trees of Leyland cypress were found everywhere irrespective of geography, age, and host density. The highest incidence of the disease in that hybrid were noted in San Joaquin (34.6%) and San Mateo (32.1%) counties.

In California, larvae of cypress bark moth *Laspeyresia cupressana* (Kearf) frequently mine the border of cankers—mostly on ×*Cup. leylandii*. From our observations, *Phloeosinus cupressi* Hopkins and *Phloeosinus cristatus* (LeConte)—both larger than the European beetles of the same genus—have likely played a marginal role in the spread of the disease.

3. Epidemiology of CCD in the Mediterranean (*S. cardinale* Invasive Population)

3.1. CCD in Italy

As described by Moriondo and Bonifacio [40], in 1968, cypress canker had already spread throughout Italy, with the exception of Sicily. First reported in Florence (1951), CCD appeared few years later in Latina (1954), and then the disease rapidly spread throughout Italy: to north and north-east (Treviso 1958), (Friuli-Venezia Giulia and Lombardy, both in 1961); while southward CCD was first reported in Rome (1961), Basilicata (Matera, 1964), and Calabria (1965). In 10–15 years, the disease spread in distances of 300 km northward and 700 km southward, and also crossed the Tyrrhenian Sea (Sardinia in 1961), suggesting an important role for human activity in transporting the pathogen.

The first picture of disease incidence available at the national level was based on a map reported by Panconesi [10] (Figure 3). According to this map, the most affected region was Tuscany, with an incidence higher than 30% around Florence, and from 21% to 30% in the surrounding provinces of Pistoia, Arezzo, and Siena. The map indicated that the incidence of cypress canker was highest in the areas of the countries where cypress was most widespread (Tuscany, Liguria, Lazio, Campania, around the pre-Alpine lakes), while the impact of the disease remained low (1%–10% incidence) in regions where cypress was scattered or sporadically distributed.

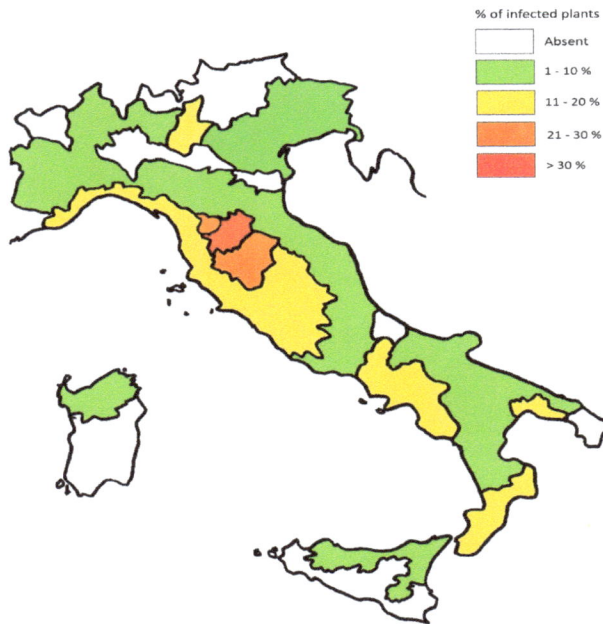

Figure 3. Incidence of CCD in Italy in 1988 according to Panconesi ([10], modified).

Despite many reports of the disease all around the world, a long-term epidemiological story can be approached only in Tuscany (central Italy). Common cypress is a representative element of the Italian cultural landscape, and plays an iconic role in Tuscany. It has traditionally been of primary importance in historical parks and gardens, villas, and boulevards, around farmhouses and sacred spots, along hillsides, and as a landmark in the countryside. Common cypress has also been successfully used for reforestation, aimed at soil protection in degraded and drought-stressed sites, and for the establishment of windbreaks and hedges. In Tuscany, the most important national nursery district located in Pistoia was the likely entrance of the disease. Then, abundance and spatial continuity of the pathogen's main host in the surrounding forests and ornamental plantations promoted its spread.

In the few years after its introduction, the pathogen became a real threat to cypress. During the 1960s, cypress canker disease started to show its destructiveness, causing high mortality. From then on, the developing epidemics attracted the attention of many specialists. The spread of the disease was recorded by a series of surveys aimed at planning proper control measures throughout the regional territory.

A geographical distribution of CCD from 1967 to 1975 in Tuscany was reported by Bartoloni et al. [41]. The maps in Figure 4 show the evolution of the disease over a period of 8 years, showing that from initial nuclei of the disease—mainly located in the provinces of Florence, Livorno and Lucca—the disease quickly spread south-eastward to the provinces of Arezzo, Siena, Grosseto, and Pistoia.

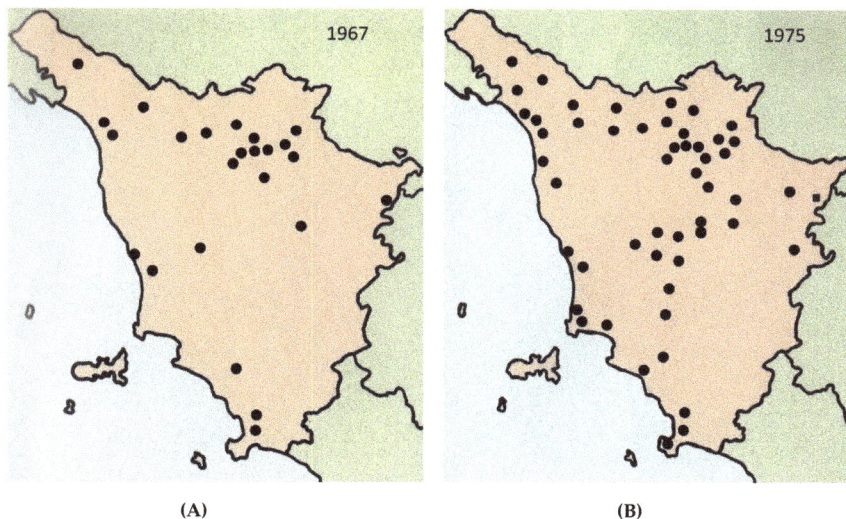

Figure 4. Maps of the incidence distribution of CCD in Tuscany in 1967 (**A**) and 1975 (**B**) according to Bartoloni et al. [41] (modified).

The first systematic survey of the disease was carried out in the province of Florence in 1978 [42]. A total of 235 cypress stands and 3,884,050 trees were examined, covering an area of 3575 ha. The mean incidence of the disease in the province of Florence was 14.5%, ranging from 1% to 30%. In woodlands, the incidence was a little lower than in ornamental plantations (12.4% and 16%, respectively). A negative correlation between elevation and incidence of CCD was found for the ornamental plantations ($r = -0.560$, $n = 49$ sites). A higher mean incidence (20.4%) of CCD was found in the sites located between 50 m and 200 m, while a lower mean value (9.9%) was recorded in sites located above 350 m Moreover, a positive correlation ($r = 0.286$, $n = 50$ municipalities) was observed between the abundance of cypress trees in each site and the recorded incidence of the disease.

This survey followed a series of extensive sanitation campaigns that had been conducted since 1973 in an attempt to control the disease, with more than 40,000 trees felled, about 20,000 that were pruned (to remove the cankered tissues from the crown), and almost 200,000 that were preventively treated with Bordeaux mixture (the poor effectiveness of copper compounds for protecting cypress from bark canker was later revealed in [43]) [42,44,45].

Based on the available data, a time course curve of the incidence of the disease in the Province of Florence from its appearance in 1951 to 1991 was reconstructed by [46] (Figure 5). The plotted data formed a saturation curve that is typical of the epidemic spread of polyetic diseases. This suggests an initial phase of spread in which the disease incidence increased relatively slowly (up to the mid-1970s), followed by a second phase of rapid increase of the rate of diseased plants (until the mid-1980s), followed by a third phase when the proportion of diseased trees reached a plateau. The threshold value at which the number of new infections started to increase at a higher rate seems to be around 15%–20%. A mean incidence of the disease of 40% for the entire provincial territory (Figure 5) appeared to be an overestimate, and could be due to the monitoring network being unevenly distributed throughout the territory.

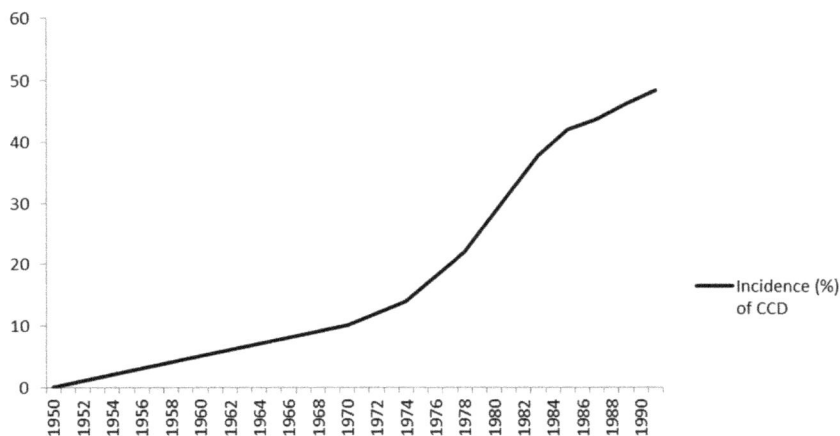

Figure 5. Evolution of the incidence of CCD in the Florence district from its first report in 1951 to 1991 according to Panconesi and Raddi (adapted from [46]).

Among the vectors of the pathogen, the bark beetles *Phloeosinus aubei* (Perris), *P. thuje* (Perris), and *P. armatus* (Reitter), the cypress seed bug *Orsillus maculatus* (Fieber) and the cypress bark moth *Cydia cupressana* Kearfott may have played a significant role in the spread of the disease, although their quantitative effect has not yet been defined. In particular, the bark beetle that breeds on infected crowns and then moves to surrounding healthy trees for feeding is well-known to be a carrier of the fungal inoculum contributing to the spread of infections to new host trees. This role is presumed to have been more important during the epidemic phase of the disease, as the number of infected dead or dying plants was rapidly increasing. The extensive survey carried out in 1994 across the region [47] by the National Forest Service of Italy (3,577,000 cypress trees in 2657 sampled areas) showed a mean incidence of 26.4%, ranging from 12.7% to 53.1% in the provinces of Siena and Massa Carrara, respectively (Table 2). The CCD was found in 93% of the sampled areas, showing that the disease had spread throughout the entire region, despite the control measures that had been adopted.

Table 2. Mean CCD disease incidence per province in Tuscany in 1994 resulting from a detailed survey carried out by the Italian National Forest Service, according to Pivi [47].

Province	Bark Canker Incidence (%)
Arezzo	32.4
Florence	27.6
Grosseto	26.4
Livorno	24.8
Lucca	25.3
Massa Carrara	53.1
Pisa	18.2
Pistoia	17.1
Siena	12.7
Regional mean	26.4

From 2000 to 2009, the META service (Regional Monitoring Network of Forest Health) of the Tuscan Regional Administration that is focussed on the monitoring of forests to plan proper protection measures included *S. cardinale* as a major forest pathogen. A network of 150 sampling areas distributed to cover all the provinces of the Tuscan region were used to monitor the evolution of the pathogen (Table 3) (2009). During this period, a slow decrease in disease incidence was reported in most of the provinces, and the regional mean value substantially dropped from 28.0% in 2002 to 20.6% in 2009. When the single provinces were considered, a lower mean value of 8.2% was recorded for Massa Carrara, whereas the highest mean value of 38.9% was recorded for Arezzo (Figures 6 and 7). Annual variations of the regional means of incidence can be also ascribed to some changes in the monitoring network. Since 2009, no other extended monitoring campaigns were conducted in Tuscany.

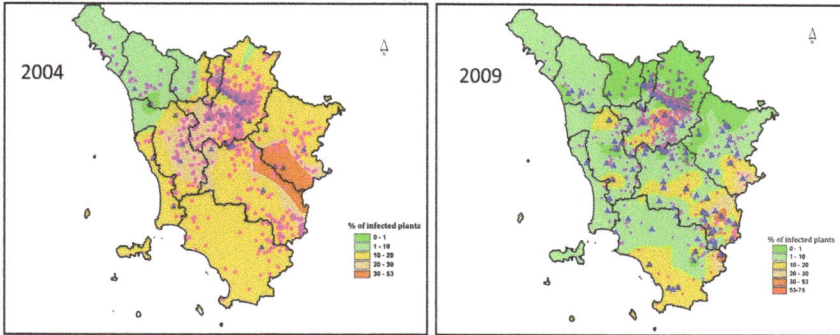

Figure 6. Maps of CCD incidence in Tuscany according to the Regional Monitoring Network of Forest Health (META) service in 2004 and 2009. The discrepancy is due to a decrease of new attacks, but also to a change of the monitoring network (courtesy of Paolo Capretti).

Figure 7. Severe attacks of CCD on *Cupressus sempervirens* in Guarniente (Arezzo province, Italy) (2003).

Table 3. Bark canker incidence (%).

	2002	2004	2005	2006	2007	2009	Mean for Province (2002–2009)
Arezzo	36.8	30.3 (7.3–77.3)	55.8 (7.3–100)	34.9 (6.3–94)	40.2 (3.1–85.2)	35.6 (1.8–58.8)	38.9
Florence	20.0 (0–52.9)	18.9 (7.5–29.5)	30.4 (4.8–84)	32.2 (4.8–62)	-	21.2 (2.7–90.5)	24.5
Grosseto	-	19.4	-	-	28.2 (15.6–47)	19.9 (0–40.7)	22.5
Livorno	18.5	16.7	-	-	-	14.8	16.7
Lucca	-	21.0 (0–38.6)	-	8.9	5.3 (3.1–7.9)	18.1 (5.7–32)	13.3
Massa Carrara	-	-	8.2	-	-	-	8.2
Pisa	-	-	28.6 (13–42.6)	-	28.2	-	28.4
Prato*	42.9	-	25.6	6.1	7.9	13.4 (0–25)	19.2
Siena	21.8 (3.1–32.8)	17.5	28.8	34.1 (12.5–83.9)	23.8 (6–41.2)	21.3 (1.9–74)	24.5
Regional mean	28.0	20.6	29.6	23.2	22.3	20.6	-

Data from META Service (modified). The surveys with maximum and minimum incidence of the disease are in brackets.

A study using G.I.S. applications was conducted at the local level in Tuscany to relate the presence of the disease to climatic and environmental parameters [48]. In a north-to-south 70 km transect along a main road through Florence, 5000 cypress plants were monitored and georeferenced. The main results showed that the incidence of the disease was generally higher in the southern and western slopes (22%–24% and 19%–21%, respectively). At higher altitudes (400–500 m), most of the diseased trees were observed in the south-east, south, and south-west slopes; while at lower altitudes (less than 300 m), the incidence was higher in the north-west, north, and north-east slopes. Excluding the effect of other local factors, this confirmed that the fungus prefers mild temperatures (its optimum in vitro is 25 °C) and that it is less adapted to both colder and warmer sites.

In the same study [48], a higher incidence was recorded in adult rather than young trees (87.3% and 12.3%, respectively), and the proximity to other cypress plants was significant: with 46.4% of diseased plants growing in groups, 43.8% in rows, and only 9.6% of the infected trees growing in isolation. This supports the suggestion that at landscape level, the pattern of distribution of the host may be a feature driving the epidemiology of CCD.

3.2. CCD in Greece

In a manner similar to California and central Italy, in Greece, cypress canker assumed epidemic proportions a few years after its introduction in 1962 [18]. In 1975, the disease was reported to be rapidly spreading throughout the country, affecting trees planted outside the natural range of *C. sempervirens*. The incidence was again related to the climatic conditions of the different districts. In 1985, the most affected areas were the western coast (including the Ionian islands) exposed to humid winds, where incidences of CCD ranged from 1%–25%, to over 50% recorded in the Peloponnese. In the eastern part of the country (characterized by a drier climate), the disease was sporadically reported, with the exception of the island of Eubea. Incidence jumped dramatically in sites where microclimatic conditions were particularly favorable (small humid valleys or higher elevation exposed to cold winter winds), reaching values as high as 80% in southern Peloponnese and 95% in Eubea [49] (Figure 8). The same authors observed in Greece what had already been reported in Tuscany: once an incidence threshold of 20% had been exceeded, the rate of the spread of the disease increased more rapidly according to a saturation curve, suggesting the existence of a kind of ecological quorum over which cypress canker acquires a true epidemic behavior.

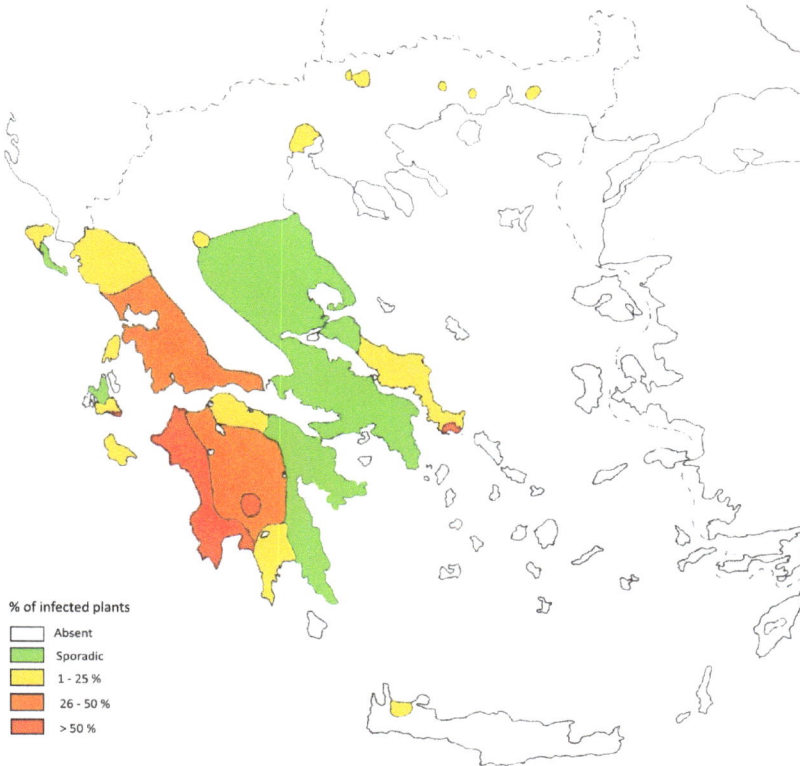

Figure 8. Map of CCD incidence in Greece in 1984 according to Xenopoulos and Diamandis (adapted from [49]).

4. Landscape Properties Influencing the CCD Epidemic

Landscape ecology focuses on the influence of habitat heterogeneity in space and time on ecological processes, and hence the impact of the landscape properties on disease dynamics. A reduction of disease risk (and incidence) is generally empirically observed in highly fragmented and diversified landscapes.

In some cases, fragmentation may have positive effects on disease dynamics by increasing the edge-to-surface ratio, thus exposing the host plants that are close to the edge to the inoculum of the pathogen present in the environment (edge effect) [50,51].

Introduction of *S. cardinale* in the Mediterranean resulted in the rebuilding of a plant–pathogen system that had already been established in California in a different geographical area. The fungus came into contact with a huge population of a susceptible host (four million cypresses in the province of Florence). The frequent distribution of cypress in patches represented by woods and groves with the presence of corridors of cypresses planted in rows along the roads acting as strips of habitat for the fungus likely promoted the movement of the pathogen and favored a large gene flow at a landscape level [52].

The frequent presence of cypresses planted in rows—an element that strongly characterizes the Tuscan landscape—has probably contributed to the spread of disease by favoring the exposure of trees to the inoculum of the fungus, maximizing the edge effect (Figure 9). Host fragmentation represented by frequent patches connected by strips of habitat (corridors), and the local abundance of source habitats (plantations of the hyper-susceptible Leyland cypress) may favor a large gene flow that facilitates the spread of virulent strains of *S. cardinale*.

Figure 9. Patches and rows of cypress are key element of the Tuscan landscape: exposure of trees to the inoculum of the fungus is maximized and favored the spread of the fungus.

Inoculum density is considered by plant pathologists as a major factor determining the probability of occurrence and the severity of disease epidemics. The local abundance of source habitats and refuges has been proven to strongly influence the prevalence of a disease. This suggests the importance of landscape composition on the dynamics of at least some diseases [51]. Currently, Leyland cypress plantations may act as a reservoir of inoculum that may influence the propagule pressure of *S. cardinale* in central Italy (Figure 10). Besides being used as an ornamental, in groups or in hedges, Leyland cypress has been extensively planted throughout the country to screen highways, main roads, railway lines, infrastructures and industrial buildings. In the last decade, from north to south, severe diebacks and mortality due to *S. cardinale* canker have been observed with increasing frequency on disfigured Leyland cypress plantations. In a survey recently carried out in a representative sample area around Florence, incidence of bark canker on Leyland cypress was higher than 54%, with more than 11% of trees killed [28]. Leyland cypress is typically planted in long rows, which could act as continuous strips of habitat (corridors) facilitating the movement and dispersal of the pathogen. Leyland cypress was found to be severely affected by bark canker throughout California when planted at a distance from the coast, and worldwide reports have shown that this tree is prone to attack and disfigurement by cypress canker in conducive areas. Vulnerability of Leyland cypress is due to its marked susceptibility to cypress canker and to the genetic uniformity which is perpetuated by the agamic propagation of few genotypes.

The occurrence of virulent and spreading infections on Leyland cypress is in contrast to the endemic (or post-epidemic) phase that *S. cardinale* canker is currently exhibiting on common cypress in Italy. Since any kind of host specialization or different pathogenicity between *S. cardinale* isolates from common and Leyland cypress has been excluded [28], the inoculum of the fungus that is progressively increasing in the diseased Leyland cypress plantations might favor the spread of the disease to the local common cypress and foster new outbreaks.

Figure 10. The hybrid ×*Cup. leylandii* is a hyper-susceptible host of *S. cardinale* which could represent a reservoir of inoculum that may influence the propagule pressure of the pathogen (Rosignano, Tuscany, 2013).

5. Disease Management of CCD in California and in the Mediterranean

Different approaches are generally required depending on the context in which the host plants are to be protected from CCD. Plantations with a landscape value, which also may have an historical and monumental importance, require a more conservative approach to preserve their ornamental and aesthetic value as much as possible (Figure 11). In cypress woods, the only possibility is the felling and removal of infected trees due to obvious operational limitations (Figure 12).

Figure 11. Sanitation of plantations with a historical and monumental value required a conservative approach to preserve their ornamental value, but only repeated interventions were effective to control the CCD (Monteriggioni, Tuscany, 2012).

Figure 12. The incidence of CCD in cypress stands was severe in the Florence surroundings in the 1990s (Monte Morello, Tuscany, 1995).

5.1. CCD Control in California: A Drastic Strategy

In California, following the rapid development of epidemics that in less than 10 years had severely injured most of the Monterey cypress plantations in inland districts, drastic measures were implemented to control the disease. All the plantations in sensitive areas were dismantled, and as a preventive measure, planting of susceptible hosts over large surfaces was discouraged. This choice has enabled optimization of the extinctive mechanical control method, minimizing the build-up of pathogen inoculum. The beneficial effects of this strategy have led to sustained results, and are still visible. Cypress canker is currently present as an endemic disease in California, and the lack of sufficient density and continuity of susceptible hosts has prevented the occurrence of new epidemic outbreaks. The policy of not planting cypress species susceptible to bark canker over large surfaces has represented a form of disease control, capable of minimizing the build-up of the pathogen inoculum; and the lack of continuity of host species has probably limited the spread of disease in the following years. Host density is recognized as a major factor driving disease epidemics, and some studies have determined a threshold below which a pathogen cannot invade a population of susceptible individuals [53,54].

As bark canker continues to be a constraint for cypress cultivation in California, effective management procedures and the use of resistant host material should help to reduce its overall impact and to use cypress successfully, although genetic resistance to CCD has been found exploitable only in *C. sempervirens*, but not in *C. macrocarpa* or ×*Cup. leylandii* [28,55,56].

5.2. CCD Control in the Mediterranean: An Integrated Disease Management Strategy

In the Mediterranean (and particularly in central Italy), the massive and extended presence of the susceptible host and the unique and irreplaceable ornamental, historic, and recreational value of cypress have required extreme caution in adopting proper control measures. Besides reducing the inoculum of the fungus, interventions had to maintain the primary role of cypress in the landscape.

The disease control strategy in Italy has been based on the integration of different methods: extinctive measures, through extensive sanitation throughout the territory; preventive chemical treatments in nurseries and young plantations, and use of selected cypress lines by conducting a breeding program for resistance.

Sanitation is the main direct method for controlling cypress canker; this consists of the extinction on a vast scale of sources of the pathogen inoculum through the removal of all infection sources. This is practically attained by felling all of the compromised or killed trees and pruning the partially affected crowns, providing that all the resultant infected material is destroyed by fire.

The effectiveness of sanitation is related to the reduction of the inoculum load that careful and extended interventions are able to attain by removing all sources of infection over a wide area. In Tuscany, execution of sanitation throughout the entire territory was considered to be difficult, due to the huge economic and human resources that had to be employed. This discouraged the start of large-scale sanitation campaigns, so the first interventions supported by public funds were performed only in 1974, more than 20 years after the first report of the disease.

As it was impossible to protect all the cypress trees in the whole region with sanitation, in the following decades, a priority program was planned. Sanitation was therefore preferentially focused on the cypress plantations which had the highest landscape, historical, and biological value.

Sanitation has been performed following two different approaches: in the woods, all affected cypresses have been felled, irrespective of how they were diseased, as pruning of crowns would involve obvious operational difficulties. Instead, in plantations that have an ornamental and historic role (rows, groups, hedges, single trees), only the compromised or killed trees were felled, while crowns that were affected to a limited extent were subjected to pruning. This has been done in an attempt to preserve the aesthetic and ornamental appearance of these vegetal structures (conservative sanitation).

These two different approaches to sanitation have shown a quite different effectiveness over the years. Sanitation has produced good results in the woods, where the removal of all the infected trees led to a significant and enduring reduction of the incidence of the disease. In ornamental plantations, however, the results of the pruning interventions were often not as good as expected due to the need to preserve the affected trees. Indeed, the identification of all current infections is not an easy task, particularly on large stems and branches, and the proper execution of cuts is not obvious, because it is necessary to remove all infected tissues and not just the drying organs (branches or stems).

Herein are reported two different cases, outlining the different results obtained by sanitation following the two different approaches. The adoption of a conservative sanitation in ornamental plantings has often been moderately effective. Although useful in reducing the inoculum sources of the pathogen, it has rarely been shown to remove all of the developing infections. The rows of cypresses have then often continued to be corridors for the spread of the disease, even after being subjected to sanitation.

5.2.1. Ornamental Plantations: The Case of the "Viale di Bolgheri"

The case of the famous cypresses of the Viale di Bolgheri (Tuscany, Italy) [57] is representative of how difficult controlling CCD through conservative sanitation may be—particularly in highly conducive sites. The Viale di Bolgheri is a straight, 5 km long road flanked by two rows of (originally) 2400 century-old cypresses located in the coastal area of Tuscany, south of Leghorn, considered part of the legally protected artistic and cultural heritage. Here the impact of the disease has always been severe due to microclimatic conditions particularly favorable to *S. cardinale* development (high humidity of the coastal area) and also due to the trees being arranged close to one another in rows that have promoted the spread of infections throughout the plantation. The Viale di Bolgheri is the only case (among ornamental plantations) in which a detailed constant monitoring of the spread of cypress canker has been carried out for years in the course of a long-term project for the preservation of the Viale. Sanitation here was conducted following a strict conservative approach due to the great historical and cultural importance of the plantation.

Since 1999, 12 surveys have been conducted (the last in 2015) to monitor the health of each cypress of the Viale. During this 16-year period, seven sanitation campaigns were carried out (Table 4).

Based on the extent of diseased crown, sanitation was intended to prune partially affected trees and to remove all trees where recovery was not possible. As an additional measure, sanitation was also extended to surrounding areas (within 2 km from the Viale) to create a protective belt by lowering the fungus inoculum load.

Table 4. Time series of the interventions aimed at the control of CCD in the monumental cypress tree-lined road "Viale di Bolgheri".

	Number of Trees in the "Viale"	Total Infected Trees (%)	Trees Felled (%)	Trees Pruned (%)	Trees Pruned More than one Time (%)	Extent of Damaged Crown (%)	Newly Infected Trees (%)
January 1999	2374	21.7	4.0	17.7	-	-	-
July 2001	2374	22.0	5.1	16.9	-	-	-
April 2002	2254	11.4	2.3	9.1	-	9.8	-
January 2005	2203	14.3	3.2	8.5	10.9	6.0	-
September 2006	2172	11.8	1.0	10.8	39.5	10.9	60.5
March 2007	2146	5.6	0.4	5.2	63.9	5.5	36.1
December 2007	2138	6.5	0.1	6.3	68.1	6.5	31.9
November 2008	2135	11.4	0.6	10.9	77.7	8.6	22.3
September 2010	2123	12.8	1.6	11.2	74.6	8.3	25.4
May 2013	2088	12.2	1.4	10.7	80.2	14.1	19.7
November 2013	2058	1.9	0.1	1.8	89.1	8.4	10.9
November 2015	2056	8.2	0.9	7.3	87.0	15.2	5.2
Since January 2016	2027 (−14.6%)	-	-	-	-	-	-

The repeated sanitation campaigns led to a gradual reduction of disease incidence which decreased from 22% in 2001 to 8.2% in 2015. Due to the numerous felling, the number of trees in 2015 was 14.6% lower than in 1999 (more than 347 out of 2374). Both the amount of trees to be felled and the number of trees to be pruned decreased progressively due to the parallel reduction of disease incidence and the decrease of the inoculum load in the Viale (Table 4). Similarly, the number of newly infected trees decreased to 5.2% in 2015, while the number of previously pruned trees showing the recurrence of new symptoms increased to 87%. These results underline the success of recurrent sanitation in reducing the spread of infections to healthy trees, and simultaneously, the difficulty of eradicating the disease by conservative pruning interventions in old and large trees, also evidencing how *S. cardinale* infections can remain hidden for years before producing symptoms in large trees.

Generally, a positive correlation was found between the time lapse between two consecutive sanitation campaigns and the number of diseased trees ($r = 0.48$), and it clearly shows that regular surveys and sanitation allowed disease incidence to remain below a "safety" threshold of around 10% (Figure 13), but that eradication of the disease is difficult. A gap of several years in the sanitation interventions could lead to a resurgence of CCD, frustrating the positive results previously achieved.

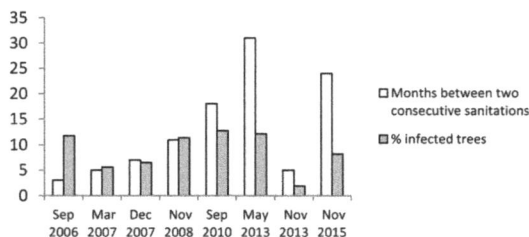

Figure 13. Effect of repeated sanitation campaigns on the CCD incidence in the "Viale di Bolgheri" (Tuscany, Italy).

5.2.2. Sanitation in Cypress Woods: The Case of the Florence Surroundings

The effectiveness of sanitation in cypress woods was evident in a trial performed in three pure cypress stands in the Florence area [58]. In 1982 the mean disease incidence in the three stands was 37.5%. In 1983 a portion in each stand was subjected to sanitation (each infected tree was felled and removed) while the remaining part was left untreated. Every two years, from 1985 to 1993, the number of newly infected trees was evaluated in both treated and untreated portions of the three cypress stands, separately recording the incidence of diseased trees (Figure 14). Results indicated a high degree of effectiveness of sanitation in all three stands. Ten years after sanitation, the incidence of disease in the treated plots ranged between 3.4% and 6.2%, while in the untreated area, the percentage of newly diseased trees ranged between 17.5%–23.5%. A clear difference in the mean annual spread rate of CCD (as a percentage of newly infected trees) emerged between the untreated stands (4.5%) and the plots subjected to sanitation (1%). The effect of reduction of the inoculum load through sanitation on the occurrence of new infections in the following years is clearly demonstrated.

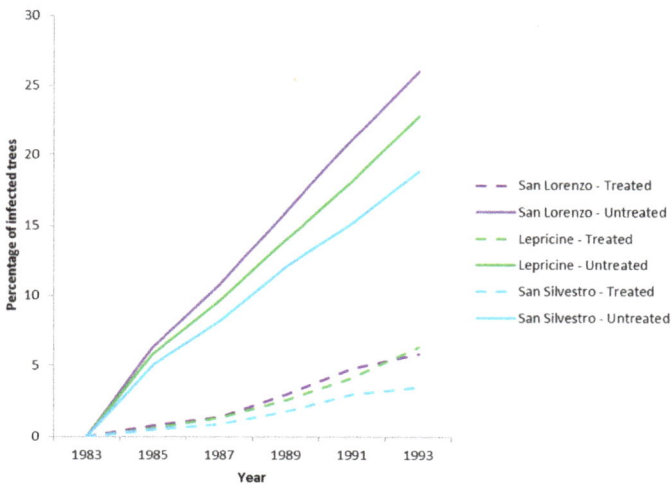

Figure 14. Effect of sanitation on the annual spread rate of CCD (% of newly infected trees) in three cypress woods (San Lorenzo, Lepricine, and San Silvestro) located in the surroundings of Florence compared with untreated neighboring areas. The disease incidence was normalized to zero at the beginning of the experiment (1983) (adapted from [58]).

Sanitation is also capable of increasing the genetic value of forest stands—particularly seed woods—by the removal of susceptible trees, thus accelerating natural selection and increasing the proportion of resistant trees.

From the above, it appears that sanitation in Tuscany could not be as drastic and incisive as the severity of epidemics would have required. Nevertheless, along with the other control measures adopted on a large scale, in the long-term, it contributed to a reduction in the inoculum of the fungus, gradually slowing the spread and intensity of epidemics, promoting natural evolution towards an endemic equilibrium.

5.3. Breeding for Resistance

As a complement to sanitation and chemical prevention, a genetic improvement program of cypress for resistance to *S. cardinale* bark canker was begun in the 1970s to tackle the rapidly spreading epidemics that were threatening woods, windbreaks, and ornamental plantings, and were spoiling the landscape in Tuscany and in other Mediterranean regions.

At first, extensive inoculation trials on plants derived from commercial seed (gathered from the seed woods) and on the half-sib progenies derived from healthy candidates sampled in the foci of infection of the disease in central Italy revealed that 88% of the trees were susceptible, 10% were tolerant, and only 1%–2% were resistant [55]. No differences were found between the populations of individuals derived from commercial seeds and the half-sib progenies of (about 200 families) derived from seeds gathered from healthy candidates selected in the foci of infection of the disease in Tuscany. Similar results were obtained from inoculation trials on different provenances sampled in the native range of *C. sempervirens* in Greece [59].

Instead, a distinct response to inoculations was clearly evident among the different cypress species. *Cupressus macrocarpa* and the other North American species included in the same phylogenetic group, such as *C. ambramsiana* Wolf, *C. goveniana* (Gordon) Bartel, *C. pygmaea* (Lemmon) Sarg, were found to be very susceptible. *Cupressus sempervirens*, *C. arizonica* Greene, *C. sargentii* Jeps, *C. macnabiana*, *C. stephensonii* Wolf were found to be susceptible. Instead, *C. glabra* (Sudw.) Little and *C. bakeri*, in addition to the Asian and North-African species, showed a good level of resistance to bark canker [56–60].

The cypress defense response to *S. cardinale* infections is known to be based on the formation of the necrophylactic periderm activated by the host to compartmentalize the damaged tissues, and is not a specific response to the pathogen. The vigor of this reaction determines the resistance level of a cypress tree to bark canker, as only plants that are able of setting up effective barriers and then compartmentalizing the infected tissues can succeed in repairing the lesion. So, the mechanism which governs the reaction of cypress to *S. cardinale* infections is under polygenic control, and its phenotypic expression can be influenced by both the host genotype and environment. On the other hand, this polygenic resistance based on the additive effects of several genes is more difficult for the pathogen to overcome [61].

The variability of the response to bark canker estimated in *C. sempervirens* was thought to be sufficient for undertaking a genetic improvement program aimed at selecting a wide genetic base of canker-resistant lines without the need to introduce resistant genes from other cypress species through interspecific crosses.

Currently, more than 100,000 accessions are included in the Institute for Sustainable Plant Protection-National Research Council of Italy (IPSP-CNR) germplasm collection (covering more than 60 hectares) which has been set up by Institute for Sustainable Plant Protection over four decades in various sites in central Italy and other Mediterranean countries. During this period, more than 400 canker-resistant *C. sempervirens* genotypes were selected, including eight patented clonal varieties [28,62,63]. Four of them have been marketed since 1995, with about 30,000 patented cypresses sold each year by the nurseries which hold the license for production. Indeed, some hundred thousand resistant cypress trees have been purportedly planted in central Italy in the last two decades. It is difficult to infer the actual contribution to reducing the reproduction and spread of the pathogen that these resistant genotypes might have played, but their role in decreasing the inoculum load of the fungus has certainly not only been theoretical.

6. Genotypic and Phenotypic Variation of California and Mediterranean *S. cardinale* Populations

A comparative analysis of the genetic diversity of *S. cardinale* populations from different areas was conducted to determine the relationships among these populations and to infer the source of the pathogen [33]. Source populations are hypothetically more diverse, while derived (introduced) populations are less diverse and nested within source populations due to bottleneck and founder effects.

Based on 7 simple sequence repeat (SSR), indices of genotypic and genetic diversity were found to be always higher for the Californian (Cal) than for the Mediterranean (Med) populations. The higher genetic diversity of the Cal population is also supported by sequencing of the β-tubulin region, which revealed the presence of two distinct haplotypes within the *S. cardinale* population in Cal, whereas only one of the two haplotypes was present among isolates from the Med. According to amplified fragment length polymorphism (AFLP) data, the two β-tubulin haplotype groups of the Cal population clustered separately in a PCA analysis [34]. This suggests that the two haplotypes are not simply two different alleles at one locus, but may be associated with two distinct groups of isolates potentially characterized by genotypic and phenotypic differences.

SSR genotyping also revealed that alleles appeared to be randomly associated in the Cal population, whereas a significant linkage disequilibrium was found for the Med population, suggesting that sexual reproduction may be ongoing in California, while a clonal reproduction characterizes the Mediterranean population [33]. A different way of reproduction seems to be confirmed by the fact that in Cal, the same genotype was found only twice at significant distances compared to 48 times in the Mediterranean. It is worth noting that *S. cardinale* is still considered to be an anamorphic fungus reproducing agamically, as its perfect stage—ascribed to the ascomycete *Leptosphaeria*—was reported on dying and dead cypress trees in north central California only once, but was not fully recognized because it was only partially described [64].

Network analysis (MSN) based on SSR multilocus genotypes and Neighbour-Joining analysis of AFLP both suggest that the Cal population of the fungus represented a source for the Med invasion [34]. In fact, isolates from California and the Mediterranean formed two distinct clusters of genotypes. A single connection between the two clusters through a series of three genotypes exclusively present in central Italy leads us to infer that a single introduction event from Cal into central Italy likely gave rise to the Med population of the fungus.

Structure analysis based on AFLP loci modeled the Mediterranean population as containing two clusters (called Med 1 and Med 2). The Med 1 population would include the founder Mediterranean population (directly derived from California), while the Med 2 population would derive from Med 1 and may be the result of a genetic process which led to a distinct population that then expanded in the Mediterranean [34].

Actually, the genotypic identity of isolates coming from locations hundreds kilometers apart suggests that the movement of infected plant material by humans is mainly responsible for pathogen expansion in the Mediterranean. As no clustering of Med isolates based on geography has been observed by PCA analysis, fungal individuals are likely to be moved frequently within this region. Central Italy as the source of the Mediterranean epidemics is also supported by identical AFLP genotypes found in Italy and Greece as well as in Italy and Spain, and by a substantial migration from Italy into Greece. Since no identical genotypes or PCA clustering of isolates from the Mediterranean and California were detected, intercontinental movement of inoculum is rather infrequent, as also confirmed by the lack of significant migration between the two areas [33].

Variation of phenotypic traits of the introduced Mediterranean population of *S. cardinale* derived from the movement and settlement of the fungus from the California source population was studied to evaluate the adaptive evolution of the pathogen during the invasive process [35]. Variability (phenotypic plasticity) of a set of characteristics considered important for establishment, infection, and spread of the pathogenic fungus were studied to determine which characteristics may favor the invasion and which are in rapid evolution. In the invasive Mediterranean population Med 1, traits associated with increased transmission and spread seemed to be favored in comparison to traits such as virulence and greater competitiveness. Thus, a smaller size of conidia and abundant sporulation are favored, while other traits such as virulence, rapid spore production, and germination are seemingly not selected during the invasive process. The source population (California) and the invasive (Mediterranean) population were phenotypically distinct (based on PCA analysis). Isolates of California showed greater aggressiveness (in terms of lesion length), a higher growth rate, a greater sporulation and germination capacity, and produced conidia of larger size. Nevertheless, in spite of the reduced genetic variability exhibited by the Med population compared to the Cal population, all of the considered characteristics showed a greater plasticity (wider range of variation) in the Med 1 population (i.e., during the first phase of invasion) than in the source Cal population. So, the phenotypic plasticity would have had a significant role in the first phase of invasion, while it would have been less important as the invasive population adapted to the new environment, as found in the isolates of the Med 2 population derived from Med 1.

7. Impacts, Risks of CCD, and Future Research Needs

The impacts attributable to the CCD where it attained an epidemic behavior have been: (a) effects on growth, survival, and abundance of the host species; (b) impacts on ecological functions of cypress woods as CCD altered disturbance regimes such as flammability, changes in hydrology, and soil erosion; (c) impacts on human social life because the presence of trees disfigured by the disease caused unpleasant feelings of degradation that were highly perceptible in environments of high landscape, historical and cultural, or recreational value.

As previously described, after its introduction from California, *S. cardinale* invasion in the Mediterranean has been subject to an evolution process over a 60-year period. A first epidemic phase occurred in the first two decades following its introduction, and had central Italy as center of gravity. Then, starting from the 1980s, epidemics expanded to cover the whole Mediterranean region. From the beginning of the 21st century, cypress canker gradually reached an endemic equilibrium (based on high disease incidence), also helped by the control measures extensively adopted in some regions.

Genetic information points to a significant isolation of the two populations, and identifies California as the center of diversity of the pathogen. Isolation and a separate evolution is also supported by different reproductive strategies and distinct phenotypic traits of the two populations. In view of these differences, plant quarantine regulations should be directed toward preventing further introductions of new strains or different lineages of the pathogen from California.

Introduction of new genetically different genotypes of the fungus in the Mediterranean could change the course of the epidemic, increasing the level of genetic variability available for selection in the new environment. Introduction and establishment of genotypes with novel mating alleles might give rise to sexual reproduction by bringing together different genetic combinations, thus increasing the pathogenicity and the level of adaptive evolution of the invasive population.

Isolates from California have also demonstrated a higher virulence, an increased growth rate, a greater adaptability to a wider range of temperatures, and greater reproductive capacity (sporulation and germination) compared to the Mediterranean isolates [35]. So, new introductions from California to the Mediterranean could have serious consequences for the local common cypress populations that survived the previous epidemics, causing a higher mortality rate and eventually new disease outbreaks. Introduction of genotypes adapted to a wider range of temperatures would cause a further expansion of the population of the pathogen to warmer and colder areas of Europe.

Shifts in the course of epidemics have been reported in other pathosystems due to intraspecific genetic recombination or gene flow among different lineages, as occurred in *Phytophthora ramorum* [65] and *P. infestans* [66] via interspecific hybridization, as was observed in the case of *Ophiostoma novo-ulmi* [67] and *Phytophthora alni* [68]. Alternatively, faster evolution and adaptation of newly introduced isolates to different environments may increase the spread of epidemics, as reported by Robin et al. [69] for *Cryphonectria parasitica*.

The regulatory systems of most countries generally do not easily allow for a distinction between groups of genotypes within a same species, and control regulations are relaxed when the same species is reported in two different countries. This underlines a considerable weakness of current international policies aimed at limiting the introduction of invasive species.

Introduction of new *S. cardinale* genotypes from California to the Mediterranean could also frustrate the results of a long-term breeding program which has led to the selection of a series of canker-resistant cypress genotypes. The selected resistance could be overcome by more aggressive genotypes of the pathogen. The durability of the selected resistant lines is crucial to the prevention or minimisation of the introduction of different genotypes of the fungus from California.

It could be useful to evaluate whether the cypress lines selected in Italy which have shown resistance against the Mediterranean isolates of *S. cardinale* are able to maintain their resistance to the most aggressive isolates from California. Actually, in California, a trial has already been started to assay the response of some canker-resistant cypress genotypes selected in Italy. If results are successful (i.e., if the tested cypress clones are confirmed to be resistant even towards the most aggressive isolates of California), their use for new plantations will be practicable, provided they are well adapted to the environmental conditions of California. If the greater aggressiveness of the isolates from California will be able to overcome the resistance of the selected clones, prevention of new introductions from this area in the Mediterranean will be imperative. This would also mean that the basic level of resistance of the cypress populations that experienced previous outbreaks in the Mediterranean will not able to prevent the development of new outbreaks of the disease. The increased aggressiveness and reproductive capacity exhibited by the isolates from California could enhance their intraspecific competitive ability, allowing them to establish within the invasive Mediterranean population of the fungus.

In the context of the control of cypress canker, management of hyper-sensitive host species which may represent a reservoir of fungal inoculum is not a minor issue. The marked worldwide vulnerability to cypress canker exhibited by Leyland cypress underlines the fact that this tree is unsuited to areas where conditions are conducive to the spread of *S. cardinale*. Leyland cypress can be used successfully in cooler climates not suited for the pathogen. In central Italy, there is the real risk that the severely cankered plantations of Leyland cypress may represent a source habitat for the inoculum of *S. cardinale* and favor a resurgence of epidemics on the local common cypress. Thus, in Italy and other Mediterranean regions, the use of this tree should be discouraged. Leyland cypress should be replaced with some autochthonous resistant cypress varieties that also have a crown habit suitable for barriers and windbreaks, which have shown a good vigor and adaptability to Mediterranean climatic conditions in the course of previous projects.

Future research should be directed toward monitoring the genetic structure of the pathogen population to highlight as early as possible the evolution of new clonal lineages with greater fitness and adaptability [69] that in a climate change scenario could lead to a geographic shift of the disease [70]. How the expected changes of climatic conditions could influence phenology of cypress and determine phenotypic changes that might alter its susceptibility [71] to CCD is another major topic that requires investigation.

To prevent further introductions of *S. cardinale*, it is important to fully understand through which routes human activities are promoting the long-range dispersal of the pathogen. Equally important is verification of whether California actually represents the worldwide source of *S. cardinale* by further comparison with populations the fungus from other areas of the world.

Acknowledgments: The authors wish to thank all the colleagues at IPSP-CNR who have worked diligently to study and protect Cypress, guided by a deep respect for this tree. In particular: Paolo Raddi, Alberto Panconesi, Vincenzo Di Lonardo, Giovanni Torraca, Anna Romagnoli. We also thank Matthew Haworth for his critical reading of the manuscript.

Author Contributions: R.D. is the primary author of this paper and conducted the literature review. G.D.R. is the corresponding author and contributed equally to the writing, editing and review of the relevant literature.

Conflicts of Interest: The authors declare no conflict of interest.

References

1. Keller, R.P.; Cadotte, M.W.; Sandiford, G. *Invasive Species in a Globalized World: Ecological, Social, and Legal Perspectives on Policy*; Keller, R.P., Cadotte, M.W., Sandiford, G., Eds.; The University of Chicago Press: Chicago, IL, USA, 2014; pp. 1–20.
2. Luque, G.M.; Bellard, C.; Bertelsmeier, C.; Bonnaud, E.; Genovesi, P.; Simberloff, D.; Courchamp, F. The 100th of the world's worst invasive alien species. *Biol. Invasions* **2014**, *16*, 981–985. [CrossRef]

3. Gonthier, P.; Faccoli, M.; Garbelotto, M.; Capretti, P. Invasioni biologiche ed effetti sulla biodiversità forestale–Bioinvasions and their effects on the forest biodiversity. In Proceedings of the Second International Congress of Silviculture, Florence, Italy, 26–29 November 2014; Ciancio, O., Ed.; Atti Accademia Italiana di Scienze Forestali: Florence, Italy, 2015; Volume II, pp. 155–160.

4. Gonthier, P.; Garbelotto, M. Reducing the threat of emerging infectious diseases of forest trees—Mini Review. *CAB Rev.* **2013**. [CrossRef]

5. Santini, A.; Ghelardini, L.; Pace, C.D.; Desprez-Loustau, M.L.; Capretti, P.; Chandelier, A.; Hantula, J. Biogeographical patterns and determinants of invasion by forest pathogens in Europe. *New Phytol.* **2013**, *197*, 238–250. [CrossRef] [PubMed]

6. Pimentel, D.; Lach, L.; Zuniga, R.; Morrison, D. Environmental and economic costs of nonindigenous species in the United States. *BioScience* **2000**, *50*, 53–65. [CrossRef]

7. Pimentel, D.; McNair, S.; Janecka, J.; Wightman, J.; Simmonds, C.; O'Connell, C.; Wong, E.; Russel, L.; Zern, J.; Aquino, T.; et al. Economic and environmental threats of alien plant, animal, and microbe invasions. *Agric. Ecosyst. Environ.* **2001**, *84*, 1–20. [CrossRef]

8. Pimentel, D.; Zuniga, R.; Morrison, D. Update on the environmental and economic costs associated with alien-invasive species in the United States. *Ecol. Econ.* **2005**, *52*, 273–288. [CrossRef]

9. Nentwig, W.; Bacher, S.; Pyšek, P.; Vilà, M.; Kumschick, S. The generic impact scoring system (GISS): A standardized tool to quantify the impacts of alien species. *Environ. Monit. Assess.* **2016**, *188*, 1–13. [CrossRef] [PubMed]

10. Panconesi, A. Pathological disorder in the Mediterranean basin. In *Agrimed. Reserach. Programme., Progress in EEC Research on Cypress Disease*; Ponchet, J., Ed.; Commission of the European Communities: Brussels, Belgium; Luxembourg, 1990; pp. 54–81.

11. Graniti, A. Cypress canker: A pandemic in progress. *Annu. Rev. Phytopathol.* **1998**, *36*, 91–114. [CrossRef] [PubMed]

12. Danti, R.; Della Rocca, G.; Panconesi, A. Cypress Canker. In *Infectious Forest Diseases*; Gonthier, P., Nicolotti, G., Eds.; CABI: Wallingford, CT, USA; Oxfordshire, UK; Boston, MA, USA, 2013; pp. 359–375.

13. Wagener, W.W. *Coryneum* canker of cypress. *Science* **1928**, *67*, 584. [CrossRef] [PubMed]

14. Birch, T.T.C. Gummosis diseases of *Cupressus. macrocarpa*. *N. Z. J. For.* **1933**, *3*, 108–113.

15. Barthelet, J.; Vinot, M. Notes sur les maladies des cultures méridionales. *Annu. Epiphyt.* **1944**, *10*, 18–20.

16. Grasso, V. Un nuovo agente patogeno del *Cupressus macrocarpa* Hartw. in Italia. *Ital. For. Mont.* **1951**, *6*, 63–65.

17. Saravì Cisneros, R. 1953 Cancrosis de los Cipreses provocada por *Coryneum cardinale* Wag. en la provincia de Buenos Aires (Argentina). *Revista de la Facultad de Agronomia de La Plata (Tercera Epoca)* **1953**, *29*, 107–119.

18. Anastassiadis, B. A new for Greece disease of the Cypress. *Ann. BPI* **1963**, *5*, 164–166.

19. Mujica, F.; Vergara, C.; Oehrens, E. *Flora Fungosa Chilena*, 2nd ed.; Ciencias Agricolas 5, Universidad de Chile: Santiago, Chile, 1980.

20. Torres, J. Grave enfermedad de los cipreses en España. *Bol. Serv. Plagas For.* **1969**, *12*, 97–99. (In Spanish)

21. Strouts, R.G. *Coryneum* canker of *Cupressus*. *J. Plant Pathol.* **1970**, *19*, 149–150.

22. Sutton, B.C. *The Coelomycetes*; CABI: Eastbourne, UK, 1980; pp. 374–379.

23. Funk, A. Microfungi associated with dieback of native *Cupressaceae*. in British Columbia. *CPDS* **1974**, *54*, 166–168.

24. Caetano, M.F.; Ramos, P.; Pinto-Ganhão, J. The phytosanitary situation of cypress in Portugal and the new prospects. In *Il Cipresso: Proposte di Valorizzazione Ambientale e Produttiva nei Paesi Mediterranei della Comunità Economica Europea*; Panconesi, A., Ed.; CNR, Regione Toscana, CEE: Florence, Italy, 1991; pp. 81–88.

25. Solel, Z.; Messinger, R.; Golan, Y.; Madar, Z. *Coryneum* canker of cypress in Israel. *Plant Dis.* **1983**, *67*, 550–551. [CrossRef]

26. Wingfield, M.J.; Swart, W.J. Cypress canker in South Africa. In Proceedings of the 5th International Congress of Plant Pathology, International Conference Hall, Kyoto, Japan, 20–27 August 1988; Harvey, I.C., Ed.; p. 361.

27. Danti, R.; Della Rocca, G.; El Wahidi, F. *Seiridium cardinale* newly reported on *Cupressus sempervirens* in Morocco. *Plant Pathol.* **2009**, *58*, 1174. [CrossRef]

28. Danti, R.; Barberini, S.; Pecchioli, A.; Di Lonardo, V.; Della Rocca, G. The epidemic spread of *Seiridium. cardinale* on Leyland cypress severely limits its use in the Mediterranean. *Plant Dis.* **2014**, *98*, 1081–1087. [CrossRef]

29. Urbasch, I. Natural occurrence of *Seiridium. cardinale* on *Thuja.* in Germany. *J. Phytopathol.* **1993**, *137*, 189–194. [CrossRef]
30. Hennon, P.E. Fungi on *Chamaecyparis nootkatensis.* *Mycologia* **1990**, *82*, 59–66. [CrossRef]
31. Faddoul, J. *Contribution à l'étude du Coryneum. cardinale Wag. Morphologie, Biologie, Physiologie*; Thèse No. 390; Université Paul Sabatier: Toulouse, France, 1973.
32. Hood, I.A.; Gardner, J.F.; Kimberley, M.O.; Gatenby, S.J.; Cox, J.C. A survey of cypress canker disease. *N. Z. Tree Grow.* **2001**, *22*, 38–41.
33. Della Rocca, G.; Eyre, C.A.; Danti, R.; Garbelotto, M. Sequence and simple-sequence repeat analyses of the fungal pathogen *Seiridium. cardinale* indicate California is the most likely source of the cypress canker epidemic for the Mediterranean region. *Phytopathology* **2011**, *101*, 1408–1417. [CrossRef] [PubMed]
34. Della Rocca, G.; Osmundson, T.; Danti, R.; Doulis, A.; Pecchioli, A.; Donnarumma, F.; Casalone, E.; Garbelotto, M. AFLP analyses of California and Mediterranean populations of *Seiridium. cardinale* provide insights on its origin, biology and spread pathways. *For. Pathol.* **2013**, *43*, 211–221.
35. Garbelotto, M.; Della Rocca, G.; Osmundson, T.; Di Lonardo, V.; Danti, R. An increase in transmission-related traits and in phenotypic plasticity is documented during a fungal invasion. *Ecosphere* **2015**, *6*, 1–16. [CrossRef]
36. Grasso, V.; Ponchet, J. Historique, distribution géographique et hôtes du Coryneum cardinale Wag. In Proceedings of the 'Seminario Il Cipresso: Malattie e Difesa', Florence, Italy, 23–24 Novembre 1979; Grasso, V., Raddi, P., Eds.; Commission of EC AGRIMED, tipografia l'Artigiano Firenze: Florence, Italy, 1980; pp. 119–126.
37. Wagener, W.W. The canker of *Cupressus.* induced by *Coryneum. cardinale* n. sp. *J. Agric. Res.* **1939**, *58*, 1–46.
38. Wolf, C.B.; Wagener, W.W. *The New World Cypresses*; El Aliso, Rancho Santa Ana Botanic Garden: Anaheim, CA, USA, 1948; p. 444.
39. Della Rocca, G.; Danti, R.; Garbelotto, M. First report of *Seiridium. unicorne* causing bark cankers on a Monterey cypress in California. *Plant Dis.* **2011**, *95*, 691.
40. Moriondo, F.; Bonifacio, A. Osservazioni preliminari sul Corineo del cipresso. *Atti dell'Accademia dei Georgofili Dispense I-II-III-IV* **1968**, *XV*, 211–221.
41. Bartoloni, P.; Panconesi, A.; Intini, M. Il Coryneum Cardinale: Notizie Biologiche e Prospettive di Lotta. In *Del Cipresso*; Cassa di Risparmio di Firenze, Ed.; Stabilimenti grafici Giunti Marzocco: Florence, Italy, 1976; pp. 49–53.
42. Poggesi, A. Intensità e ripercussioni economiche degli attacchi parassitari di Corinneum (Seiridium) cardinale Wag. e da Cinara cupressi Bckt. sul cipresso comune, con particolare riferimento alla provincia di Firenze. In Proceedings of the 'Seminario Il Cipresso: Malattie e difesa', Florence, Italy, 23–24 November 1979; Grasso, V., Raddi, P., Eds.; Commission of EC AGRIMED, tipografia l'Artigiano Firenze: Florence, Italy, 1980; pp. 135–147.
43. Della Rocca, G.; Di Lonardo, V.; Danti, R. Newly-assessed fungicides for the control of cypress canker caused by *Seiridium. cardinale.* *Phytopathol. Mediterr.* **2011**, *50*, 65–73.
44. Uzielli, A. Il consorzio per la difesa del cipresso. In *Del Cipresso*; Cassa di Risparmio di Firenze, Ed.; Stabilimenti grafici Giunti Marzocco: Florence, Italy, 1976; pp. 57–59.
45. Accolti Gil, F.; Funaioli, U. Consuntivo della prima campagna d'interventi effettuati nel 1974 in difesa del cipresso in provincia di Firenze. In *Del Cipresso*; Cassa di Risparmio di Firenze, Ed.; Stabilimenti grafici Giunti Marzocco: Florence, Italy, 1976; pp. 63–66.
46. Panconesi, A.; Raddi, P. Cancro del cipresso aspetti biologici ed epidemiologici . In *Il Cipresso: Proposte di Valorizzazione Ambientale e Produttiva nei Paesi Mediterranei della Comunità Economica Europea*; Panconesi, A., Ed.; CNR, Regione Toscana, CEE: Florence, Italy, 1991; pp. 49–60.
47. Pivi, R. Primi risultati di un'indagine epidemiologica sul cancro del cipresso in Toscana. In Proceedings of 'Il Recupero Del Cipresso Nel Paesaggio E Nel Giardino Storico' Collodi, Pistoia, Italy, 15 March 1995; Regione Toscana Giunta Regionale Dipartimento Agricoltura e Foreste: Florence, Italy, 1995; pp. 37–43.
48. Feducci, M. *Relazioni Tra Parametri Climatico–Ambientali E Diffusione Del Cancro Del Cipresso in Toscana Mediante Applicazioni G.I.S.–G.P.S. Tesi di Laurea di I livello in Scienze Forestali e Ambientali. Facoltà di Agraria*; Università Degli Studi di Florence: Florence, Italy, 2003–2004.
49. Xenopoulos, S.; Diamandis, S. A distribution map for *Seiridium. cardinale* causing the cypress canker disease in Greece. *Eur. J. For. Path.* **1985**, *15*, 223–226. [CrossRef]

50. Rizzo, D.M.; Garbelotto, M. Sudden oak death: Endangering California and Oregon forest ecosystems. *Front. Ecol. Environ.* **2003**, *1*, 197–204. [CrossRef]

51. Plantegenest, M.; Le May, C.; Fabre, F. Landscape epidemiology of plant diseases. *J. R. Soc. Interface* **2007**, *4*, 963–972. [CrossRef] [PubMed]

52. Tischendorf, L.; Fahrig, L. On the usage and measurement of landscape connectivity. *Oikos* **2000**, *90*, 7–19. [CrossRef]

53. McCallum, H.; Barlow, N.; Hone, J. How should pathogen transmission be modelled? *Trends Ecol. Evol.* **2001**, *16*, 295–300. [CrossRef]

54. Otten, W.; Gilligan, C.A. Soil structure and soil-borne diseases: Using epidemiological concepts to scale from fungal spread to plant epidemics. *Eur. J. Soil Sci.* **2006**, *57*, 26–37. [CrossRef]

55. Raddi, P. Variabilità della resistenza al cancro nell'ambito del cipresso comune. In Proceedings of the 'Seminario Il Cipresso: Malattie E Difesa', Florence, Italy, 23–24 Novembre 1979; Grasso, V., Raddi, P., Eds.; Commission of EC AGRIMED, tipografia l'Artigiano Firenze: Florence, Italy, 1980; pp. 185–193.

56. Andreoli, C.; Ponchet, J. Potential uses of exotic cypress species resistant to canker disease. In *Il Cipresso: Proposte di Valorizzazione Ambientale e Produttiva nei Paesi Mediterranei della Comunità Economica Europea*; Regione Toscana, CEE, Eds.; CNR: Florence, Italy, 1991; pp. 150–167.

57. Danti, R.; Della Rocca, G.; Barberini, S.; Raddi, P. The historic cypresses of the Viale di Bolgheri: Need of an organic and non-stop action to preserve a living monument. In Proceedings of the 6th International Congress on "Science and Technology for the Safeguard of Cultural Heritage in the Mediterranean Basin" Vol I, Sessions A, C Resource of the Territory, Biological Diversity, Athens, Greece, 22–25 October 2013; pp. 367–374.

58. Panconesi, A.; Danti, R. Esperienze tecnico-scientifiche nella bonifica del cipresso. In Proceedings of 'Il Recupero Del Cipresso Nel Paesaggio E Nel Giardino Storico' Collodi, Pistoia, Italy, 15 March 1995; Regione Toscana Giunta Regionale Dipartimento Agricoltura e Foreste: Florence, Italy, 1995; pp. 9–21.

59. Raddi, P.; Panconesi, A.; Xenopoulos, S.; Ferrandes, P.; Andreoli, C. Genetic improvement for resistance to cypress canker. In *Agrimed Reserach Programme, Progress in EEC Research on Cypress Disease*; Ponchet, J., Ed.; Report EUR 12493 EN: Brussels, Belgium; Luxembourg, 1990; pp. 127–134.

60. Allemand, P. Relations phylogéniques dans le genre Cupressus (Cupressaceae). In Proceedings of the 'Seminario Il Cipresso: Malattie e difesa', Florence, Italy, 23–24 November 1979; Grasso, V., Raddi, P., Eds.; Commission of EC AGRIMED, tipografia l'Artigiano Firenze: Florence, Italy, 1980; pp. 51–67.

61. Parlevliet, J.E. Durability of resistance against fungal, bacterial and viral pathogens; present situation. *Euphytica* **2002**, *124*, 147–156. [CrossRef]

62. Danti, R.; Raddi, P.; Panconesi, A.; Di Lonardo, V.; Della Rocca, G. "Italico" and "Mediterraneo": Two *Seiridium cardinale* canker-resistant cypress cultivars of *Cupressus sempervirens*. *HortScience* **2006**, *41*, 1357–1359.

63. Danti, R.; Di Lonardo, V.; Pecchioli, A.; Della Rocca, G. 'Le Crete 1' and 'Le Crete 2': Two new *Seiridium cardinale* canker-resistant cultivars of *Cupressus sempervirens*. *For. Pathol.* **2013**, *43*, 204–210.

64. Hansen, H. The perfect stage of *Coryneum. cardinale*. *Phytopathology* **1956**, *46*, 636–637.

65. Ivors, K.; Garbelotto, M.; Vries, I.D.E.; Ruyter-Spira, C.; Hekkert, B.T.E.; Rosenzweig, N.; Bonants, P. Microsatellite markers identify three lineages of *Phytophthora. ramorum* in US nurseries, yet single lineages in US forest and European nursery populations. *Mol. Ecol.* **2006**, *15*, 1493–1505. [CrossRef] [PubMed]

66. Goodwin, S.B.; Sujkowski, L.S.; Dyer, A.T.; Fry, B.A.; Fry, W.E. Direct detection of gene flow and probable sexual reproduction of Phytophthora infestans in Northern North America. *Phytopathology* **1995**, *85*, 473–479. [CrossRef]

67. Paoletti, M.; Buck, K.W.; Brasier, C.M. Selective acquisition of novel mating type and vegetative incompatibility genes via interspecies gene transfer in the globally invading eukaryote *Ophiostoma. novo-ulmi*. *Mol. Ecol.* **2006**, *15*, 249–262. [CrossRef] [PubMed]

68. Brasier, C.M.; Cooke, D.E.L.; Duncan, J.M. Orgin of a new *phytophthora* pathogen through interpecific hybridization. *Proc. Natl. Acad. Sci. USA* **1999**, *96*, 5878–5883. [CrossRef] [PubMed]

69. Robin, C.; Andanson, A.; Saint-Jean, G.; Fabreguettes, O.; Dutech, C. What was old is new again: Thermal adaptation within clonal lineages during range expansion in a fungal pathogen. *Mol. Ecol.* **2017**. [CrossRef] [PubMed]

70. Bebber, D.P. Range-expanding pests and pathogens in a warming world. *Annu. Rev. Phytopathol.* **2015**, *53*, 335–356. [CrossRef] [PubMed]
71. Anacker, B.L.; Rank, N.E.; Hüberli, D.; Garbelotto, M.; Gordon, S.; Harnik, T.; Meentemeyer, R. Susceptibility to *Phytophthora. ramorum* in a key infectious host: Landscape variation in host genotype, host phenotype, and environmental factors. *New Phytol.* **2008**, *177*, 756–766. [CrossRef] [PubMed]

![forests logo] *forests*

MDPI

Article

Laurel Wilt in Natural and Agricultural Ecosystems: Understanding the Drivers and Scales of Complex Pathosystems

Randy C. Ploetz [1,*], Paul E. Kendra [2], Robin Alan Choudhury [3], Jeffrey A. Rollins [3], Alina Campbell [2], Karen Garrett [3], Marc Hughes [4] and Tyler Dreaden [5]

[1] Tropical Research & Education Center, University of Florida, 18905 SW 280th Street, Homestead, FL 33031-3314, USA
[2] USDA-ARS Subtropical Horticulture Research Station, Miami, FL 33158-1857, USA; paul.kendra@ars.usda.gov (P.E.K.); Alina.Campbell@ars.usda.gov (A.C.)
[3] Plant Pathology Department, University of Florida, Gainesville, FL 32611, USA; ra.choudhury@ufl.edu (R.A.C.); rollinsj@ufl.edu (J.A.R.); karengarrett@ufl.edu (K.G.)
[4] College of Tropical Agriculture and Human Resources, University of Hawaii at Manoa, 875 Komohana Street, Hilo, HI 96720, USA; Mhughes7@hawaii.edu
[5] Forest Health Research and Education Center, Southern Research Station, USDA-Forest Service, Lexington, KY 40546, USA; tdreaden@fs.fed.us
* Correspondence: kelly12@ufl.edu; Tel.: +1-786-217-9278

Academic Editors: Matteo Garbelotto and Paolo Gonthier
Received: 22 December 2016; Accepted: 13 February 2017; Published: 18 February 2017

Abstract: Laurel wilt kills members of the Lauraceae plant family in the southeastern United States. It is caused by *Raffaelea lauricola* T.C. Harr., Fraedrich and Aghayeva, a nutritional fungal symbiont of an invasive Asian ambrosia beetle, *Xyleborus glabratus* Eichhoff, which was detected in Port Wentworth, Georgia, in 2002. The beetle is the primary vector of *R. lauricola* in forests along the southeastern coastal plain of the United States, but other ambrosia beetle species that obtained the pathogen after the initial introduction may play a role in the avocado (*Persea americana* Miller) pathosystem. Susceptible taxa are naïve (new-encounter) hosts that originated outside Asia. In the southeastern United States, over 300 million trees of redbay (*P. borbonia* (L.) Spreng.) have been lost, and other North American endemics, non-Asian ornamentals and avocado—an important crop that originated in MesoAmerica—are also affected. However, there are no reports of laurel wilt on the significant number of lauraceous endemics that occur in the Asian homeland of *R. lauricola* and *X. glabratus*; coevolved resistance to the disease in the region has been hypothesized. The rapid spread of laurel wilt in the United States is due to an efficient vector, *X. glabratus*, and the movement of wood infested with the insect and pathogen. These factors, the absence of fully resistant genotypes, and the paucity of effective control measures severely constrain the disease's management in forest ecosystems and avocado production areas.

Keywords: laurel wilt; Lauraceae; redbay; avocado; *Raffaelea lauricola*; *Xyleborus glabratus*; ambrosia beetles; coevolution

1. Introduction

New diseases are developing at an alarming rate on the world's trees. Diverse forest communities, pulp and timber plantations, and agricultural production are impacted in tropical, temperate and boreal environments [1–4]. Host jumps, pathogen hybridization and climate change have been associated with some of the new diseases [5–8], but other outbreaks have resulted from the invasion of naïve ecosystems by exotic pathogens [9–12]. The occasionally catastrophic responses of new-encounter hosts to alien pathogens are poorly understood, unpredictable and warrant further study [1].

In 2004, stands of redbay, *Persea borbonia* (L.) Spreng., began to die on Hilton Head Island, South Carolina, United States [13] (Figure 1A). At the time, the problem was not widespread, and abiotic factors, such as drought or salt-water damage, were invoked as possible causes for losses of this native tree. However, as the problem spread a new disease, laurel wilt, was recognized [13].

Figure 1. (**A**) Devastation caused by laurel wilt in a former stand of redbay, *Persea borbonia*, in Georgia in 2009 (photo: R. Ploetz). (**B**) Adult female of *Xyleborus glabratus* (PaDIL photo 33462, Justin Bartlett, http://www.padil.gov.au/). (**C**) Galleries of *X. glabratus* in a redbay tree that was killed by laurel wilt (photo: P. Kendra). (**D**) Internal staining of the sapwood of a redbay tree that is affected by laurel wilt (photo: R. Ploetz). Note inconspicuous entrance to a gallery of *X. glabratus* denoted with an arrow. During the early development of this disease, evidence for *X. glabratus* is limited. (**E**) Laurel wilt-induced death of redbay trees in Florida (photo: R. Ploetz). Laurel wilt-affected trees of native *Persea* spp. in the southeastern United States defoliate, as in **A**), only several months after dying.

Laurel wilt affects trees and shrubs in the Lauraceae (Laurales, Magnoliid complex). It is caused by a fungal symbiont, *Raffaelea lauricola* T.C. Harr., Fraedrich and Aghayeva (Ophiostomatales), of an invasive Asian ambrosia beetle, *Xyleborus glabratus* Eichhoff (Curculionidae: Scolytinae) [13] (Figure 1B,C). Starting in its putative epicenter of Port Wentworth, Georgia, United States (i.e., where *X. glabratus* was first detected in 2002) [14,15], the disease had spread by August 2016 as far west as 95° W, as far east as 78.5° W, as far north as 35° N and as far south as 25.5° N [16].

An estimated 300 million redbay trees have been killed by laurel wilt [17], and damage to the ecosystems that are associated with this important species has been documented or predicted [18–23]. Other common species in the southeastern United States, such as sassafras (*Sassafras albidum* (Nutt.) Nees), silk bay (*P. humilis* Nash), and swamp bay (*P. palustris* (Raf.) Sarg.), are decimated by the disease, as are threatened (pondspice, *Litsea aestivalis* (L.) Fernald) and endangered endemics (pondberry, *Lindera melissifolia* (Walter) Blume) [24–26]. Commercial production of a crop from MesoAmerica, avocado (*P. americana* Miller), was affected in Florida beginning in 2012, and that outbreak poses an increasingly serious threat to commercial production there and in other, currently unaffected areas [27,28]. Other lauraceous species from Europe and the United States are also susceptible, including bay laurel (*Laurus nobilis* L.) and, after artificial inoculation, California laurel (*Umbellularia californica* (Hook. and Arn.) Nutt.), gulf licaria (*Licaria triandra* (Sw.) Kosterm.), lancewood (*Nectandra coriacea* (Sw.) Griseb.), Northern spicebush (*Lindera benzoin* (L.) Blume) and Viñátigo (*P. indica* (L.) Spreng.) [13,26,29–32]. An Asian endemic, camphortree (*Cinnamomum camphora* (L.) J. Presl.), generally tolerates the disease [33].

2. Origins

Rabaglia et al. [15] indicated that *X. glabratus* was native to Asia and had been recorded from Bangladesh, India, Japan, Myanmar and Taiwan. Recently, Hulcr and Lou [34] reported the insect in mainland China. They confirmed that *X. glabratus* preferred lauraceous hosts (*Phoebe zhennan* S. Lee and F.N. Wei, *Machilus nanmu* Nees, *C. camphora* and *Phoebe neurantha* (Hemsl.) Gamble in China), and that taxa in other families were rarely colonized (only two of 40 collections).

Harrington et al. [35] recovered *R. lauricola* from specimens of *X. glabratus* from Japan and Taiwan. Since there are no reports of laurel wilt in the United States prior to 2004, *X. glabratus* probably carried the pathogen when it was first detected in Port Wentworth [14,15]. The focal and temporal spread of laurel wilt from that area (see [16]) and the genetically uniform populations of *R. lauricola* and *X. glabratus* that are found throughout the southeastern United States suggest that a single founding event, in or before 2002, may be responsible for the laurel wilt epidemic [17,36,37].

Non-Asian suscepts in the Lauraceae are all naïve (new-encounter) hosts [1,26]. Despite the wide geographic range of *X. glabratus* and the large number of species in the Lauraceae that are endemic to Asia (see below), there is only one report of laurel wilt in Asia, and that was on the introduced non-Asian host, avocado [38]. Fraedrich et al. [33] indicated that "there are no reports that indicate *R. lauricola* causes a plant disease in Asia", and Hulcr and Lou [34] doubted that (sic) "*X. glabratus* displays tree-killing behavior in its native range." If laurel wilt occurs in Asia on Asian members of the Lauraceae, it must be inconspicuous.

3. Coevolution

The term "coevolution" was coined to describe butterfly x plant interactions [39]. However, the idea that reciprocal evolution occurred between sympatric species was discussed by Darwin [40] and described in a plant-pathological context in the 1950s. In describing results from his classic research on flax rust, Flor [41] suggested that " . . . obligate parasites, such as the rust fungi, must have evolved in association with their hosts" and that " . . . during their parallel evolution, host and parasite developed complementary genic systems". Gene-for-gene and genetically quantitative/multi-gene systems have now been identified in many other pathosystems, and the specific adaptation of pathogens to host taxa, such as those described as formae speciales, is generally accepted as "the outcome of coevolution" [42]. These relationships can be conceived of as arms races in which increased disease resistance develops in a host in response to increased virulence in a pathogen [43,44].

Although these interactions can be difficult to document [43,45,46], coevolution appears to be an important factor in the development of many pathosystems [47]. Several criteria can be used to identify possible coevolved pathosystems [41–43,45,46,48–52]; they include:

1. A limited, often specific host range for the pathogen;
2. An original geographic distribution of the pathogen that overlaps with that of the host;
3. The occurrence of significant disease resistance in the host's primary center of origin;
4. Regional overlap of resistance and pathogenicity factors and phenotypes in the respective host and pathogen populations (i.e., geographic evidence for reciprocal selection);
5. Gene-for-gene relationships (although quantitative, non-gene-for-gene interactions can also co-evolve); and
6. Tandem speciation (also known as parallel cladogenesis).

Due to the rigorous criteria that are needed to confirm these relationships, it is not surprising that relatively few unequivocal examples of host x pathogen coevolution exist [51,52]. Most proposed coevolved pathosystems possess some, but not all, of the above attributes [10,42,53,54]. For example, although the first three of the above criteria are met for *R. lauricola* in Asia, there are no data for criteria 4–6. Future work may provide additional support for the idea that *R. lauricola* coevolved with endemic laurels in Asia. In the meantime, studies of Asian species in the Lauraceae, such as camphortree, could provide valuable insight into how tolerant hosts respond to this pathogen, and which attributes should be sought during the development of laurel wilt-resistant genotypes.

4. Ambrosial Symbioses

Xyleborine ambrosia beetles (Curculionidae:Scolytinae:Xyleborini) exhibit a haplodiploid, sibling mating system, which is also known as arrhenotoky [55–57]. Females are diploid and establish colonies after dispersion to uncolonized portions of the same or new host trees. Fertilized females can establish new colonies of females and haploid males, whereas nonfertilized females lay haploid eggs that become males. Males are flightless and rarely leave their natal galleries [58], where they mate with their mother and sisters.

Ambrosia beetles carry their fungal symbionts in specialized structures called mycangia [59–61]. In the Xyleborini, paired pre-oral (also known as mandibular) mycangia are small invaginations at mandible bases, mesothoracic mycangia are single, large invaginations between the meso- and metanotum, and elytral mycangia are small cavities at elytra bases [61,62]. In the species that have been examined, only one type of mycangium is present. However, mycangia are absent in some species that plunder fungal gardens of other species [63].

When adult females disperse to new trees, they bore brood galleries into host tree xylem, in which they cultivate gardens of the fungal symbionts. The developing colony feeds on these fungi (not wood), and as the colony matures new females are eventually produced. They then perpetuate the species by dispersing and establishing new colonies. Although some ambrosia beetle species (e.g., *Xylosandrus compactus* Eichhoff) can attack and colonize (establish brood colonies in) healthy trees [64,65], most reproduce only in stressed or dead trees.

During the early stages of the laurel wilt epidemic, Fraedrich et al. [13,33] examined *X. glabratus* x redbay interactions. They indicated that initial attacks by the insect in healthy trees were aborted and that reproduction by the insect was not observed in such trees. Nonetheless, aborted attacks were sufficient to infect trees with *R. lauricola*. Only after laurel wilt began to develop in infected trees was brood development by *X. glabratus* observed [13,33]. If this sequence is typical, *X. glabratus* may resemble other ambrosia beetle species in that it preferentially colonizes and reproduces in compromised or dead trees.

Most ambrosia beetles are generalists with wide host ranges. Thus, *X. glabratus* is unusual as it displays a strong preference for trees in the Lauraceae. Sesquiterpenes, rather than ethanol, the stress metabolite to which ambrosia beetles are usually attracted, appear to be a significant component of the attraction signature of these trees [66]. This difference has been cited when indicating that the *X. glabratus* vector relationship is exceptional, but there appears to be no evidence that trees identified with these signatures then support colonization and brood development before disease develops. Thus, indications that the interaction of *X. glabratus* with its host trees is atypical [61,67,68] should be clarified. Clearly, better information on, and a distinction between, the early (attack phase) and mid-stage interactions of *X. glabratus* with its host trees (colonization and brood development) are needed.

In discussing the appearance of laurel wilt in the United States and its absence in the Asian homeland of *X. glabratus*, Hulcr and Dunn [68] proposed that "the sudden emergence of pathogenicity" was due to "a new evolutionary phenomenon." A more parsimonious explanation for the emergence of laurel wilt is that coevolution with *R. lauricola* eliminated susceptible species in Asia, but not in the United States. The idea that American strains of *R. lauricola* became pathogenic after their move from Asia is not supported by data from studies that detected no differences in pathogenicity to avocado and swamp bay between isolates of *R. lauricola* from Asia and the United States [36,38]. Hulcr and Dunn [68] also suggested that an "olfactory mismatch" may be responsible for the identification by *X. glabratus* of nondiseased host trees as suitable for colonization. Since *X. glabratus* propagates in compromised host trees, as do other ambrosia beetle species, the attractive sesquiterpenes may simply enable *X. glabratus* to identify trees in which its ambrosial symbiont, *R. lauricola*, will establish, regardless of the host tree's health status. In summary, pathological differences have not been evident between *R. lauricola* in Asia and the United States, and *X. glabratus* and other ambrosia beetles appear to have similar reproductive preferences for compromised or dead trees. However, no other ambrosial symbiont is known to be a systemic, lethal pathogen. Even when ambrosial symbionts kill trees (e.g., *R. quercus-mongolicae* K.H. Kim, Y.J. Choi and H.D. Shin, *R. quercivora* Kubono and Shin. Ito, and *Fusarium euwallaceae* S. Freeman, Z. Mendel, T. Aoki et O'Donnell), mortality is due to localized (nonsystemic) necrosis and multiple attacks by the associated ambrosia beetle vectors [1].

More work is needed to understand the impact of *R. lauricola* on naïve American hosts. Host tree colonization by *X. glabratus* is incompletely understood, and discerning the role that other ambrosia beetle species may play in the laurel wilt epidemic has only begun. Several other species are associated with redbay, avocado and other lauraceous hosts, but they have been considered as vectors of *R. lauricola* only recently. The general absence of *X. glabratus* in avocado orchards that are affected by laurel wilt, the pathogen's presence in other species of ambrosia beetle that are recovered from avocado and other host trees, and the experimental demonstration of pathogen transmission and subsequent laurel wilt development in redbay and avocado suggest that species other than *X. glabratus* could play roles in the epidemiology of this disease [1,69,70].

5. Vectors of *Raffaelea lauricola*

Although ambrosia beetles have an obligate association with nutritional fungi [59,61,71,72], these can be promiscuous relationships wherein a given beetle species carries more than a single symbiont, and the same fungus species is present in more than one species of beetle. The movement of symbionts amongst ambrosia beetle species had been recognized previously [69,71,73,74]. However, the magnitude and speed with which this has occurred for *R. lauricola* is unprecedented [70]. Since its introduction into the United States in or before 2002, *R. lauricola* has been horizontally transferred from *X. glabratus* to nine additional ambrosia beetle species [1,69,70,75,76].

Reared (from laurel wilt-affected host trees) or trapped individuals (in the proximity of laurel wilt-affected trees) of 14 species of ambrosia beetle (*Ambrosiodmus*, *Euwallacea*, *Premnobius*, *Xyleborus*, *Xyleborinus* and *Xylosandrus* spp.) were assayed for *R. lauricola* by Ploetz et al. [70]. During 10 experiments, the pathogen was recovered from 34% (246 of 726) of the individuals that were associated with *Persea* spp. that are native to the southeastern United States, but only 6% (58 of 931) of those that were associated with avocado. *Raffaelea lauricola* was recovered from 10 of the ambrosia beetle species that were assayed, including *X. glabratus*, but was most prevalent in *Xyleborus* congeners [70]. Previous reports had suggested that *Raffaelea* spp. were the primary symbionts of *Xyleborus* spp. [74,75]. In general, mycangia of *X. glabratus* contained 10–1000 times more colony forming units (CFUs) of *R. lauricola* than the other assayed species. From native *Persea* spp. and avocado, *R. lauricola* was recovered from a respective 91% and 60% of the live specimens of *X. glabratus* that were assayed [70].

Although little is known about symbioses that are established between *R. lauricola* and different ambrosia beetle species, some of these insects may be involved in the ongoing epidemic in the southeastern United States [28,70]. In no-choice experiments, Carrillo et al. [69] reported that six and two species other than *X. glabratus* transmitted *R. lauricola* to potted redbay and avocado trees, and that laurel wilt developed in six and one of these interactions, respectively.

Kostovcik et al. [74] indicated that different types of mycangia may "support functionally and taxonomically distinct" symbioses. In summarizing a study of microbial communities in mycangia of *Xyleborus affinis* Eichhoff, *Xyleborus ferrugineus* Fabricius and *Xylosandrus crassiusculus* Motschulsky, they concluded that the mandibular (pre-oral) mycangium found in *Xyleborus* enabled the establishment of a broader array of symbionts than the larger and more exposed mesonotal mycangium possessed by *X. crassiusculus* (and *Xyleborinus saxesenii* Ratzeburg).

Hulcr et al. [77] studied how different ambrosia beetle species responded to *R. lauricola*. In olfactometer assays, *X. glabratus* was significantly attracted to *R. lauricola* in 54 of 84 assays ($p = 0.004$). In contrast, adult females of *X. saxesenii* and *X. crassiusculus* were significantly repelled by the fungus [77], which corresponds with its uncommon recovery from these species (1%–4% of all assayed individuals in [70]). Interestingly, another beetle that carried *R. lauricola* more frequently than *X. saxesenii* and *X. crassiusculus*, *X. ferrugineus* (11%–57% of the individuals in [70]), had a net nonresponse to the fungus; i.e., was repelled about as often (156 assays) as it was attracted (132 assays) ($p = 0.16$) [77].

Although beetle attraction to, or avoidance of, *R. lauricola* may impact whether it is a factor in the epidemiology of this disease [28,70], it is unclear whether species that are repelled by *R. lauricola* in olfactometers could still be occasional vectors of the pathogen and whether attracted or neutral species would disseminate the pathogen more frequently. Clearly, the involvement of other ambrosia beetle species in the laurel wilt epidemic requires more study. Nonetheless, we are beginning to understand the vector portion of this puzzle. By virtue of their affinity for *Raffaelea* spp. [70,74,75], species of *Xyleborus* might be expected to foster symbioses with *R. lauricola* more readily than species in other genera. In contrast, species with mesonotal mycangia, such as *X. crassiusculus* and *X. saxesenii*, should be able to carry greater quantities of the pathogen, based on the larger size of this organ compared to pre-oral mycangia. Even though they are infested with *R. lauricola* infrequently and are not attracted to the fungus in olfactometers, highly infested individuals with mesonotal mycangia could be vectors of *R. lauricola*.

6. Vector Chemical Ecology and Host Location

Host-based attractants (kairomones) have been studied most extensively for the primary vector of *R. lauricola*, dispersing females of *X. glabratus*. This beetle is not attracted to ethanol [78], the standard lure for ambrosia beetle detection [79]. The strongest female attractants identified to date are terpenoid kairomones, specifically volatile sesquiterpenes emitted from host wood [80–82]. In comparative studies with nine lauraceous species, emissions of four sesquiterpenes (α-copaene, α-cubebene, α-humulene, and calamenene) were positively correlated with in-flight attraction of *X. glabratus*, and electroantennography has confirmed olfactory chemoreception of these compounds [66]. A succession of field lures has been developed using essential oils naturally high in sesquiterpenes, including manuka oil (derived from *Leptospermum scoparium* J.R. Forst. and G. Forst., Myrtaceae; [80]), phoebe oil (*Phoebe porosa* (Nees and Martius) Barroso, Lauraceae; [80]), and cubeb oil (*Piper cubeba* L. f., Piperaceae; [83–86]). Laboratory bioassays with fractionated cubeb oil identified the negative enantiomer of α-copaene as a primary attractant, sufficient to evoke positive chemotaxis. Currently, the most effective lure for *X. glabratus* is an essential oil product enriched to 50% ($-$)-α-copaene content [87,88].

Exiting the natal tree to locate and colonize new resources is critical for the reproductive success of ambrosia beetles, but this brief dispersal event potentially exposes them to predation and adverse environmental conditions. To minimize risks associated with dispersion, it would be adaptive for females to have efficient host-seeking behaviors, guided by reliable cues. Kendra et al. [66] proposed that host location and acceptance is a multistep process directed by a series of cues presented in sequential order.

Initiation of dispersal flight in scolytine beetles is determined by light intensity, temperature, relative humidity and other environmental cues ([89] and references therein). With *X. glabratus*, females engage in host-seeking flight during the late afternoon and early evening, several hours earlier than other species of *Xyleborus* in Florida [90]. While in flight, females orient initially toward long-range olfactory cues; α-copaene appears to be the primary kairomone, but other terpenoids likely contribute to generate an attractive "signature bouquet" of the Lauraceae [66,82,87]. In addition to sesquiterpenes, several monoterpenes have been reported as kairomones, including eucalyptol (1,8 cineole) [91] and a blend of redbay leaf volatiles [92]. There is no evidence that *X. glabratus* utilizes sex or aggregation pheromones [78].

As females approach the focal source of host kairomones, visual cues assist in directing flight toward individual trees. Mayfield and Brownie [93] demonstrated that stem silhouette diameter aids in host location, but only when presented in an appropriate chemical context (i.e., presented concurrently with host odors). These experimental data support field observations that the oldest, largest-diameter trees are typically the first to be attacked by *X. glabratus*. In addition, there is a higher density of beetle entrance holes on the trunk and large diameter branches of redbay and swampbay, compared to smaller diameter branches [10,58,84]. In mature avocado trees, the highest emissions of α-copaene and α-cubebene are detected on the trunk and larger branches [94]. This chemical gradient may further assist with location of optimal sites for landing and initiation of a gallery. Even though large hosts are preferred, once those resources are depleted, smaller diameter trees are utilized, thereby enabling low populations of *X. glabratus* to persist for many years [95].

The short-range cues that prompt a shift from host-seeking to host acceptance and boring behaviors have received scant study. Knowledge in this area could facilitate the development of effective repellents or boring deterrents. Once a female makes contact with a potential host, she likely integrates a variety of stimuli, including olfactory, gustatory, contact chemosensory, tactile and visual cues, all of which must reinforce the message that an appropriate host has been found before a reproductive effort is initiated. In short-range laboratory bioassays, *X. glabratus* is attracted to volatiles emitted from its fungal symbiont [77], and in field tests these volatiles increased beetle captures when combined with host-based lures [96]. The ability to detect food-based odors may be adaptive in the host-choice process, confirming that the tree under evaluation is capable of supporting growth of required (nutritional) fungal resources.

7. Pathogen Attributes

The fungal symbionts of ambrosia beetles typically colonize only the lining of the natal gallery. In rare cases, these fungi cause serious damage in host trees, for example, *R. quercus-mongolicae* and *R. quercivora* on *Quercus* spp., and *F. euwallaceae* on diverse host species [1,97,98]. However, in these and other cases in which significant damage occurs, the fungal symbiont does not move systemically in host xylem and causes only localized damage; in these situations, tree mortality is associated with mass attack by the beetle vectors. *Raffaelea lauricola* is a unique symbiont, in that it systemically infects tree xylem and a single infection of susceptible trees can be fatal.

The population of *R. lauricola* in the United States is genetically uniform, apparently resulting from a single founding event (perhaps the importation of *X. glabratus* to Port Wentworth, Georgia). Little to no genetic variation was detected in isolates of the pathogen across the southeastern United States when Amplified Fragment Length Polymorphisms (AFLPs) and microsatellite markers were used [17]. With the same microsatellite markers and sequences of the large subunit of ribosomal (LSU) DNA, Wuest et al. [37] also concluded that the population of *R. lauricola* in the United States was quite uniform, in contrast to far greater diversity that they detected in the pathogen in Japan and Taiwan. When isolates from Asia were tested on avocado and swamp bay, they were as pathogenic as isolates from the United States [36,38]. Thus, although greater genetic diversity in *R. lauricola* has been found in Asia, differences in pathogenicity have not been apparent when isolates from the two regions have been compared.

The role of symbiont pathogenicity in the ecology of ambrosia beetles is not well understood [1,99]. Evident by comparison with other introductions is that the aggressiveness by which the laurel wilt pathogen systemically colonizes its North American tree hosts and causes lethal vascular wilt is an extreme scenario [1]. The range of fungal-specific damage observed among ambrosia symbionts in their native habitats varies from asymptomatic (nonpathogenic) to localized damage and host recovery (mildly virulent).

From the perspective of fungal pathogenicity, two hypotheses may help to understand the extreme aggressiveness of *R. lauricola* on North American Lauraceae. The first of these is that *R. lauricola* is an "accidental pathogen". Adaptations that enable *R. lauricola* to colonize natal galleries may, when encountered beyond the coevolved range of hosts, result in a massive host defense response that ultimately leads to xylem dysfunction and host death. The second hypothesis is that pathogenicity is a selective force in Asia that is kept in balance by coevolutionary processes; susceptible hosts, such as the naïve laurels in the southeastern United States, are eliminated, whereas those that tolerate infection persist and eventually replace their susceptible relatives.

Tests of these hypotheses are underway by comparing genomes of *R. lauricola* and its close relatives [100]. These analyses are aided by recently published phylogenies of *Raffaelea* [101,102], which have identified the closest, extant relatives of *R. lauricola*; notably, all other members of the phylogenetic clade in which *R. lauricola* resides (*Raffaelea sensu stricto*) are not plant pathogens. Of particular interest are genome-wide comparisons between *R. lauricola* and its close relative *R. aguacate* D. R. Simmons, Dreaden and Ploetz. *Raffaelea aguacate* has only been recovered from avocado [102], but it is neither systemic nor pathogenic on avocado or redbay [103]. Evidence for the accidental versus adapted pathogen hypotheses is being sought in comparative analyses between these two genomes and among genomes of related species. Virulence gene content, including those for putative effectors, elicitors and toxins, will be assessed, as will the expansion of these pathogenicity-associated gene families, and evidence for diversifying selection among candidate virulence factors.

Unraveling these fundamental questions may provide significant insight and predictive value for what might be expected in the future. If all ambrosial symbionts have the potential to be plant pathogens, is it a matter of time before another beetle–fungus–tree combination facilitates the emergence of another laurel wilt-like pathogen ("symbiont roulette")? If, on the other hand, pathogenicity is a derived character, what are the attributes required for weakly pathogenic symbionts to be aggressive pathogens when introduced outside their native range?

8. Hosts of Laurel Wilt

The Lauraceae is a large family that includes over 50 genera and 2500 to 3000 species [104]. The family represents some of the earliest angiosperms and has a fossil record dating back to the Mid-Cretaceous [105]. It is well represented on both sides of the Pacific Basin, and the so-called "amphi-Pacific tropical disjunction" of the *Persea* and *Cinnamomum* groups in the family has been examined to understand the presence, and the origins and relatedness, of family members in the Eastern and Western Hemispheres [101,106,107].

Although there are notable exceptions, such as naturalized populations of camphortree in the southeastern USA and the globally cultivated crop, avocado [108,109], most species in the family have restricted distributions. In the American tropics, lauraceous taxa comprise significant portions of lowland forests and montane environments [110,111]. Considering their ecological importance, laurel wilt could have an even greater impact as it spreads to new areas in the Western Hemisphere.

Of economic concern is the potential impact of the disease in currently unaffected avocado-production areas [28]. Avocado is a subtropical/tropical tree, and a significant fruit crop. In 2014, 5 million metric tonnes (MMT) of fruit were harvested worldwide, and Mexico was the leading producer (1.5 MMT) and exporter (in 2013, 0.6 MMT valued at ca. US$1.1 billion) [112]. Depending on the cultivars that are grown, which vary considerably in cold tolerance, the crop is grown commercially from United States Department of Agriculture Hardiness Zones 10 to 11, with moderate urban production of some cultivars occurring into Zone 9. Florida's avocado industry is the state's second-largest fruit industry after citrus, and 85% of the producers have orchards of less than 13 hectares. In the United States, more than 6600 growers on mostly small farms (less than 45 hectares) produce avocado; annual production worth more $1.6 billion is at risk [112].

Laurel wilt has affected avocado in Florida for at least a decade, but so far only ca. 3% of the avocado trees in commercial production have been killed [113]. However, the short time laurel wilt has been in South Florida (where most commercial production occurs), the great susceptibility of avocado to this disease, and experience with other lethal, invasive diseases [11,114] suggests that laurel wilt will, in the future, cause increasing damage on avocado in Florida. Laurel wilt imperils avocado production in other states (e.g., California and Hawaii), USA protectorates (Puerto Rico) and neighboring countries such as Mexico [28].

Persea americana is divided into Mexican (M), Guatemalan (G) and West Indian (WI) (also known as Lowland or Antillean) botanical races, respectively vars. *drymifolia* (Schltdl. and Cham.) S.F. Blake, *guatemalensis* (L. O.Williams) Scora, and *americana* Mill. [115]. Hybridization occurs freely among the races and is associated with a range of responses to calcareous soil, salinity, high and low temperatures, and other stresses [109,116]. Different responses to laurel wilt have also been noted among the races and racial combinations of the species; in general, greater susceptibility occurs in WI cultivars [117,118].

Although the most serious outbreaks of laurel wilt have occurred on lauraceous natives from the southeastern United States, a few natives from the region, gulf licaria and lancewood, display moderate tolerance after artificial inoculation [32,119]. They develop vascular symptoms, but do not die, which is similar to the response of camphortree [33]. Notably, rare tolerance to laurel wilt has been selected in redbay, and there is hope that these selections could be used to re-establish this tree in the southeastern coastal plain [36,120].

More data are needed on the susceptibility of American and non-American species in the family, and how susceptible and resistant hosts respond to infection by *R. lauricola*. To date, we have only rudimentary understandings of these processes [33,121–123]. Although natural selection against susceptibility has probably occurred in camphortree and other Asian Lauraceae [1,33], host attributes that are associated with laurel wilt tolerance have been studied only recently [121].

9. Host Responses to Infection by *Raffaelea lauricola*

Vascular wilt diseases typically exhibit wilting, sapwood discoloration (Figure 1D) and vascular dysfunction associated with physical and histological changes in the host [124]. The production of gels [125] and tyloses [126] in the xylem are two of the most common attributes of affected woody plants [127–131].

Tyloses are formed in xylem lumena in response to pathogen infection, embolism, aging, and injury [131]. They are outgrowths from adjacent parenchyma cells, and even though they can prevent desiccation, damage, and infection of adjacent cells [131], they also reduce hydraulic conductivity (water conductance) [132]. Gels induced by pathogens generally arise from host perforation plates, end walls, and pit membranes of the primary wall and middle lamella [131,133,134]. Breakdown of these cellular components by the pathogen results in the accumulation of gels [135].

Symptoms of laurel wilt on avocado, redbay, and swampbay include rapid wilting of foliage and vascular discoloration (Figures 1D,E and 2). Xylem blockage associated with tylose and gel formation appears to be at least partially responsible for the wilting symptoms [122,123]. Xylem function (the ability to conduct water) in avocado is impaired as soon as 3 days after inoculation, before the development of external or internal symptoms of the disease are apparent [123] (Figure 3). Tree mortality is associated with vascular functionalities of less than 10%.

Artificial inoculations with as few as 100 conidia of *R. lauricola* can kill avocado and swamp bay [136]. After laurel wilt-susceptible (avocado and swamp bay) and tolerant (camphortree) hosts were inoculated with a green fluorescent protein (GFP)-labelled strain of *R. lauricola*, Campbell et al. [120] reported that the pathogen was scarcely visible in microscopic cross sections, even in dead or dying plants. Although they observed that a maximum of 0.9% of the xylem lumena of avocado were colonized by the GFP-labelled strain 30 days after inoculation (dai) (Figure 4), about 40% of the lumena of avocado were occluded by tyloses 21 dai in another study [123] (Figure 5). Mobility of the pathogen or its metabolites in the xylem seems to be related to susceptibility.

Symptom development has been associated with reduced water transport in other, similar tree diseases. Wych elm affected by Dutch elm disease (caused by *Ophiostoma novo-ulmi* Brasier) [137], bitternut hickory affected by hickory decline (*Ceratocystis smalleyi* J. A. Johnson and T. C. Harr. [138], *Quercus* spp. affected by Japanese oak wilt (*R. quercivora*) [139], and *Notholithocarpus* (formerly *Lithocarpus*) *densiflorus* (Hook. and Arn.) Manos, Cannon and S.H.Oh affected by sudden oak death (*Phytophthora ramorum* Werres et al.) [140] all exhibited reduced xylem function which was, in turn, correlated with symptom development. For example, Park et al. [138] detected an inverse relationship between xylem sap flow and sapwood infection by *C. smalleyi*. They proposed that tylose formation induced by infection was responsible for reduced water transport in affected trees; *C. smalleyi* caused multiple cankers and reduced xylem function, resulting in crown wilt and decline of bitternut hickory.

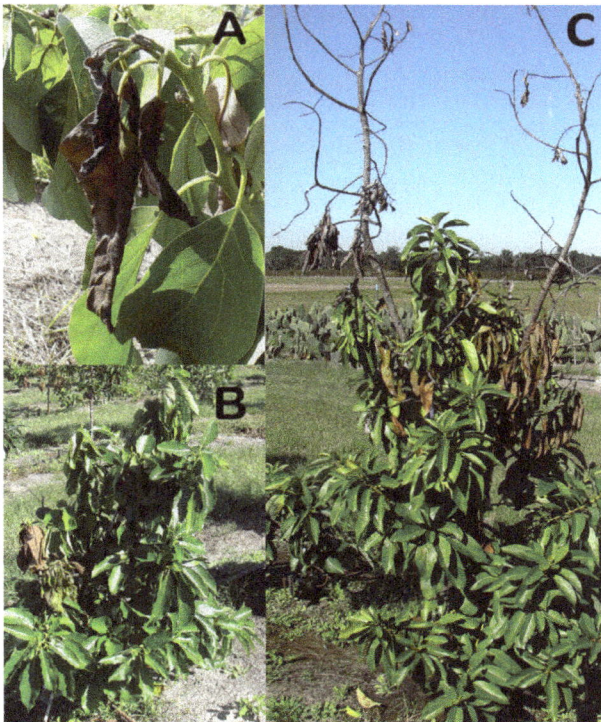

Figure 2. Foliar symptoms of laurel wilt on avocado include (**A**) wilting and an oily grey discoloration that rapidly progresses to necrosis; (**B**) initial symptom development in only a portion of a tree's canopy, involving a limited number of vascular traces; and (**C**) involvement of greater portions of the canopy followed by defoliation within a month or two of symptom onset (far more rapidly than occurs in redbay and other native *Persea* spp.) (photos: R. Ploetz).

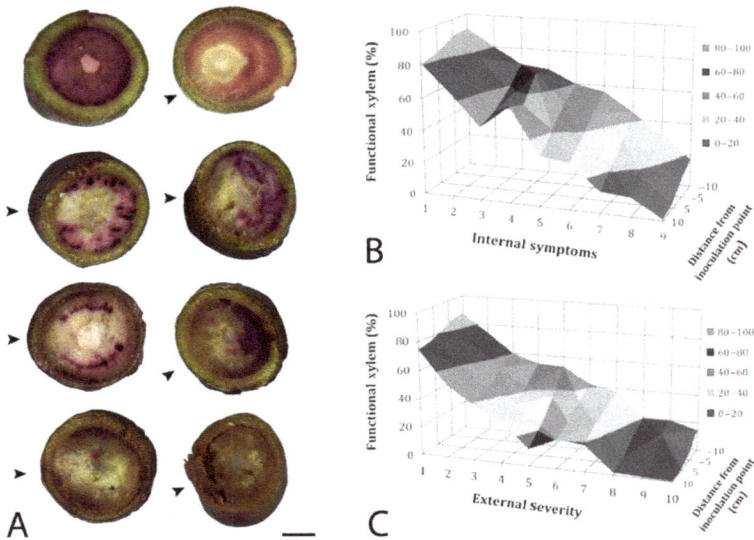

Figure 3. "Simmonds" avocado trees were inoculated with *Raffaelea lauricola* and xylem function was assessed with an acid fuchsin stain [122]. In (**A**), burgundy to pink staining in stem sections indicates functional xylem, which was quantified in scanned, digital images. From left to right and top to bottom, xylem in cross sections are: 98% functional (water control, internal symptoms (is) = 1; 86% (3 days after inoculation (dai)), is = 1; 76% (7 dai, is = 2); 71% (14 dai, is = 3); 32% (21 dai, is = 5); 30% (21 dai, is = 6; 5% (42 dai, is = 9; and 1% (42 dai, is = 9). Arrows (^) indicate the position on a stem that was inoculated, and the scale bar = 0.5 cm. In the response surfaces, % of functional xylem is graphed on the y-axes against the distance from the inoculation point on the z-axes and (**B**) internal and (**C**) external symptom development on the x-axes, which was rated on a 1–10 scale, where 1 = healthy, no symptoms, and 10 = dead, totally symptomatic.

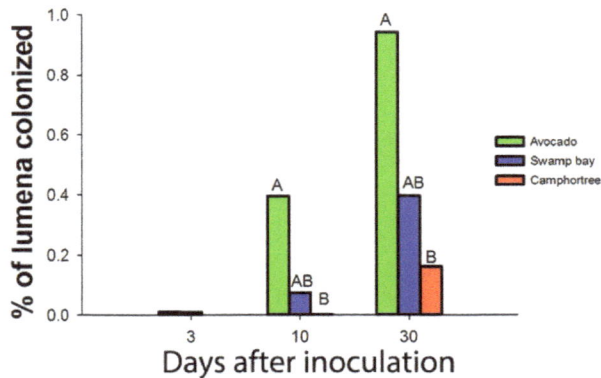

Figure 4. Campbell et al. [121] observed that less than 1% of the lumena of avocado, swamp bay and camphortree were colonized with a GFP-labelled strain of *Raffaelea lauricola*, even 30 days after inoculation (dai) (when most plants of avocado and swamp bay had died from laurel wilt).

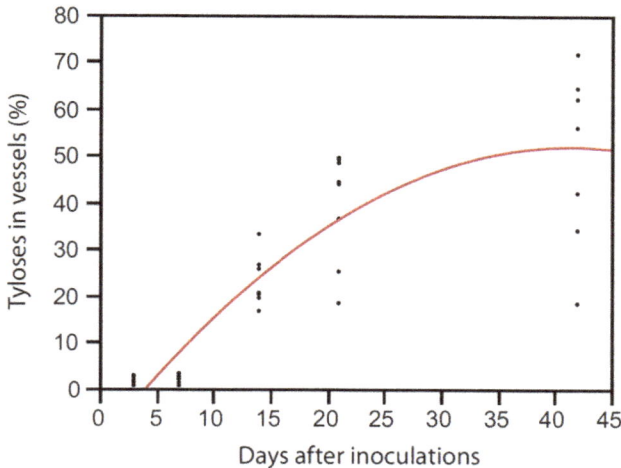

Figure 5. "Simmonds" avocado was artificially inoculated with *Raffaelea lauricola* and xylem lumena were examined in stem cross sections for tylose formation 7, 14 and 42 days after inoculation ($R^2 = 0.78$, $y = -0.69 + 1.79x - 0.04 (x - 17.4)^2$, $p < 0.0001$) [123].

Ploetz et al. [118] hypothesized that an avocado scion's susceptibility to laurel wilt is related to its ability to conduct water. They observed that pre-inoculation sap flow rates were greater ($p = 0.05$) in a susceptible cultivar, "Russell" WI, compared to the less susceptible cultivars "Brogdon" M × G × WI and "Marcus Pumpkin" G; however, sap flow plummeted in "Russell" soon after inoculation. Recently, Campbell et al. [121] reported that cross sections of xylem lumena of "Simmonds", another WI avocado cultivar, were nearly 2.5 times larger (3063 μm^2 ± 226) than those of camphortree (1250 μm^2 ± 221). Although they indicated that wood chemistry might impede the development of *R. lauricola* in camphortree (it contains antifungal compounds [141]), they also observed that smaller lumena in camphortree could hinder the movement of *R. lauricola* compared to host trees with larger lumena.

In long-term studies to develop elms that resist Dutch elm disease, vessel size and sap flow were reported to be key factors associated with the susceptibility of *Ulmus minor* Mill. and *Ulmus* hybrids [142–144]. Susceptible trees had significantly wider and longer vessels than those that resisted the disease [144]. Solla and Gil [142] hypothesized that vessel diameters and sap flow affected the dispersal of *O. novo-ulmi* in the elm host, and Venturas et al. [144] suggested that xylem structure restricted the pathogen's spread in resistant elms.

Additional work is needed to understand the relationship between susceptibility to laurel wilt and sap flow and other xylem-associated attributes, as it may facilitate the identification of laurel wilt-tolerant genotypes of host trees. Whether sap flow rates and xylem attributes are general predictors of laurel wilt susceptibility and tolerance in avocado are examined in ongoing work.

10. Ecology and Epidemiology

To date, native *Persea* spp. (redbay, swamp bay and silk bay, which some refer to collectively as redbay) have been most severely impacted by laurel wilt. Within a few years of affecting a stand, most mature trees of these species succumb to this disease [13,21,22,145,146]. Symptoms of laurel wilt develop soon after host trees are infected by *R. lauricola* (usually within 2 to 3 weeks of artificial inoculation) [117]. As the disease develops, host trees increase the production of volatiles that attract *X. glabratus* [92], which then promotes mass attack, colonization and brood development by *X. glabratus*.

The explosive nature of the laurel wilt epidemic is due to the great susceptibility of the native *Persea* spp. [26], their attractiveness to *X. glabratus* [66,78], the rapid increase in *X. glabratus* numbers in affected stands of these trees [58], and the substantial amounts of inoculum that most females of *X. glabratus* carry [70]. Furthermore, tree-to-tree spread via interconnected root systems, which has been suspected in avocado [28] and sassafras [147], and demonstrated in pondberry [148], enables movement of the pathogen without vector assistance. It is hard to imagine a more efficient means for killing trees than this disease.

Raffaelea lauricola typically infects avocado only via ambrosia beetle vectors and interconnecting root grafts (see below), although other avenues of infection have been studied experimentally. Mechanical transmission of the pathogen was possible with artificially infested handsaws on potted plants, but the pathogen did not survive on circular saws that are used to prune avocado trees in commercial groves, presumably due to the high heat that was generated on these blades during use [149]. Seed and scion transmission of *R. lauricola* have also been discounted. After potted, fruit-bearing trees were artificially inoculated with *R. lauricola* plants were systemically colonized by the fungus, but in no instance did infection progress further than the hilum of fruit (87 fruit and their associated pedicles) [150]. Furthermore, when scions from artificially infected trees were used as grafting material, they did not establish on recipient rootstocks [151].

Most commercial avocado production in the eastern United States is located in Miami-Dade County in southern Florida; it is bordered on the north and east by metropolitan Miami and on the west and south by Everglades National Park (ENP) [152,153]. The proximity of a densely populated urban area and a protected natural area creates significant challenges for managing this disease.

Residential avocado trees are common in Miami. In other pathosystems, residential trees can act as reservoirs for pathogens from which neighboring agricultural areas are colonized. For example, residential trees played a significant role in epidemics of citrus canker in Florida and plum pox in Canada [154,155]. Residential trees are also associated with the regional spread of pathogens via the nursery trade, as infected asymptomatic plants can be sold and distributed through this network [156,157]. In this case, the short latent period of laurel wilt on susceptible hosts may be advantageous, as asymptomatic infection of these species would probably play a negligible role in spreading the disease. In contrast, tolerant species, such as camphortree, could serve as symptomless reservoirs of *R. lauricola* and *X. glabratus* in the landscape as well as a means by which the disease could spread; greater understanding of their potential role in the regional dissemination of this disease is needed.

Natural areas that border avocado production areas in southern Florida may present risks similar to those of residential avocado trees. Prior to the laurel wilt epidemic, swamp bay occurred throughout the ENP. Swamp bay is highly susceptible to laurel wilt [26] and is attractive to, and supports significant egg production of, *X. glabratus* [158]. An aerial survey of the Everglades in 2011 and 2013 detected rapid spread of laurel wilt on this host tree [25]. Since source-to-sink dynamics from unmanaged areas to agricultural areas have played an important role in other diseases with insect vectors [159], there was concern that swamp bay trees might act as disease and vector refugia in the ENP. However, recent spatial analyses suggested that laurel wilt outbreaks in the avocado production area are concentrated there, and that there is little connectivity between laurel wilt in the avocado and natural areas [160].

Once laurel wilt has established in an avocado orchard, it apparently moves among trees through root grafts [28]. Root grafting (i.e., the establishment of functional unions between the roots of different plants), is a common phenomenon in trees [161,162], and it plays a significant role in the movement of similar plant pathogens, such as those that cause oak wilt (*Ceratocystis fagacearum* (Bretz) Hunt) and Dutch elm disease (*Ophiostoma* spp.) [163–165]. In avocado orchards, high densities of trees are planted in rows in which root grafting occurs among adjacent trees. Previously, the root-graft movement of herbicides and another avocado pathogen, *Avocado sunblotch viroid* (ASBVd), were recognized [117,166–168]. Subsequent evidence that root-graft transmission of *R. lauricola* occurs has included the natural infection of roots by the pathogen, rapid expansion of disease foci, and the fact that prompt fungicide treatment of trees adjacent to diseased trees impedes spread and expansion of foci [28,117].

Preventing root-graft spread in avocado orchards is difficult. Since trenching to sever root grafts stops the spread of oak wilt and Dutch elm disease, it would presumably be useful in laurel wilt-affected avocado orchards. However, trenching would damage irrigation systems that are typically in place in avocado orchards. Other techniques to limit root graft spread are being explored, such as establishing barriers to pathogen movement with fungicide or herbicide treatments [28,169]. Since root grafts tend to develop in older plants, maintaining younger orchards and smaller trees might also inhibit transmission among trees.

While root-graft transmission enables rapid movement of the pathogen among trees, transmission also occurs among and within orchards via ambrosia beetles. As discussed above, *X. glabratus* is most important in natural settings, as it predominates in both trapping and rearing assays with these host trees [26,28,66], carries the greatest amounts of the pathogen [70], and was most consistently associated with transmission in no-choice assays [69]. However, the role that this species and its relatives play in the avocado system requires further study. The above spatial analyses [160] suggest that laurel wilt in swamp bay populations in the Everglades (on which *X. glabratus* would predominate) plays an insignificant role in the avocado epidemic. All of these data suggest that a complex of beetles, which may or may not include *X. glabratus*, are involved in laurel wilt dissemination in avocado. These beetles are understudied, and relatively little is known about their habits and roles in the environment. Understanding the ecological drivers of these insects is critical for future assessments of disease risk.

An early epidemiological model suggested that *X. glabratus* would not reach the southern tip of Florida until approximately 2015 [170]. However, *X. glabratus* was observed in Miami-Dade County in early 2010, and laurel wilt was reported in 2011 [153]. While this model accounted for host density and climatic features that benefit *X. glabratus*, it likely underestimated the effects of anthropogenic spread [170].

Although assigning an anthropogenic role to an outbreak of laurel wilt can be difficult, large geographic jumps in the disease's distribution and the close proximity of an outbreak to parks or other places where firewood is used, or where wood or wood products are handled, are attributes that suggest human agency. For example, Cameron et al. [171] reported outbreaks of laurel wilt in Georgia near a pulp mill and mulch plant. They also noted an apparent firewood-related outbreak near a campsite 120 km from the nearest outbreak. As this disease has spread, others have noted large jumps in its distribution. Riggins et al [172] reported a laurel wilt outbreak in Mississippi, over 500 km from the nearest known outbreak, and a recent finding in Texas was 300 km from the nearest known occurrence [16]. The above findings have been on redbay, but sassafras has also been impacted. Bates et al. [173] noted an outbreak on sassafras in Alabama, 160 km from the nearest outbreak, Fraedrich et al. [174] described an outbreak in Louisiana, 400 km from the nearest outbreak, and Olatinwo et al. [175] described a subsequent finding in Arkansas, 134 km north of the Louisiana outbreak. These range expansions suggest that anthropogenic spread is an important factor in the epidemiology of this disease. In addition, the increased prevalence of laurel wilt on sassafras, a common species that extends into Canada and which supports propagation of *X. glabratus*, indicates that the disease has the potential to continue its northward movement.

Effective landscape modeling of laurel wilt will need to account for biological (e.g., host range, host density and vector competency) and social factors (e.g., proximity to campgrounds and areas in which wood or wood products are handled). Integrating these factors is critical for understanding long-term disease prevention and control. Shearman et al. [176] used forest inventory and analysis data to estimate risk factors for redbay. They determined that the presence of laurel wilt in a county increased the odds that a given tree would be killed by approximately 154%, for every year after the initial outbreak, and that mortality rates increased by 5% for increases of 1 cm diameter breast height (DBH) in a tree's trunk diameter.

Epidemic network analysis is becoming a common part of the epidemiological toolbox [177,178]. Harwood et al. [179] used network analysis to evaluate likely outcomes for epidemics caused by

P. ramorum and *P. kernoviae* in the United Kingdom, as influenced by inspection strategies. These pathogens can spread through natural systems, as well as the nursery trade. They concluded that if the rates at which these pathogens were introduced increased, inspection activities could not keep up with the influx. Network analysis has also been used to identify key sampling locations for tracking epidemic progress [180,181]. For laurel wilt, spread via infested wood is a key vulnerability for long-distance movement, but this activity is difficult to monitor. To the extent that epidemic network structures can be characterized for a disease such as laurel wilt, strategies may be implemented to make inspection more efficient.

Impact Network Analysis [182] is a multilayer network analysis that uses agent-based modeling [183] to evaluate the likely system-level effects of regional management implementation. Multilayer networks integrate linked processes, such as networks of the spread of information about disease and linked networks for the spread of the disease and/or its vectors [180,184–186]. Multilayer networks can provide analyses about how networks interact to produce system level outcomes.

In Impact Network Analysis, three components are emphasized [182]. The first is a management concept or tool, such as a recommended cultural practice, a resistant variety, or effective pesticide, which is associated with a level of efficacy, a level of confidence in the estimated efficacy, and the cost of implementation. The second component is the network of communication and influence among managers who decide whether to implement the management tool [187], as well as the management landscape (e.g., natural or agro-ecosystem) in which management is or is not applied. The third component incorporates information on the epidemic network. Notably, establishment of a given disease in a new location may be strongly influenced by the management landscape. Impact Network Analysis is a platform for evaluating how a proposed management type or portfolio of management types is likely to influence regional epidemics, including the effects of manager decision-making [182].

An Impact Network Analysis is being developed for the laurel wilt epidemic in Florida and beyond. When the first component of the analysis is considered, the tools that are available for management are somewhat disheartening (see below). Currently, in natural systems that are populated by highly susceptible host species, there are no viable measures. Furthermore, although there are effective tools for avocado production, they are not consistently adopted by producers. The second component of the analysis, the network of communication and influence among managers, presents additional challenges. For example, some avocado managers may not be convinced of the importance of management, or may be motivated to manage only minimally until land is diverted for other purposes. The third component, the epidemic network, indicates that unmanaged outbreaks of laurel wilt are important risk factors in the avocado system. Growers who do not effectively manage laurel wilt may increase the difficulty with which the disease is managed by other avocado producers. Additionally, the impact of residential avocado production is a poorly understood, but potentially important facet of the epidemic network.

Serious challenges exist for managing laurel wilt in most situations. However, if laurel wilt epidemics in natural systems are shown to exert little or no influence on the avocado system, regional avocado management strategies could be simplified. Understanding the regional system will be a first step toward scaling up [188] risk assessment and management strategies in Cuba, Mexico and neighboring avocado-production areas.

Early detection of laurel wilt (ideally *R. lauricola*-infected trees before they develop symptoms) is critical to the containment of laurel wilt and the success of disease management efforts in avocado orchards. Since symptoms of laurel wilt can be confused with those associated with other biotic and abiotic factors, the presence of *R. lauricola* should be confirmed when a laurel wilt diagnosis is in doubt [26,28]. Currently, a taxon-specific method can be used to identify the pathogen and distinguish it from its closest relative, *R. aguacate* [102,189].

Visible–near infrared spectroscopy has been tested for the nondestructive detection of laurel wilt on avocado [190]. Classification studies were conducted with visible near infrared spectra of asymptomatic and symptomatic leaves from plants artificially infected with *R. lauricola*, as well as leaves from noninfected freeze-damaged and healthy plants. Scores from principal component analyses were used as input features in four classifiers: linear discriminant analysis, quadratic discriminant analysis, Naïve-Bayes classifier, and bagged decision trees. All of the classifiers were able to discriminate leaves from plants with laurel wilt from freeze-damaged leaves. False negatives resulted mainly from asymptomatic leaves from infected plants being classified as healthy.

In subsequent work, spectral data were used to distinguish healthy, laurel wilt-affected, and Phytophthora root rot-affected avocado trees [191]. With a modified camera, spectral images were taken during helicopter surveys of commercial avocado orchards [192]. RmodGB digital data were used to calculate vegetation indices (VIs), band ratios, and VI combinations for healthy and laurel wilt-affected trees. Significant differences were observed in all VIs calculated among laurel wilt affected and healthy trees, although the best results were achieved with Excess Red, (Red–Green) and Combination 1. These results were used to modify a MCA-6 Tetracam camera with different spectral filters (580–10 nm, 650–10 nm, 740–10 nm, 750–10 nm, 760–10 nm and 850–40 nm), which was then used to take multispectral images of avocado trees at early, intermediate and late stages of laurel wilt development at three altitudes (180, 250 and 300 m) [191]. Inexpensive devices that use this technology need to be developed.

Canine detection was recently promoted for detecting avocado trees infected with *R. lauricola*, prior to the development of laurel wilt symptoms [193]. Although there has been some success in detecting other diseases under controlled settings [194], data for canine detection of laurel wilt has only been presented in seminars (no peer reviewed publications are available). Recently, canine detection was reported in avocado trees that displayed no obvious symptoms of laurel wilt [195]. Although *R. lauricola* was not isolated from 42% of the "detected" trees, in another avocado orchard, laurel wilt subsequently developed in most of the detected trees. Thus, trees from which the pathogen had not been isolated were, presumably, infected with *R. lauricola* when detected by dogs.

The reliability of canine detection requires additional evaluations. For example, the ability of dogs to distinguish infection in root-grafted trees, detect the pathogen prior to symptom development (see [196]), and distinguish *R. lauricola* from its close relatives, many of which are prevalent in ambrosia beetles that infest avocado [102,197], should be determined. Nonetheless, the available evidence suggests that canine detection could be a useful tool for managing laurel wilt in avocado orchards.

As alluded to above, the spread of laurel wilt is affected by poorly understood social factors. For example, even though early detection could help stem its spread the avocado system, producers might not adopt the available techniques (the second component in the above Impact Network Analysis). Furthermore, early detection would only be effective if detected trees were then promptly removed (sanitation) or treated with fungicide. Agricultural producers respond to uncertainty in different ways, and only a subset of all producers will adopt a given technology, no matter how effective it might be. Better understandings are needed for the impacts of different social, economic, and cognitive factors on the decision making process [183,198].

Laurel wilt management in commercial avocado orchards is possible, but difficult [199]. Effective fungicide treatments have been described, but they are expensive and their use may not be sustainable. Annual retreatment is indicated for even the best fungicide treatments. It becomes increasingly difficult to retreat trees as they heal from previous applications (wound tissue impedes fungicide uptake in previously treated sites). Subsequent applications are forced to move above the root collar, which is the most effective application site. Early detection and the rapid removal and destruction of affected trees is a more effective measure for managing this disease, but its adoption varies widely among producers. Retaining infected trees in an orchard is a dangerous, but common practice, as many producers are averse to removing trees that might produce fruit. Once focal development of the disease (root-graft movement of the pathogen) begins, laurel wilt becomes incredibly difficult to manage. In general, avocado producers who have been forced out of production by laurel wilt have practiced tardy and insufficient management.

In natural environments that are impacted by laurel wilt, the above management strategies are not useful; in these situations, eradication of the disease is not possible, sanitation has been ineffective when it has been used, and fungicide treatment is far too expensive [26,200]. Nonetheless, laurel wilt-tolerant individuals of redbay have been identified [120]. Although it would be a long-term proposition, these selections might be useful for re-establishing devastated populations of this important species.

11. Outlook

Laurel wilt has changed the composition of forests throughout the southeastern United States [26]. It has eliminated redbay and swamp bay, two keystone species, from major portions of the region, and other important species, such as sassafras, will be increasingly impacted as the disease spreads. Other rare suscepts, such as pondberry, are threatened with extinction [24].

Time will tell when and whether new areas are affected by this destructive tree disease [28]. The ultimate impact of laurel wilt will be determined by which areas are invaded, the susceptibility of hosts that occur in these areas, the suite of vectors that are involved in new outbreaks and their effectiveness in transmitting *R. lauricola*. Clearly, ecosystems in the western United States and tropical America that are populated by suscepts are at risk [26,28].

Laurel wilt has begun to cause alarming losses in avocado production in Florida [28]. The state's primary commercial production area was first affected in 2012, and losses will continue to increase as the disease consolidates and spreads in the area. The economic impact of laurel wilt could increase dramatically when and if it spreads to California, Mexico and other major production areas.

Much remains to be learned about this enigmatic disease. Basic information is lacking on vector x host x pathogen interactions, vector identity and ecology in the avocado system and the disease's epidemiology in natural and agricultural environments. More data are needed on the disease's host range and the nature of resistance and susceptibility. Until better information is available in these and other areas, laurel wilt will remain a destructive disease for which we have few management options.

12. Conclusions

Laurel wilt is caused by the only known systemic and lethal ambrosia beetle symbiont, *R. lauricola*. To date, all suscepts are naïve trees in the Lauraceae plant family that do not have an evolutionary history with this pathogenic symbiont. In little more than a decade, laurel wilt spread throughout the southeastern coastal plant of the United States. Significant populations of native trees have been eliminated by the disease, and an important fruit crop, avocado, is affected in southern Florida and threatened in other production areas. Advances have been made in understanding the interactions of the pathogen with various host trees, as well as how the host responds to the disease. The disease's epidemiology is generally understood, but important gaps remain in what is known about the ambrosia beetle vector portion of the puzzle, especially in the avocado system. Although progress has been made in the management of laurel wilt, successful control in avocado production is still difficult, and in natural systems is all but impossible. An increasing impact of laurel wilt is predicted as it spreads within and beyond its present range.

Acknowledgments: This research was partially funded by USDA grants from the Florida Department of Agriculture and Consumer Services, FDACS Sponsor #019730 and #021757, the USDA-ARS National Plant Disease Recovery System, and NIFA grants 2009-51181-05915 and 2015-51181-24257.

Author Contributions: P.E.K. wrote the section on vector chemical ecology and host location, R.A.C. and K.G. wrote portions of the ecology and epidemiology section, J.A.R. wrote portions of the section on pathogen attributes, A.C. wrote portions of the section on host responses to infection, M.H. wrote portions of the sections on pathogen attributes and ecology and epidemiology, and T.D. wrote some of the section on pathogen attributes. R.C.P. contributed to the above sections, wrote all other portions of the manuscript, and conceived and edited its content.

Conflicts of Interest: The authors declare no conflict of interest.

References

1. Ploetz, R.C.; Hulcr, J.; Wingfield, M.; de Beer, Z.W. Ambrosia and bark beetle-associated tree diseases: Black Swan events in tree pathology? *Plant Dis.* **2013**, *95*, 856–872. [CrossRef]
2. Pautasso, M.; Schlegel, M.; Holdelrieder, O. Forest health in a changing world. *Microb. Ecol.* **2015**, *69*, 826–842. [CrossRef] [PubMed]
3. Wingfield, M.J.; Brockerhoff, E.G.; Wingfield, B.D.; Slippers, B. Planted forest health: The need for a global strategy. *Science* **2015**, *349*, 832–836. [CrossRef] [PubMed]
4. Wingfield, M.J.; Garnas, J.R.; Hajek, A.; Hurley, B.P.; de Beer, Z.W.; Taerum, S.J. Novel and co-evolved associations between insects and microorganisms as drivers of forest pestilence. *Biol. Invasions* **2016**, *18*, 1045–1056. [CrossRef]
5. Anderson, P.K.; Cuningham, A.A.; Patel, N.G.; Morales, F.J.; Epstein, P.R.; Daszak, P. Emerging infectious diseases of plants: Pathogen pollution, climate change and agrotechnology drivers. *Trends Ecol. Evol.* **2004**, *19*, 535–544. [CrossRef] [PubMed]
6. Lu, M.; Wingfield, M.J.; Gillette, N.; Sun, J.-H. Do novel genotypes drive the success of an invasive bark beetle–fungus complex? Implications for potential reinvasion. *Ecology* **2011**, *92*, 2013–2019. [CrossRef] [PubMed]
7. Fisher, M.C.; Henk, D.A.; Briggs, C.J.; Brownstein, J.S.; Madoff, L.C.; McCraw, S.L.; Gurr, S.J. Emerging fungal threats to animal, plant and ecosystem health. *Nature* **2012**, *484*, 186–194. [CrossRef] [PubMed]
8. Trumbore, S.; Brando, P.; Hartmann, H. Forest health and global change. *Science* **2015**, *349*, 814–818. [CrossRef] [PubMed]
9. Anagnostakis, S.L. Chestnut blight: The classical problem of an introduced pathogen. *Mycologia* **1987**, *79*, 22–37. [CrossRef]
10. Ploetz, R.C. Diseases of tropical perennial crops: Challenging problems in diverse environments. *Plant Dis.* **2007**, *91*, 644–663. [CrossRef]
11. Rackham, O. Ancient woodlands: Modern threats. *New Phytol.* **2008**, *180*, 571–586. [CrossRef] [PubMed]
12. Wingfield, M.J.; Hammerbacher, A.; Ganley, R.J.; Steenkamp, E.T.; Gordon, T.R.; Wingfield, B.D.; Coutinho, T.A. Pitch canker caused by *Fusarium circinatum*—A growing threat to pine plantations and forests worldwide. *Australas. Plant Pathol.* **2008**, *37*, 319–334. [CrossRef]
13. Fraedrich, S.W.; Harrington, T.C.; Rabaglia, R.J.; Ulyshen, M.D.; Mayfield, A.E., III; Hanula, J.L.; Eickwort, J.M.; Miller, D.R. A fungal symbiont of the redbay ambrosia beetle causes a lethal wilt in redbay and other Lauraceae in the southeastern USA. *Plant Dis.* **2008**, *92*, 215–224. [CrossRef]
14. Haack, R.A. Exotic bark- and wood-boring Coleoptera in the United States: Recent establishments and interceptions. *Can. J. For. Res.* **2006**, *36*, 269–288. [CrossRef]
15. Rabaglia, R.J.; Dole, S.A.; Cognato, A.I. Review of American Xyleborina (Coleoptera: Curculionidae: Scolytinae) Occurring North of Mexico, with an Illustrated Key. *Ann. Entomol. Soc. Am.* **2006**, *99*, 1034–1056. [CrossRef]
16. Barton, C.; Bates, C.; Cutrer, B.; Eickwort, J.; Harrington, S.; Jenkins, D.; Stones, D.M.; Reid, L.; Riggins, J.J.; Trickel, R. Distribution of Counties with Laurel wilt Disease by Year of Initial Detection. Available online: http://www.fs.usda.gov/Internet/FSE_DOCUMENTS/fseprd513913.pdf (accessed on 16 Feb 2017).
17. Hughes, M.A.; Riggins, J.J.; Koch, F.H.; Cognato, A.I.; Anderson, C.; Formby, J.P.; Dreaden, T.J.; Ploetz, R.C.; Smith, J.A. No rest for the laurels: Symbioclone invader causes unprecedented damage to southern USA forests. *Biol. Invasions* **2017**, in press.
18. Goldberg, N.; Heine, J. A comparison of arborescent vegetation pre- (1983) and post- (2008) outbreak of the invasive species the Asian ambrosia beetle *Xyleborus glabratus* in a Florida maritime hammock. *Plant Ecol. Divers.* **2009**, *2*, 77–83. [CrossRef]
19. Gramling, J.M. Potential effects of laurel wilt on the flora of North America. *Southeast Nat.* **2010**, *9*, 827–836. [CrossRef]
20. Shields, J.; Jose, S.; Freeman, J.; Bunyan, M.; Celis, G.; Hagan, D.; Morgan, M.; Pieterson, E.C.; Zak, J. Short-term impacts of laurel wilt on redbay (*Persea borbonia* L. Spreng.) in a mixed evergreen-deciduous forest in northern Florida. *J. For.* **2011**, *109*, 82–88.

21. Evans, J.P.; Scheffers, B.R.; Hess, M. Effect of laurel wilt invasion on redbay populations in a maritime forest community. *Biol. Invasions* **2013**, *16*, 1581–1588. [CrossRef]

22. Spiegel, K.S.; Leege, L.M. Impacts of laurel wilt disease on redbay (*Persea borbonia* (L.) Spreng.) population structure and forest communities in the coastal plain of Georgia, USA. *Biol. Invasions* **2013**, *15*, 2467–2487. [CrossRef]

23. Chupp, A.D.; Battaglia, L.L. Potential for host shifting in *Papilio palamedes* following invasion of laurel wilt disease. *Biol. Invasions* **2014**, *16*, 2639–2651. [CrossRef]

24. Fraedrich, S.W.; Harrington, T.C.; Bates, C.; Johnson, J.; Reid, L.; Leininger, T.; Hawkins, T. Susceptibility to laurel wilt and disease incidence in two rare plant species, pondberry and pondspice. *Plant. Dis.* **2011**, *95*, 1056–1062. [CrossRef]

25. Rodgers, L.; Derksen, A.; Pernas, T. Expansion and impact of laurel wilt in the Florida Everglades. *Fla. Entomol.* **2014**, *97*, 1247–1250. [CrossRef]

26. Hughes, M.A.; Smith, J.A.; Ploetz, R.C.; Kendra, P.E.; Mayfield, A.E., III; Hanula, J.L.; Hulcr, J.; Stelinski, L.L.; Cameron, S.; Riggins, J.J.; et al. Recovery plan for laurel wilt on redbay and other forest species caused by *Raffaelea lauricola* and disseminated by *Xyleborus glabratus*. *Plant Health Progr.* **2015**, *16*, 173–210.

27. Mosquera, M.; Evans, E.A.; Ploetz, R. Assessing the profitability of avocado production in south Florida in the presence of laurel wilt. *Theor. Econ. Lett.* **2015**, *5*, 343–356. [CrossRef]

28. Ploetz, R.C.; Hughes, M.A.; Kendra, P.E.; Fraedrich, S.W.; Carrillo, D.; Stelinski, L.L.; Hulcr, J.; Mayfield, A.E., III; Dreaden, T.L.; Crane, J.H.; et al. Recovery Plan for Laurel Wilt of Avocado, caused by *Raffaelea lauricola*. *Plant Health Progr.* **2017**, *18*, in press.

29. Hughes, M.A.; Shin, K.; Eickwort, J.; Smith, J.A. First report of laurel wilt disease caused by *Raffaelea lauricola* on silk bay in Florida. *Plant Dis.* **2012**, *96*, 910. [CrossRef]

30. Hughes, M.A.; Brar, G.; Ploetz, R.C.; Smith, J.A. Field and growth chamber inoculations demonstrate *Persea indica* as a newly recognized host for the laurel wilt pathogen, *Raffaelea laurciola*. *Plant Health Progr.* **2013**. [CrossRef]

31. Hughes, M.A.; Black, A.; Smith, J.A. First report of laurel wilt, caused by *Raffaelea lauricola*, on bay laurel (*Laurus nobilis*) in the United States. *Plant Dis.* **2014**, *98*, 1159. [CrossRef]

32. Ploetz, R.C.; Konkol, J. First report of gulf licaria, *Licaria trianda*, as a suscept of laurel wilt. *Plant Dis.* **2013**, *97*, 1248. [CrossRef]

33. Fraedrich, S.W.; Harrington, T.C.; Best, G.S. *Xyleborus glabratus* attacks and systemic colonization by *Raffaelea lauricola* associated with dieback of *Cinnamomum camphora* in the southeastern United States. *For. Pathol.* **2015**, *45*, 60–70.

34. Hulcr, J.; Lou, Q.-Z. The redbay ambrosia beetle (Coleoptera: Curculionidae) prefers Lauraceae in its native range: Records from the Chinese national insect collection. *Fla. Entomol.* **2013**, *96*, 1595–1596. [CrossRef]

35. Harrington, T.C.; Yun, H.Y.; Lu, S.S.; Goto, H.; Aghayeva, D.N.; Fraedrich, S.W. Isolations from the redbay ambrosia beetle, *Xyleborus glabratus*, confirm that the laurel wilt pathogen, *Raffaelea lauricola*, originated in Asia. *Mycologia* **2011**, *103*, 1028–1036. [CrossRef] [PubMed]

36. Hughes, M.A. The Evaluation of Natural Resistance to Laurel wilt Disease in Redbay (*Persea borbonia*). Ph.D. Thesis, University of Florida, Gainesville FL, USA, 2013.

37. Wuest, C.E.; Harrington, T.C.; Fraedrich, S.W.; Yun, H.-Y.; Lu, S.-S. Genetic variation in native populations of the laurel wilt pathogen, *Raffaelea lauricola*, in Taiwan and Japan and the introduced population in the USA. *Plant Dis.* **2017**, *101*, in press.

38. Ploetz, R.C.; Thant, Y.Y.; Hughes, M.A.; Dreaden, T.J.; Konkol, J.L.; Kyaw, A.T.; Smith, J.A.; Harmon, C.L. Laurel wilt, caused by *Raffaelea lauricola*, is detected for the first time outside the southeastern USA. *Plant Dis.* **2016**, *100*, 2166. [CrossRef]

39. Erlich, P.R.; Raven, P.H. Butterflies and plants: A study in coevolution. *Evolution* **1964**, *18*, 586–608. [CrossRef]

40. Darwin, C.R. *On the Origin of Species by Means of Natural Selection, or the Preservation of Favoured Races in the Struggle for Life*; John Murray: London, UK, 1859.

41. Flor, H.H. Host-parasite interaction in flax rust—Its genetics and other implications. *Phytopathology* **1955**, *45*, 680–685.

42. Crute, I.R. The elucidation and exploitation of gene-for-gene recognition. *Plant Pathol.* **1998**, *47*, 107–113. [CrossRef]

43. Bergelson, J.; Dwyer, G.; Emerson, J.J. Models and data on plant-enemy coevolution. *Annu. Rev. Genet.* **2001**, *35*, 469–499. [CrossRef] [PubMed]

44. Kareiva, P. Coevolutionary arms races: Is victory possible? *Proc. Natl. Acad. Sci. USA* **1999**, *96*, 8–10. [CrossRef] [PubMed]

45. Burdon, J.J.; Thrall, P.H. Spatial and temporal patterns in coevolving plant and pathogen associations. *Am. Nat.* **1999**, *153*, S15–S33. [CrossRef]

46. Thompson, J.N. Specific hypotheses on the geographic mosaic of coevolution. *Am. Nat.* **1999**, *153*, S1–S14. [CrossRef]

47. Burdon, J.J.; Thrall, P.H.; Ericson, L. The current and future dynamics of disease in plant communities. *Annu. Rev. Phytopathol.* **2006**, *44*, 19–39. [CrossRef] [PubMed]

48. Haldane, J.B.S. Disease and evolution. *La Ric. Sci. (Suppl.)* **1949**, *19*, 68–76.

49. Harlan, J.R. Diseases as a factor in plant evolution. *Annu. Rev. Phytopathol.* **1976**, *14*, 31–51. [CrossRef]

50. Clay, K.; Kover, P.X. The red queen hypothesis and plant/pathogen interactions. *Annu. Rev. Phytopathol.* **1996**, *34*, 29–50. [CrossRef] [PubMed]

51. Schardl, C.L.; Leuchtmann, A.; Chung, K.-R. Coevolution by common descent of fungal symbionts (*Epichloë* spp.) and grass hosts. *Mol. Biol. Evol.* **1997**, *14*, 133–143. [CrossRef]

52. Holst-Jensen, A.; Kohn, L.M.; Jakobsen, K.S.; Schumacher, T. Molecular phylogeny and evolution of *Monilinia* (Sclerotiniaceae) based on coding and noncoding rDNA sequences. *Amer. J. Bot.* **1997**, *84*, 686–701. [CrossRef]

53. Evans, H.C. Invasive neotropical pathogens of tree crops. In *Tropical Mycology: Volume 2, Micromycetes*; CABI Publishing: Wallingford, UK, 2002; pp. 83–112.

54. O'Donnell, K.; Sink, S.; Libeskind-Hadas, R.; Ploetz, R.C.; Konkol, J.L.; Ploetz, J.N.; Carrillo, D.; Campbell, A.; Duncan, R.E.; Kasson, M.T.; et al. Cophylogenetic analysis of the *Fusarium–Euwallacea* (Coleoptera: Scolytinae) mutualism suggests their discordant phylogenies are due to repeated host shifts. *Fungal Genet. Biol.* **2015**, *82*, 277–290. [CrossRef] [PubMed]

55. Kirkendall, L.R. Ecology and evolution of biased sex ratios in bark and ambrosia beetles. In *Evolution and Diversity of Sex Ratio: Insects and Mites*; Wrensch, D.L., Ebbert, M.A., Eds.; Chapman and Hall: New York, NY, USA, 1993; pp. 235–345.

56. Jordal, B.H.; Normark, B.B.; Farrell, B.D. Evolutionary radiation of a haplodiploid beetle lineage (Curculionidae, Scolytinae). *Biol. J. Linn. Soc.* **2000**, *71*, 483–499. [CrossRef]

57. Jordal, B.H.; Cognato, A.I. Molecular phylogeny of bark and ambrosia beetles reveals multiple origins of fungus farming during periods of global warming. *BMC Evol. Biol.* **2012**, *12*, 133. [CrossRef] [PubMed]

58. Maner, M.L.; Hanula, J.L.; Braman, S.K. Gallery productivity, emergence, and flight activity of the redbay ambrosia beetle (Coleoptera: Curculionidae: Scolytinae). *Environ. Entomol.* **2013**, *42*, 642–647. [CrossRef] [PubMed]

59. Farrell, B.D.; Sequeira, A.; O'Meara, B.; Normark, B.B.; Chung, J.; Jordal, B. The evolution of agriculture in beetles (Curculionidae: Scolytinae and Platypodinae). *Evolution* **2001**, *55*, 2011–2027. [CrossRef] [PubMed]

60. Six, D.L. Bark beetle-fungus symbioses. In *Insect Symbiosis*; Bourtzis, K., Miller, T., Eds.; CRC Press: Boca Raton, FL, USA, 2003; pp. 97–114.

61. Hulcr, J.; Stelinski, L.L. The ambrosia symbiosis: From evolutionary ecology to practical management. *Annu. Rev. Entomol.* **2017**, *62*, 285–303. [CrossRef] [PubMed]

62. Francke-Grosmann, H. Ectosymbiosis in wood-inhabiting insects. In *Symbiosis (Volume 2—Associations of Invertebrates, Birds, Ruminants and Other Biota*; Henry, S.M., Ed.; Academic Press: New York, NY, USA, 1967; pp. 141–206.

63. Hulcr, J.; Cognato, A.I. Repeated evolution of crop theft in fungus-farming ambrosia beetles. *Evolution* **2010**, *64*, 3205–3212. [CrossRef] [PubMed]

64. Ngoan, N.D.; Wilkinson, R.C.; Short, D.E.; Moses, C.S.; Mangold, J.R. Biology of an introduced ambrosia beetle, *Xylosandrus compactus*, in Florida. *Ann. Entomol. Soc. Am.* **1976**, *69*, 872–876. [CrossRef]

65. Ranger, C.M.; Reding, M.E.; Persad, A.B.; Herms, D.A. Ability of stress-related volatiles to attract and induce attacks by *Xylosandrus germanus* and other ambrosia beetles. *Agric. For. Entomol.* **2010**, *12*, 177–185. [CrossRef]

66. Kendra, P.E.; Montgomery, W.S.; Niogret, J.; Pruett, G.E.; Mayfield, A.E., III; MacKenzie, M.; Deyrup, M.A.; Bauchan, G.R.; Ploetz, R.C.; Epsky, N.D. North American Lauraceae: Terpenoid emissions, relative attraction

and boring preferences of redbay ambrosia beetle, *Xyleborus glabratus* (Coleoptera: Curculionidae: Scolytinae). *PLoS ONE* **2014**, *9*, e102086. [CrossRef] [PubMed]

67. Kuhnholz, S.; Borden, J.H.; Uzunovic, A. Secondary ambrosia beetles in apparently healthy trees: Adaptions, potential causes and suggested research. *Integr. Pest Manag. Rev.* **2002**, *6*, 209–219. [CrossRef]

68. Hulcr, J.; Dunn, R.R. The sudden emergence of pathogenicity in insect–fungus symbioses threatens naive forest ecosystems. *Proc. Roy. Soc. B Sci.* **2011**, *278*, 2866–2873. [CrossRef] [PubMed]

69. Carrillo, D.; Duncan, R.E.; Ploetz, J.N.; Campbell, A.; Ploetz, R.C.; Peña, J.E. Lateral transfer of a phytopathogenic symbiont among native and exotic ambrosia beetles. *Plant Pathol.* **2014**, *63*, 54–62. [CrossRef]

70. Ploetz, R.C.; Konkol, J.L.; Narvaez, T.; Duncan, R.E.; Saucedo, R.J.; Campbell, A.; Mantilla, J.; Carrillo, D.; Kendra, P.E. Presence and prevalence of *Raffaelea lauricola*, cause of laurel wilt, in different species of ambrosia beetle in Florida USA. *J. Econ. Entomol.* **2017**. [CrossRef] [PubMed]

71. Batra, L.R. Ambrosia fungi: Extent of specificity to ambrosia beetles. *Science* **1966**, *173*, 193–195. [CrossRef] [PubMed]

72. Harrington, T.C. Ecology and evolution of mycophagous bark beetles and their fungal partners. In *Insect-Fungal Associations. Ecology and Evolution*; Vega, F.E., Blackwell, M., Eds.; Oxford University Press: New York, NY, USA, 2005; pp. 257–291.

73. Gebhardt, H.; Begerow, D.; Oberwinkler, F. Identification of the ambrosia fungus of *Xyleborus monographus* and *X. dryographus* (Coleoptera: Curculionidae, Scolytinae). *Mycol. Prog.* **2004**, *3*, 95–102. [CrossRef]

74. Kostovcik, M.; Bateman, C.C.; Kolarik, M.; Stelinski, L.L.; Jordal, B.H.; Hulcr, J. The ambrosia symbiosis is specific in some species and promiscuous in others: Evidence from community pyrosequencing. *ISME J.* **2015**, *9*, 126–138. [CrossRef] [PubMed]

75. Harrington, T.C.; Aghayeva, D.N.; Fraedrich, S.W. New combinations in *Raffaelea, Ambrosiella*, and *Hyalorhinocladiela*, and four new species from the redbay ambrosia beetle, *Xyleborus glabratus*. *Mycotaxon* **2010**, *111*, 337–361. [CrossRef]

76. Harrington, T.C.; Fraedrich, S.W. Quantification of propagules of the laurel wilt fungus and other mycangial fungi from the redbay ambrosia beetle, *Xyleborus glabratus*. *Phytopathology* **2010**, *100*, 1118–1123. [CrossRef] [PubMed]

77. Hulcr, J.; Mann, R.; Stelinski, L.L. The scent of a partner: Ambrosia beetles are attracted to volatiles from their fungal symbionts. *J. Chem. Ecol.* **2011**, *37*, 1374–1377. [CrossRef] [PubMed]

78. Hanula, J.L.; Mayfield, A.E., III; Fraedrich, S.W.; Rabaglia, R.J. Biology and host associations of redbay ambrosia beetle (Coleoptera: Curculionidae: Scolytinae), exotic vector of laurel wilt killing redbay trees in the southeastern United States. *J. Econ. Entomol.* **2008**, *101*, 1276–1286. [CrossRef] [PubMed]

79. Miller, D.R.; Rabaglia, R.J. Ethanol and (−)-α-pinene: Attractant kairomones for bark and ambrosia beetles in the southeastern U.S. *J. Chem. Ecol.* **2009**, *35*, 435–448. [CrossRef] [PubMed]

80. Hanula, J.L.; Sullivan, B. Manuka oil and phoebe oil are attractive baits for *Xyleborus glabratus* (Coleoptera: Curculionidae: Scolytinae), the vector of laurel wilt. *Environ. Entomol.* **2008**, *37*, 1403–1409. [CrossRef] [PubMed]

81. Kendra, P.E.; Montgomery, W.S.; Niogret, J.; Peña, J.E.; Capinera, J.L.; Brar, G.; Epsky, N.D.; Heath, R.R. Attraction of the redbay ambrosia beetle, *Xyleborus glabratus*, to avocado, lychee, and essential oil lures. *J. Chem. Ecol.* **2011**, *37*, 932–942. [CrossRef] [PubMed]

82. Niogret, J.; Kendra, P.E.; Epsky, N.D.; Heath, R.R. Comparative analysis of terpenoid emissions from Florida host trees of the redbay ambrosia beetle, *Xyleborus glabratus* (Coleoptera: Curculionidae: Scolytinae). *Fla. Entomol.* **2011**, *94*, 1010–1017. [CrossRef]

83. Hanula, J.L.; Sullivan, B.T.; Wakarchuk, D. Variation in manuka oil lure efficacy for capturing *Xyleborus glabratus* (Coleoptera: Curculionidae: Scolytinae), and cubeb oil as an alternative attractant. *Environ. Entomol.* **2013**, *42*, 333–340. [CrossRef] [PubMed]

84. Kendra, P.E.; Montgomery, W.S.; Niogret, J.; Epsky, N.D. An uncertain future for American Lauraceae: A lethal threat from redbay ambrosia beetle and laurel wilt disease. *Amer. J. Plant Sci.* **2013**, *4*, 727–738. [CrossRef]

85. Kendra, P.E.; Montgomery, W.S.; Niogret, J.; Schnell, E.Q.; Deyrup, M.A.; Epsky, N.D. Evaluation of seven essential oils identifies cubeb oil as most effective attractant for detection of *Xyleborus glabratus*. *J. Pest Sci.* **2014**, *87*, 681–689. [CrossRef]

86.   Kendra, P.E.; Niogret, J.; Montgomery, W.S.; Deyrup, M.A.; Epsky, N.D. Cubeb oil lures: Terpenoid emissions, trapping efficacy, and longevity for attraction of redbay ambrosia beetle (Coleoptera: Curculionidae: Scolytinae). *J. Econ. Entomol.* **2015**, *108*, 350–361. [CrossRef] [PubMed]

87.   Kendra, P.E.; Montgomery, W.S.; Deyrup, M.A.; Wakarchuk, D. Improved lure for redbay ambrosia beetle developed by enrichment of α-copaene content. *J. Pest Sci.* **2016**, *89*, 427–438. [CrossRef]

88.   Kendra, P.E.; Montgomery, W.S.; Schnell, E.Q.; Deyrup, M.A.; Epsky, N.D. Efficacy of α-copaene, cubeb, and eucalyptol lures for detection of redbay ambrosia beetle (Coleoptera: Curculionidae: Scolytinae). *J. Econ. Entomol.* **2016**, *109*, 2428–2435. [CrossRef] [PubMed]

89.   Chen, H.; Li, Z.; Tang, M. Laboratory evaluation of flight activity of *Dendroctonus armando* (Coleoptera: Curculionidae: Scolytinae). *Canad. Entomol.* **2010**, *142*, 378–387. [CrossRef]

90.   Kendra, P.E.; Montgomery, W.S.; Niogret, J.; Deyrup, M.A.; Guillén, L.; Epsky, N.D. *Xyleborus glabratus, X. affinis*, and *X. ferrugineus* (Coleoptera: Curculionidae: Scolytinae): Electroantennogram responses to host-based attractants and temporal patterns in host-seeking flight. *Environ. Entomol.* **2012**, *41*, 1597–1605. [CrossRef] [PubMed]

91.   Kuhns, E.H.; Martini, X.; Tribuiani, Y.; Coy, M.; Gibbard, C.; Peña, J.; Hulcr, J.; Stelinski, L.L. Eucalyptol is an attractant of the redbay ambrosia beetle, *Xyleborus glabratus*. *J. Chem. Ecol.* **2014**, *40*, 355–362. [CrossRef] [PubMed]

92.   Martini, X.; Hughes, M.A.; Smith, J.A.; Stelinski, L.L. Attraction of redbay ambrosia beetle, *Xyleborus glabratus*, to leaf volatiles of its host plants in North America. *J. Chem. Ecol.* **2015**, *41*, 613–621. [CrossRef] [PubMed]

93.   Mayfield, A.E., III; Brownie, C. The redbay ambrosia beetle (Coleoptera: Curculionidae: Scolytinae) uses stem silhouette diameter as a visual host-finding cue. *Environ. Entomol.* **2013**, *42*, 743–750. [CrossRef] [PubMed]

94.   Niogret, J.; Epsky, N.D.; Schnell, R.J.; Boza, E.J.; Kendra, P.E.; Heath, R.R. Terpenoid variations within and among half-sibling avocado trees, *Persea americana* Mill. (Lauraceae). *PLoS ONE* **2013**, *8*, e73601. [CrossRef] [PubMed]

95.   Maner, M.L.; Hanula, J.L.; Horn, S. Population trends of the redbay ambrosia beetle (Coleoptera: Curculionidae: Scolytinae): Does utilization of small diameter redbay trees allow populations to persist? *Fla. Entomol.* **2014**, *97*, 208–216. [CrossRef]

96.   Kuhns, E.H.; Tribuiani, Y.; Martini, X.; Meyer, W.L.; Peña, J.; Hulcr, J.; Stelinski, L.L. Volatiles from the symbiotic fungus *Raffaelea lauricola* are synergistic with manuka lures for increased capture of the redbay ambrosia beetle *Xyleborus glabratus*. *Agric. For. Entomol.* **2014**, *16*, 87–94. [CrossRef]

97.   Kim, K.-H.; Choi, Y.J.; Seo, S.-T.; Shin, H.-D. *Raffaelea quercus-mongolicae* sp. nov. associated with *Platypus koryoensis* on oak in Korea. *Mycotaxon* **2009**, *110*, 189–197. [CrossRef]

98.   Matsuda, Y.; Kimura, K.; Ito, S.-I. Genetic characterization of *Raffaelea quercivora* isolates collected from areas of oak wilt in Japan. *Mycoscience* **2010**, *51*, 310–316. [CrossRef]

99.   Six, D.L.; Wingfield, M.J. The role of phytopathogenicity in bark beetle-fungus symbioses: A challenge to the classic paradigm. *Annu. Rev. Entomol.* **2011**, *56*, 255–272. [CrossRef] [PubMed]

100. Rollins, J.A.; Ploetz, R.C.; Zhang, Y. Genomic insights into the mechanisms of pathogenesis in *Raffaelea lauricola*, causal agent of laurel wilt disease. University of Florida: Gainesville, FL, USA, Unpublished data. 2017.

101. Dreaden, T.J.; Davis, J.M.; de Beer, W.Z.; Ploetz, R.C.; Soltis, P.S.; Wingfield, M.J.; Smith, J.A. Phylogeny of ambrosia beetle symbionts in the genus *Raffaelea*. *Fungal Biol.* **2014**. [CrossRef] [PubMed]

102. Simmons, D.R.; de Beer, Z.W.; Huang, Y.-T.; Bateman, C.C.; Campbell, A.; Dreaden, T.J.; Li, Y.; Ploetz, R.C.; Li, H.-F.; Chen, C.-Y.; et al. New *Raffaelea* species (Ophiostomataceae) from the United States and Taiwan associated with ambrosia beetles and plant hosts. *IMA Fungus* **2016**, *7*, 265–273. [PubMed]

103. Dreaden, T.J.; Campbell, A.S.; Gonzalez-Benecke, C.A.; Ploetz, R.C.; Smith, J.A. Response of swamp bay, *Persea palustris*, and redbay, *P. borbonia*, to *Raffaelea* spp. isolated from *Xyleborus glabratus*. *For. Pathol.* **2016**. [CrossRef]

104. Chanderbali, A.S.; van der Werff, H.; Renner, S.S. Phylogeny and historical biogeography of Lauraceae: Evidence from the chloroplast and nuclear genomes. *Ann. Missouri Bot. Garden* **2001**, *88*, 104–134. [CrossRef]

105. Drinnan, A.; Crane, P.; Friis, E.; Pedersen, K. Lauraceous flowers form the Potomac Group (Mid-Cretaceous) of Eastern North America. *Bot. Gaz.* **1990**, *151*, 370–384. [CrossRef]

106. Li, L.; Li, J.; Rohwer, J.G.; van der Werff, H.; Wang, Z.-H.; Li, H.-W. Molecular phylogenetic analysis of the *Persea* group (Lauraceae) and its biogeographic implications on the evolution of tropical and subtropical amphi-Pacific disjunctions. *Am. J. Bot.* **2011**, *98*, 1520–1536. [CrossRef] [PubMed]

107. Huang, J.-F.; Li, L.; van der Werff, H.; Li, H.-W.; Rohwer, J.G.; Crayn, D.M.; Meng, H.H.; van der Merwe, M.; Conran, J.G.; Li, J. Origins and evolution of cinnamon and camphor: A phylogenetic and historical biogeographical analysis of the *Cinnamomum* group (Lauraceae). *Molec. Phylogenet. Evol.* **2016**, *96*, 33–44. [CrossRef] [PubMed]

108. Langeland, K.A.; Cherry, H.M.; McCormick, C.M.; Burks, K.A.C. *Identification and Biology of Nonnative Plants in Florida's Natural Areas*; IFAS Communication Services, University of Florida: Gainesville, FL, USA, 2008.

109. Schaffer, B.; Gil, P.M.; Mickelbart, M.V.; Whiley, A.W. Ecophysiology. In *The Avocado: Botany, Production and Uses*, 2nd ed.; Schaffer, B., Wolstenholme, B.N., Whiley, A.W., Eds.; CAB International Publishing: Wallingford, UK, 2013; pp. 168–199.

110. Gentry, A. Neotropical floristic diversity: Phytogeographical connections between Central and South America, Pleistocene climate fluctuations, or an accident of the Andean orogeny? *Ann. Mo. Bot. Gard.* **1982**, *69*, 557–593. [CrossRef]

111. Bosque, C.; Ramírez, R.; Rodríguez, D. The diet of the oilbird in Venezuela. *Ornitol. Neotropical* **1995**, *6*, 67–80.

112. Food and Agriculture Organization of the United Nations. FAOSTAT. Available online: http://www.fao.org/faostat/en/#data (accessed on 21 December 2016).

113. Evans, E.A.; Current losses of avocado to laurel wilt. University of Florida, Homestead, FL, USA. Personal communication, 2017.

114. Podger, F.D. *Phytophthora cinnamomi*, a cause of lethal disease in indigenous plant communities in Western Australia. *Phytopathology* **1972**, *62*, 972–981. [CrossRef]

115. Lahav, E.; Lavi, U. Genetics and breeding. In *The Avocado: Botany, Production and Uses*, 2nd ed.; Schaffer, B., Wolstenholme, B.N., Whiley, A.W., Eds.; CAB International Publishing: Wallingford, UK, 2013; pp. 51–85.

116. Whiley, A.W.; Schaffer, B. Avocado. In *Environmental Physiology of Fruit Crops, Vol. 2, Subtropical and Tropical Crops*; Schaffer, B., Andersen, P.C., Eds.; CRC Press: Boca Raton, FL, USA, 1994; pp. 165–197.

117. Ploetz, R.C.; Pérez-Martínez, J.M.; Smith, J.A.; Hughes, M.; Dreaden, T.J.; Inch, S.A.; Fu, Y. Responses of avocado to laurel wilt, caused by *Raffaelea lauricola*. *Plant Pathol.* **2012**, *61*, 801–808. [CrossRef]

118. Ploetz, R.C.; Schaffer, B.; Vargas, A.I.; Konkol, J.L.; Salvatierra, J.; Wideman, R. Impact of laurel wilt, caused by *Raffaelea lauricola*, on leaf gas exchange and xylem sap flow in avocado, *Persea americana*. *Phytopathology* **2015**, *105*, 433–440. [CrossRef] [PubMed]

119. Ploetz, R.C.; Hughes, M.A. Susceptibility of gulf licaria and lancewood to laurel wilt. University of Florida: Homestead, FL, USA; University of Hawaii: Hilo, HI, USA, Unpublished data. 2017.

120. Hughes, M.A.; Smith, J.A. Vegetative propagation of putatively laurel wilt-resistant redbay (*Persea borbonia*). *Native Plants* **2014**, *115*, 42–50. [CrossRef]

121. Campbell, A.S.; Ploetz, R.C.; Rollins, J.A. Comparing avocado, swamp bay, and camphortree as hosts of *Raffaelea lauricola* using a green fluorescent protein (GFP)-labeled strain of the pathogen. *Phytopathology* **2016**, *107*, 70–74. [CrossRef] [PubMed]

122. Inch, S.A.; Ploetz, R.C. Impact of laurel wilt, caused by *Raffaelea lauricola*, on xylem function in avocado. *Forest Pathol.* **2012**, *42*, 239–245. [CrossRef]

123. Inch, S.A.; Ploetz, R.C.; Held, B.; Blanchette, R. Histological and anatomical responses in avocado, *Persea americana*, induced by the vascular wilt pathogen, *Raffaelea lauricola*. *Botany* **2012**, *90*, 627–635. [CrossRef]

124. Beckman, C.H. Host response to vascular infection. *Annu. Rev. Phytopathol.* **1964**, *2*, 231–252. [CrossRef]

125. Bonsen, K.J.M.; Kucera, L.J. Vessel occlusions in plants: morphological functional and evolutionary aspects. *IAWA J.* **1990**, *11*, 393–399. [CrossRef]

126. Bishop, C.D.; Cooper, R.M. Ultrastructure of vascular colonization by fungal wilt pathogens. II. Invasion of resistant cultivars. *Physiol. Plant Pathol.* **1984**, *24*, 277–289. [CrossRef]

127. Struckmeyer, B.E.; Beckman, C.H.; Kuntz, J.E.; Ricker, A.J. Plugging of vessels by tyloses and gums in wilting oaks. *Phytopathology* **1954**, *44*, 148–153.

128. Jacobi, W.R.; MacDonald, W.L. Colonization of resistant and susceptible oaks by *Ceratocystis fagacearum*. *Phytopathology* **1980**, *70*, 618–623. [CrossRef]

129. Rioux, D.; Ouellette, G.B. Light microscope observations of histological changes induced by *Ophiostoma ulmi* in various nonhost trees and shrubs. *Can. J. Bot.* **1989**, *67*, 2335–2351. [CrossRef]

130. Fry, S.M.; Milholland, R.D. Response of resistant, tolerant, and susceptible grapevine tissues to invasion by pierce's disease bacterium, *Xylella fastidiosa*. *Phytopathology* **1990**, *80*, 66–69. [CrossRef]

131. Rioux, D.; Nicole, M.; Simard, M.; Ouellette, G.B. Immunocytochemical evidence that secretion of pectin occurs during gel (gum) and tylosis formation in trees. *Phytopathology* **1998**, *88*, 494–505. [CrossRef] [PubMed]

132. Collins, B.R.; Parke, J.L.; Lachenbruch, B.; Hansen, E.M. The effects of *Phytophthora ramorum* infection on hydralic conductivity and tylosis formation in tanoak sapwood. *Can. J. For. Res.* **2009**, *39*, 1766–1776. [CrossRef]

133. VanderMolen, G.E.; Beckman, C.H.; Rodehorst, E. Vascular gelation: A general response phenomenon following infection. *Physiol. Plant Pathol.* **1977**, *11*, 85–100. [CrossRef]

134. Aist, J.R. Structural responses as resistance mechanisms. In *The dynamics of host defense*; Bailey, J.A., Deverall, B.J., Eds.; Academic Press: Sydney, Australia, 1983; pp. 33–70.

135. Gagnon, C. Histochemical studies on the alteration of lignin and pectic substances in white elm infected by *Ceratocystis ulmi*. *Can. J. Bot.* **1967**, *45*, 1619–1623. [CrossRef]

136. Hughes, M.A.; Inch, S.A.; Ploetz, R.C.; Er, H.L.; van Bruggen, A.H.C.; Smith, J.A. Responses of swamp bay, *Persea palustris*, and avocado, *Persea americana*, to the laurel wilt pathogen, *Raffaelea lauricola*. *For. Pathol.* **2015**, *45*, 111–119. [CrossRef]

137. Urban, J.; Dvořák, M. Occlusion of sap flow in elm after artificial inoculation with *Ophiostoma novo-ulmi*. *Acta Hort.* **2013**, *991*, 301–306. [CrossRef]

138. Park, J.-H.; Juzwik, J.; Cavender-Bares, J. Multiple *Ceratocystis smalleyi* infections associated with reduced stem water transport in bitternut hickory. *Phytopathology* **2013**, *103*, 565–574. [CrossRef] [PubMed]

139. Murata, M.; Yamada, T.; Ito, S. Changes in water status in seedlings of six species in the Fagaceae after inoculation with *Raffaelea quercivora* Kubono et Shin-Ito. *J. For. Res.* **2005**, *10*, 251–255. [CrossRef]

140. Parke, J.L.; Oh, E.; Voelker, S.; Hansen, E.M.; Buckles, G.; Lachenbruch, B. *Phytophthora ramorum* colonizes tanoak xylem and is associated with reduced stem water transport. *Phytopathology* **2007**, *97*, 1558–1567. [CrossRef] [PubMed]

141. Li, Q.; Wang, X.X.; Lin, J.G.; Liu, J.; Jiang, M.S.; Chu, L.X. Chemical Composition and Antifungal Activity of Extracts from the Xylem of *Cinnamomum camphora*. *BioResources* **2014**, *9*, 2560–2571. [CrossRef]

142. Solla, A.; Gil, L. Xylem vessel diameter as a factor in resistance of *Ulmus minor* to *Ophiostoma novo-ulmi*. *For. Pathol.* **2002**, *32*, 123–134. [CrossRef]

143. Solla, A.; Martín, J.A.; Corral, P.; Gil, L. Seasonal changes in wood formation of *Ulmus pumila* and *U. minor* and its relation with Dutch elm disease. *New Phytol.* **2005**, *166*, 1025–1034. [CrossRef] [PubMed]

144. Venturas, M.; Lopez, R.; Martın, R.A.; Gasco, A.; Gil, L. Heritability of *Ulmus minor* resistance to Dutch elm disease and its relationship to vessel size, but not to xylem vulnerability to drought. *Plant Pathol.* **2014**, *63*, 500–509. [CrossRef]

145. Cameron, R.S.; Hanula, J.; Fraedrich, S.W.; Bates, C. Progression of laurel wilt disease within redbay and sassafras populations in southeast Georgia. *Southeast Nat.* **2015**, *14*, 650–674. [CrossRef]

146. Smith, C.K.; Landreaux, E.; Steinmann, H.; McGrath, D.; Hayes, C.; Hayes, R. Redbay survival eleven years after infection with an exotic disease on St. Catherines Island, Georgia, USA. *Environ. Nat. Resour. Res.* **2015**, *6*, 27–34. [CrossRef]

147. Cameron, R.S.; Bates, C.; Johnson, J. Progression of laurel wilt disease in Georgia: 2009–2011 (Project SC-EM-08-02). In *Forest Health Monitoring: National Status, Trends and Analysis*; Potter, K.M., Conlking, B.L., Eds.; Southern Research Station, USDA-ARS: Asheville, NC, USA, 2012; pp. 145–151.

148. Best, S.; Fraedrich, S. The impact of laurel wilt caused by *Raffaelea lauricola* on clonal populations of pondberry (*Lindera melissifolia*). *Phytopathology* **2016**, *106*, S4.124.

149. Beckman, F.C. Laurel Wilt: Assessing the Risk of Pruning Tool Transmission of *Raffaelea lauricola*. Master's Thesis, University of Florida, Gainesville, FL, USA, 2012.

150. Ploetz, R.C.; Inch, S.A.; Pérez-Martínez, J.M.; White, T.L., Jr. Systemic infection of avocado, *Persea americana*, by *Raffaelea lauricola*, does not progress into fruit pulp or seed. *J. Phytopathol.* **2012**, *160*, 491–495. [CrossRef]

151. Ploetz, R.C.; White, T.; Konkol, J. *Raffaelea lauricola* is not transmitted through scions of avocado that are used for grafting. University of Florida: Homestead, FL, USA, Unpublished data. 2017.

152. Evans, E.A.; Ballen, F.H. An econonmetric demand model for Florida green-skin avocados. *HortTechnology* **2015**, *25*, 405–411.

153. Ploetz, R.C.; Peña, J.E.; Smith, J.A.; Dreaden, T.L.; Crane, J.H.; Schubert, T.; Dixon, W. Laurel wilt is confirmed in Miami-Dade County, center of Florida's commercial avocado production. *Plant Dis.* **2011**, *95*, 1599. [CrossRef]

154. Gottwald, T.R.; Hughes, G.; Graham, J.H.; Sun, X.; Riley, T. The citrus canker epidemic in Florida: The scientific basis of regulatory eradication policy for an invasive species. *Phytopathology* **2001**, *91*, 30–34. [CrossRef] [PubMed]

155. Gottwald, T.R.; Wierenga, E.; Luo, W.; Parnell, S. Epidemiology of Plum pox 'D'strain in Canada and the USA. *Can. J. Plant Pathol.* **2013**, *35*, 442–457. [CrossRef]

156. Drew, J.; Anderson, N.; Andow, D. Conundrums of a complex vector for invasive species control: A detailed examination of the horticultural industry. *Biol. Invas.* **2010**, *12*, 2837–2851. [CrossRef]

157. Fry, W.E.; McGrath, M.T.; Seaman, A.; Zitter, T.A.; McLeod, A.; Danies, G.; Gugino, B.K. The 2009 late blight pandemic in the eastern United States-Causes and results. *Plant Dis.* **2013**, *97*, 296–306. [CrossRef]

158. Brar, G.S.; Capinera, J.L.; Kendra, P.E.; McLean, S.; Peña, J.E. Life cycle, development, and culture of *Xyleborus glabratus* (Coleoptera: Curculionidae: Scolytinae). *Fla. Entomol.* **2013**, *96*, 1158–1167. [CrossRef]

159. Plantegenest, M.; Le May, C.; Fabre, F. Landscape epidemiology of plant diseases. *J. R. Soc. Interface* **2007**, *4*, 963–972. [CrossRef] [PubMed]

160. Carrillo, D.; Choudrey, R.; Garrett, K. Spatial distribution of laurel wilt outbreaks in commercial avocado production areas in south Florida suggest no connection with outbreaks of the disease in neighboring natural areas. University of Florida: Homestead, FL, USA, Unpublished data. 2017.

161. La Rue, C.D. Root grafting in trees. *Am. J. Bot.* **1934**, *21*, 121–126. [CrossRef]

162. La Rue, C.D. Root-grafting in tropical trees. *Science* **1952**, *115*, 296. [CrossRef] [PubMed]

163. Epstein, A.H. Root graft transmission of tree pathogens. *Annu. Rev. Phytopathol.* **1978**, *16*, 181–192. [CrossRef]

164. Himelick, E.B.; Neely, D. Root grafting of city-planted American elms. *Plant Dis. Rep.* **1962**, *46*, 86–87.

165. Sinclair, W.A.; Lyon, J. *Diseases of Trees and Shrubs*, 2nd ed.; Cornell University Press: Ithaca, NY, USA, 2005; pp. 1–680.

166. Horne, W.T.; Parker, E.R. The avocado disease called sunblotch. *Phytopathology* **1931**, *21*, 235–238.

167. Wallace, J.M.; Drake, R.J. A high rate of seed transmission of avocado sun blotch virus from symptomless trees and the origin of such trees. *Phytopathology* **1962**, *52*, 237–241.

168. Dann, E.; Ploetz, R.C.; Pegg, K.G.; Coates, L. Diseases of avocado. In *The Avocado*, 2nd ed.; Schaffer, B., Ed.; CABI: Wallingford, UK, 2012; pp. 380–422.

169. Ploetz, R.C.; Konkol, J. Herbicide treatment to establish barriers around laurel wilt affected avocado trees. University of Florida: Homestead, FL, USA, Unpublished data. 2017.

170. Koch, F.H.; Smith, W.D. Spatio-temporal analysis of *Xyleborus glabratus* (Coleoptera: Circulionidae: Scolytinae) invasion in eastern US forests. *Environ. Entomol.* **2008**, *37*, 442–452. [CrossRef] [PubMed]

171. Cameron, R.S.; Bates, C.; Johnson, J. *Distribution and Spread of Laurel wilt Disease in Georgia: 2006–2008 Survey and Field Observations*; Georgia Forestry Commission Report, 2006–2008; Georgia Forestry Commission: Tifton, GA, USA, 2008.

172. Riggins, J.J.; Hughes, M.; Smith, J.A.; Mayfield, A.E., III; Layton, B.; Balbalian, C.; Campbell, R. First occurrence of laurel wilt disease caused by *Raffaelea lauricola* on redbay trees in Mississippi. *Plant Dis.* **2010**, *94*, 634. [CrossRef]

173. Bates, C.A.; Fraedrich, S.W.; Harrington, T.C.; Cameron, R.S.; Menard, R.D.; Best, G.S. First report of laurel wilt, caused by *Raffaelea lauricola*, on sassafras (*Sassafras albidum*) in Alabama. *Southeast Nat.* **2015**, *14*, 650–674.

174. Fraedrich, S.W.; Johnson, C.W.; Menard, R.D.; Harrington, T.C.; Olatinwo, R.; Best, G.S. First report of *Xyleborus glabratus* (Coleoptera: Curculionidae: Scolytinae) and laurel wilt in Louisiana, USA: The disease continues westward on sassafras. *Fla. Entomol.* **2015**, *98*, 1266–1268. [CrossRef]

175. Olatinwo, R.; Barton, C.; Fraedrich, S.W.; Johnson, W.; Hwang, J. First report of laurel wilt, caused by *Raffaelea lauricola*, on sassafras (*Sassafras albidum*) in Arkansas. *Plant Dis.* **2016**, *100*, 2331. [CrossRef]

176. Shearman, T.M.; Wang, G.G.; Bridges, W.C. Population dynamics of redbay (*Persea borbonia*) after laurel wilt disease: An assessment based on forest inventory and analysis data. *Biol. Invasions* **2015**, *17*, 1371–1382. [CrossRef]

177. Moslonka-Lefebvre, M.; Finley, A.; Dorigatti, I.; Dehnen-Schmutz, K.; Harwood, T.; Jeger, M.J.; Xu, X.M.; Holdenrieder, O.; Pautasso, M. Networks in plant epidemiology: From genes to landscapes, countries, and continents. *Phytopathology* **2011**, *101*, 392–403. [CrossRef] [PubMed]

178. Shaw, M.W.; Pautasso, M. Networks and plant disease management: Concepts and applications. *Annu. Rev. Phytopathol.* **2014**, *52*, 477–493. [CrossRef] [PubMed]

179. Harwood, T.D.; Xu, X.; Pautasso, M.; Jeger, M.J.; Shaw, M.W. Epidemiological risk assessment using linked network and grid based modelling: *Phytophthora ramorum* and *Phytophthora kernoviae* in the UK. *Ecol. Model.* **2009**, *220*, 3353–3361. [CrossRef]

180. Nopsa, J.F.H.; Daglish, G.J.; Hagstrum, D.W.; Leslie, J.F.; Phillips, T.W.; Scoglio, C.; Thomas-Sharma, S.; Walter, G.H.; Garrett, K.A. Ecological networks in stored grain: Key postharvest nodes for emerging pests, pathogens, and mycotoxins. *Bioscience* **2015**, *65*, 985–1002. [CrossRef] [PubMed]

181. Sutrave, S.; Scoglio, C.; Isard, S.A.; Hutchinson, J.M.S.; Garrett, K.A. Identifying highly connected counties compensates for resource limitations when evaluating national spread of an invasive pathogen. *PLoS ONE* **2012**, *7*, e37793. [CrossRef] [PubMed]

182. Garrett, K.A. Impact Network Analysis: A framework for evaluating the effects of information and other technologies through linked socioeconomic and biophysical networks. *bioRxiv* **2017**, in press.

183. Grimm, V.; Revilla, E.; Berger, U.; Jeltsch, F.; Mooij, W.M.; Railsback, S.F.; Thulke, H.H.; Weiner, J.; Wiegand, T.; DeAngelis, D.L. Pattern-oriented modeling of agent-based complex systems: Lessons from ecology. *Science* **2005**, *310*, 987–991. [CrossRef] [PubMed]

184. Garrett, K.A. Information networks for plant disease: Commonalities in human management networks and within-plant signaling networks. *Eur. J. Plant Pathol.* **2012**, *133*, 75–88. [CrossRef]

185. Mills, P.; Dehnen-Schmutz, K.; Ilbery, B.; Jeger, M.; Jones, G.; Little, R.; MacLeod, A.; Parker, S.; Pautasso, M.; Pietravalle, S.; et al. Integrating natural and social science perspectives on plant disease risk, management and policy formulation. *Philos. Trans. R. Soc. B Biol. Sci.* **2011**, *366*, 2035–2044. [CrossRef] [PubMed]

186. Rebaudo, F.; Dangles, O. Coupled information diffusion-pest dynamics models predict delayed benefits of farmer cooperation in pest management programs. *PLoS Comput. Biol.* **2011**, *7*, e1002222. [CrossRef] [PubMed]

187. McRoberts, N.; Hall, C.; Madden, L.V.; Hughes, G. Perceptions of disease risk: From social construction of subjective judgments to rational decision making. *Phytopathology* **2011**, *101*, 654–665. [CrossRef] [PubMed]

188. Borer, E.; Laine, A.-L.; Seabloom, E. A multiscale approach to plant disease using the metacommunity concept. *Annu. Rev. Phytopathol.* **2016**, *54*, 397–418. [CrossRef] [PubMed]

189. Dreaden, T.J.; Davis, J.M.; Harmon, C.L.; Ploetz, R.C.; Palmateer, A.J.; Soltis, P.S.; Smith, J.A. Development of multilocus PCR assays for *Raffaelea lauricola*, causal agent of laurel wilt disease. *Plant Dis.* **2014**, *98*, 379–383. [CrossRef]

190. Sankaran, S.; Ehsani, R.; Inch, S.A.; Ploetz, R.C. Evaluation of visible-near infrared reflectance spectra of avocado leaves as a non-destructive sensing tool for detection of vascular infection by the laurel wilt pathogen, *Raffaelea lauricola*. *Plant Dis.* **2012**, *96*, 1683–1689. [CrossRef]

191. De Castro, A.I.; Ehsani, R.; Ploetz, R.; Crane, J.H.; Abdulridh, J. Optimum spectral and geometric parameters for early detection of laurel wilt disease in avocado. *Rem. Sens. Environ.* **2015**, *171*, 33–44. [CrossRef]

192. De Castro, A.I.; Ehsani, R.; Ploetz, R.C.; Crane, J.H.; Buchanon, S. Detection of laurel wilt disease in avocado using low altitude aerial imaging. *PLoS ONE* **2015**, *10*, e0124642. [CrossRef] [PubMed]

193. Pero, J. Bug off my guacamole. *Mech. Engin.-CIME* **2015**, *137*, 96–97.

194. Angle, C.; Waggoner, L.P.; Ferrando, A.; Haney, P.; Passler, T. Canine detection of the volatilome: A review of implications for pathogen and disease detection. *Front. Vet. Sci.* **2016**, *3*, 47. [CrossRef] [PubMed]

195. Mills, D. *The Science of Laurel Wilt Canine Detection*; Seminar Series; Tropical Research & Education Center, University of Florida: Homestead, FL, USA, 2017.

196. Adkins, J. Canines Detect Deadly Disease in Historic Avocado Trees. Available online: https://news.fiu.edu/2015/10/canines-detect-deadly-disease-in-historic-avocado-trees/92863 (accessed on 31 January 2017).

197. Campbell, A.S.; Ploetz, R.C.; Kendra, P.E.; Montgomery, W.S.; Dreaden, T.J. Geographic variation in mycangial communities of *Xyleborus glabratus*. *Mycologia* **2016**, *108*, 657–667. [CrossRef] [PubMed]

198. An, L. Modeling human decisions in coupled human and natural systems: review of agent-based models. *Ecol. Model.* **2012**, *229*, 25–36. [CrossRef]

199. Ploetz, R.C.; Konkol, J.L.; Pérez-Martínez, J.M.; Fernandez, R. Management of laurel wilt of avocado, caused by *Raffaelea lauricola*. *Eur. J. Plant Pathol.* **2017**. [CrossRef]
200. Mayfield, A.E., III; Barnard, E.L.; Smith, J.A.; Bernick, S.C.; Eickwort, J.M.; Dreaden, T.J. Effect of propiconazole on laurel wilt disease development in redbay trees and on the pathogen *in vitro*. *Arboric. Urban For.* **2008**, *34*, 317–324.

MDPI AG
St. Alban-Anlage 66
4052 Basel, Switzerland
Tel. +41 61 683 77 34
Fax +41 61 302 89 18
http://www.mdpi.com

Forests Editorial Office
E-mail: forests@mdpi.com
http://www.mdpi.com/journal/forests

www.ingramcontent.com/pod-product-compliance
Lightning Source LLC
Chambersburg PA
CBHW051729210326
41597CB00032B/5663